Toxicological Risk Assessment

Volume II

General Criteria and Case Studies

Editors

D. B. Clayson
Chief, Toxicological Research Division
Bureau of Chemical Safety
Sir Frederick G. Banting Research Centre
Ottawa, Canada

D. Krewski
Chief, Biostatistics and Computer Applications
Environmental Health Directorate
Health Protection Branch
Health and Welfare Canada
Ottawa, Canada

I. Munro
Director, Canadian Center for Toxicology
Guelph, Ontario
Canada

CRC Press, Inc.
Boca Raton, Florida

Library of Congress Cataloging in Publication Data
Main entry under title:

Toxicological risk assessment.

Bibliography: p.
Includes index.
Contents: v. 1. Biological and statistical
criteria -- v. 2. General criteria and case studies.
3. Environmental health--Evaluation.   4. Environmental
health--Government policy--Decision making.
5. Environmental health--Government policy--United
States--Decision making.   I. Clayson, D. B.
II. Krewski, D.   III. Munro, Ian C.   IV. Title: Risk assessment.
RA1199.T68   1985   615.9   84-12679
ISBN 0-8493-5976-7 (v. 1)
ISBN 0-8493-5977-5 (v. 2)

This book represents information obtained from authentic and highly regarded sources. Reprinted material is quoted with permission, and sources are indicated. A wide variety of references are listed. Every reasonable effort has been made to give reliable data and information, but the author and the publisher cannot assume responsibility for the validity of all materials or for the consequences of their use.

Direct all inquiries to CRC Press, Inc., 2000 Corporate Blvd., N.W., Boca Raton, Florida, 33431.

© 1985 by CRC Press, Inc.
Second Printing, 1986

International Standard Book Number 0-8493-5976-7 (Volume I)
International Standard Book Number 0-8493-5977-5 (Volume II)

Library of Congress Card Number 84-12679
Printed in the United States

FOREWORD

The rise of the modern pharmaceutical and chemical industry during the first half of this century has resulted in a dramatic enhancement in the quality of life and life expectancy, particularly in the industrialized nations. Modern treatment methods have substantially influenced the pattern of human disease. Mortality has changed from many deaths at younger ages due to microbial diseases to many deaths being attributable to cancer, heart disease, or other degenerative diseases which occur usually, but not always, in older people. Although the number of cases of the latter diseases have increased, it is noteworthy that for most types of cancer (excluding those of the bronchus and lung and of the stomach and pancreas), age-adjusted rates of mortality have been relatively constant for the last 50 years.

Our society, instead of lauding the chemical and pharmaceutical industries for their major contribution to the control of disease, seemingly holds industry responsible as a perceived major cause of the increased number of cancer cases. This reaction has been fuelled by sensational public reports that certain individual substances may induce cancer in laboratory animals and, by extension, in humans as well. Concern is further heightened by discoveries that extremely toxic chemicals, such as dioxin and aflatoxin, contaminate our environment.

This is the frenetic background against which modern toxicological science is developing. Society-mediated political pressure has led to the adoption and use of toxicological tests to protect the public health, sometimes before these tests are validated and before their significance to the species of primary concern, *Homo sapiens,* has been evaluated. Thus, we find ourselves with an elaborate battery of whole animal, microorganism, cell or tissue culture tests for various aspects of toxicology that are routinely carried out by highly competent and able scientists. Often, confusion arises first about how these tests should be applied to the protection of humanity and, second, of the part science on the one hand, or political considerations on the other, should play in controlling environmental risks. The latter difficulty can only be overcome if risk assessment, the scientific estimate of risk, is clearly separated from risk management, the regulatory decision making procedure that balances the scientific data on risk against the overall significance of regulation to the whole community. This is a highly judgmental process.

The editors of the present book recognized the cardinal importance of risk assessment to reasonable regulation. They have sought contributions from scientists, statisticians, and regulators with a broad experience in this area and have assembled the work in a form that they believe will be valuable to toxicologists, statisticians, regulators, and legislators who wish to obtain a deeper understanding of the difficulties and pitfalls that may befall the unwary in this area. Additionally, authors have been asked to suggest how particular areas of toxicological science may develop in future years.

The book is presented in two volumes. The first is concerned with the biological and statistical criteria that must be regarded as the foundations of any regulatory decision. These are presented separately: Section A is concerned with the experimental findings and Section B with statistical analysis. The second volume considers more general criteria and case studies. Section A is concerned with the interpretation of the scientific data base extrapolated to likely human exposure levels. Consideration is also given to assessing the benefits of chemical usage as well as other regulatory issues. In Section B, a number of specific examples are considered in detail as illustrations of the care needed and the difficulties that can be met in the use of the data. The emphasis of these two volumes is directed mainly towards the induction of cancer, as this is the most active area of toxicological risk assessment at this time. However, the issue of how to best increase the safety of individual members of our community without at the same time destroying the apparent benefits of our current lifestyle is common to all aspects of toxicology and regulation.

In Volume I, Chapter 1, Hart and Fishbein discuss biological factors that may lead to species differences in response to particular toxic agents. The importance of xenobiotic metabolism to such differences is considered in greater detail in the following two chapters. Withey, in Chapter 2, discusses the importance of pharmacokinetic studies on the test agent and its metabolites. In Chapter 3, O'Flaherty indicates how metabolism may vary quantitatively or even qualitatively with the dose level of the agent. The concluding two chapters address the induction of cancer by chemicals. In Chapter 4, Shank and Barrows discuss what is known about the mechanisms of chemical carcinogenesis and how they may ultimately be found to be more diverse than hitherto imagined, a conclusion firmly based on these authors' observations. Clayson, in Chapter 5, concludes this section with a discussion of a possible means of deriving the strength or potency of a carcinogen from bioassay data. He speculates on whether a compatible value for potency could be derived for humans using in vitro studies.

The statistical methodology discussed in Section B serves two functions. First, methods for the design and analysis of toxicological experiments which are both valid and efficient are discussed. Second, statistical models are identified to permit the extrapolation from the high doses fed to animals to the much lower doses to which humans may be exposed. The first two chapters in this section are devoted to carcinogenesis bioassay in small rodents. Bickis and Krewski (Chapter 6) discuss the many complex design issues involved in such tests and provide an overview of currently used methods of analysis. Lagakos and Louis (Chapter 7) treat the IARC method for analyzing time to tumor data in detail. Seilken (Chapter 8) discusses the one extrapolation model that appeared to fit the data of the massive ED01 experiment on mouse carcinogenicity induced by N, 2-fluorenylacetamide. This model differs from others in considering both dose and time on test and, if sufficient data is available, may prove of great value for low dose extrapolation of carcinogenicity bioassay results. In Chapter 9, Crump and Howe review other methods for low dose extrapolation and consider the most promising avenues for future development. Hoel, in Chapter 10, emphasizes the importance of attempting to make low dose extrapolations compatible with the results of pharmacokinetic studies, a valuable approach to the consolidation of more than one type of toxicological test within the statistical data base.

Volume II is concerned with the practical implications of toxicity data. It opens with a consideration by Day of the utility and sensitivity of epidemiological methodology to identify chemicals or mixtures that may adversely affect people (Chapter 1). Without the information provided by epidemiology, toxicology would appear to be an esoteric academic exercise rather than the major practical endeavor that it is today. Kraybill (Chapter 2) outlines the immense body of knowledge we presently have on human exposures to foreign chemicals with a major emphasis on the drinking water supply as an example of possible sources of human contamination. Newberne (Chapter 3), on the other hand, discusses the difficult area of nutrition in respect to toxic response and clearly shows how different levels and types of nutrition may markedly affect the responses in man and animals. Section A concludes with three chapters discussing how the body of information on human exposure and toxic effects may be reconciled with demands for a safe environment, and with the clear need for our society to continue. In Chapter 4, Darby points out that many of the chemicals with which we as a society are most concerned pose benefits as well as risks. Thompson, Chapter 5, considers health benefits as a way to balance the apparent health hazards associated with the continued use of chemicals. In Chapter 6, Miller addresses the specific issues to food regulation in which regulatory decisions may affect each and every member of our community.

Finally, an attempt is made in Section B to determine how our societal efforts at chemical regulation are working out in practice. This is achieved by considering five chemicals in considerable depth. All of these are well known either because of their widespread use or because of recent discussions in the media. They are asbestos discussed by Meek, Shannon, and Toft (Chapter 7); formaldehyde by Starr, Gibson, and Swenberg (Chapter 9); polychlorinated biphenyls by Cordle, Locke, and Springer (Chapter 10); and finally the very controversial artificial sweetening agent, saccharin, by Arnold and Clayson (Chapter 11).

It is hoped that this work will be of use in helping society, including the regulators, toxicologists, consumers, and industrialists, to arrive at a more unified view of where we should be going in order to achieve a safer but still satisfying environment for society as a whole. The editors would each like to pay tribute to each individual author for a substantial contribution achieved at the cost of personal effort and time. The problems that remain to be solved are immense, but, if we can learn where we are presently situated, we have a far better chance of deciding where we need to go and thereafter to choose the correct path.

D. B. Clayson
D. Krewski
I. C. Munro

THE EDITORS

Dr. David B. Clayson Ph.D., is Chief, Toxicology Research Division, Bureau of Chemical Safety of the Canadian Food Directorate. He trained in Natural Sciences (Chemistry) at the University of Oxford, England and took his Ph.D. in Experimental Pathology and Cancer Research at the University of Leeds, England. Dr. Clayson's interests during the 26 years he was at the University of Leeds were in occupational bladder cancer, the mode of chemical induction of bladder tumors, the nature and mode of action of chemical carcinogens. In 1974, he was appointed Deputy Director of the Epply Institute for Cancer Research, University of Nebraska Medical Center in Omaha, and came to Ottawa in 1981. Dr. Clayson, in 1962, published *Chemical Carcinogenesis* the first comprehensive text on this subject, and has contributed over 120 publications to the Cancer Research literature ranging from the mode of action of occupational bladder carcinogens to possible approaches to the difficult problems of carcinogen regulation. Dr. Clayson was Founder Secretary of the Toxicology Forum and is a member of the Society of Toxicology and the American Association of Cancer Research. He has been a member of numerous scientific committees in the United Kingdom, the United States, and Canada, and has also served on international panels such as the International Commission for the Protection of the Environment against Mutagens and Carcinogens, and groups established by the World Health Organization and the International Agency for Research on Cancer.

Dr. Daniel Krewski Ph.D., is currently Chief of the Biostatistics and Computer Applications Division in the Environmental Health Directorate of Health and Welfare Canada and a member of the Laboratory for Research Statistics and Probability at Carleton University in Ottawa. Dr. Krewski has published over 40 articles on biostatistics and risk assessment and has participated in scientific committees established by the Toxicology Forum, International Life Sciences Institute, Institute for Environmental Studies, and International Agency for Research on Cancer. He is presently a member of the Board of Directors of the American Statistical Association and is an Associate Editor of the *Canadian Journal of Statistics* and *Risk Abstracts*.

Dr. Ian C. Munro, Ph.D., is the Director of the Canadian Centre for Toxicology. Immediately prior to accepting the position of Director of the Centre, Dr. Munro was the Director General of the Food Directorate, Health Protection Branch, Health and Welfare Canada from 1979 to 1983. He was past Chairman of the Tripartite Toxicology Committee (U.S., U.K., and Canada) and is Assistant Editor of the *Journal of the American College of Toxicology*. He has published over 60 articles in the field of toxicology.

CONTRIBUTORS

D. L. Arnold, Ph.D.
Research Scientist
Toxicology Research Division
Food Directorate
Health and Welfare Canada
Ontario, Canada

L. R. Barrows, Ph. D.
Department of Pharmacology
George Washington University
Washington, D.C.

M. Bickis
Senior Consulting Statistician
Environmental Health Directorate
Health and Welfare Canada
Ontario, Canada

F. Cordle
Chief
Epidemiology and Clinical Toxicology
Center for Food Safety and Applied
  Nutrition
Food and Drug Administration
Washington, D.C.

K. S. Crump, Ph.D.
President
K. S. Crump and Co.,
Ruston, Louisiana

W. J. Darby, M.D., Ph.D.
Professor Emeritus of Biochemistry
  (Nutrition)
Vanderbilt University School of Medicine
Nashville, Tennessee

N. E. Day, Ph.D.
Chief, Unit of Biostatistics and Field
  Studies
International Agency for Research on
  Cancer
Lyon, France

L. Fishbein
Associate Director for Scientific
  Coordination
National Center for Toxicological
  Research
Jefferson, Arkansas

J. E. Gibson, Ph.D.
Vice President and Director of Research
Chemical Industry Institute of Toxicology
Research Triangle Park, North Carolina

R. W. Hart, Ph.D.
Director, National Center for
  Toxicological Research
National Center for Toxicological
  Research
Jefferson, Arkansas

D. G. Hoel, Ph.D.
Director, Biometry and Risk Assessment
  Program
National Institute of Environmental
  Health Sciences
Research Triangle Park, North Carolina

R. B. Howe, Ph.D.
Professor of Mathematics and Statistics
Louisiana Tech University
Ruston, Louisiana

H. F. Kraybill, Ph.D.
Scientific Coordinator for Environmental
  Cancer
National Cancer Institute
Bethesda, Maryland

S. W. Lagakos, Ph.D.
Associate Professor, Biostatistics
Harvard School of Public Health
Boston, Massachusetts

R. Locke
Environmental Protection Agency
Washington, D.C.

T. A. Louis, Ph.D.
Associate Professor, Biostatistics
Harvard School of Public Health
Boston, Massachusetts

M. E. Meek, M.Sc.
Biologist
Monitoring and Criteria Division
Bureau of Chemical Hazards
Department of National Health and
  Welfare
Ottawa, Canada

S. A. Miller, Ph.D.
Director, Center for Food Safety and
  Applied Nutrition
Food and Drug Administration
Washington, D.C.

P. M. Newberne, D.V.M., Ph.D.
Professor of Nutritional Pathology
Massachusetts Institute of Technology
Cambridge, Massachusetts
  and Professor of Pathology
Boston University School of Medicine
Boston, Massachusetts

E. J. O'Flaherty, Ph.D.
Associate Professor of Environmental
  Health
Department of Environmental Health
University of Cincinnati College of
  Medicine
Cincinnati, Ohio

G. M. Paddle, Ph.D., D.I.C.
Biostatistician to Central Medical Group
Central Medical Group
Imperial Chemical Industries PLC
Cheshire, England

I. F. H. Purchase, Ph.D., FRC Path.
Director, Central Toxicology Laboratory
Imperial Chemical Industries PLC
Cheshire, England

R. C. Shank, Ph.D.
Professor, Community and Environmental
  Medicine and Pharmacology
College of Medicine
University of California
Irvine, California

H. S. Shannon, Ph.D.
Assistant Professor, Department of
  Clinical Epidemiology and Biostatistics
McMaster University
Hamilton, Canada

R. L. Sielken, Jr., Ph.D.
Professor of Statistics
Texas A and M University
College Station, Texas

K. J. Skinner, Ph.D.
Special Assistant, Center for Food Safety
  and Applied Nutrition
Food and Drug Administration
Washington, D.C.

J. Springer
Epidemiology Branch
Environmental Protection Agency
Washington, D.C.

J. Stafford, Ph.D.
Plastics and Petrochemicals Division
Imperial Chemical Industries PLC
Hertfordshire, England

T. B. Starr, Ph.D.
Scientist-Mathematician
Department of Epidemiology
Chemical Industry Institute of Toxicology
Research Triangle Park, North Carolina

J. A. Swenberg, Ph.D., DVM.
Head, Department of Biochemical
  Toxicology and Pathobiology
Chemical Industry Institute of Toxicology
Research Triangle Park, North Carolina

M. S. Thompson, Ph.D.
Associate Professor, Health Policy and
  Management
Harvard University
Boston, Massachusetts

P. Toft, D.Phil.
Chief
Monitoring and Criteria Division
Bureau of Chemical Hazards
Department of National Health and
  Welfare
Ottawa, Canada

J. R. Withey, Ph.D.
Research Scientist
Toxicology Research Division
Foods Directorate
Bureau of Chemical Safety
Health and Welfare Canada
Ontario, Canada

TABLE OF CONTENTS

Volume I

TABLE OF CONTENTS

Volume II

Section A—General Criteria

Chapter 1

EPIDEMIOLOGICAL METHODS FOR THE ASSESSMENT OF HUMAN CANCER RISK

Nicholas E. Day

TABLE OF CONTENTS

I. INTRODUCTION

To be credible, methods used for assessing risk must have observational support, and methods used for assessing human risk should be backed by human data. If the risk concerned is cancer, then these data will be derived from analytical epidemiological cancer studies. These studies are usually either of the cohort or case-control type. In the former, an identified group of individuals is followed forward in time. Information on exposures of interest is collected on individuals when they enter the study, and often during the period of follow-up as well. Information on cancer occurrence is obtained during the period of follow-up. In this way, absolute measures of risk, i.e., incidence of mortality rates, can be obtained for groups exposed at different levels. To assess whether a particular cancer is in excess, the observed risk can be compared either with some external group, often the general population of the region or country concerned, or, preferably, comparisons can be made within the cohort between different levels of exposure, including a nonexposed group if possible. The excess risk can be expressed in either absolute or relative terms.[1]

In case-control studies, by contrast, cases of disease, e.g., a particular form of cancer, and a suitable comparison group are recruited into the study; information on past exposure is obtained from the two groups, and the results expressed as the relative increase in risk for the different levels of exposure.[2] If the cases and controls can be taken as representative of a well defined population for which incidence rates are known for the cancer in question, then the estimated relative risks can be expressed as absolute increases in risk.

It was shown by Cornfield[3] that logically, similar measures of relative risk are available from case-control and cohort studies. Provided the quality of the information is the same in the two types of study, results should be in concordance. A clear example is given by the association between lung cancer and cigarette smoking. One of the first case-control studies in this area was conducted in England between 1948 and 1952.[4] The same investigators started a prospective cohort study among British doctors in 1951, and the results of 20 years of follow-up have been published.[5,6] The estimates of relative risk of lung cancer for different levels of cigarette smoking are given in Table 1. The overall trends are strikingly similar; the higher levels seen systematically for the cohort study probably arise from the very small number of nonsmoking cases in the two series. If the reference category is taken as 15 to 24 cigarettes per day, the relative risks among smokers are nearly identical.

The use of absolute excess risk, although superficially appealing, has a serious drawback, in that the absolute excess risk associated with many exposures is heavily dependent on the population in question, particularly the age structure of the population. Relative risks, on the other hand, are often fairly stable (for an extended discussion of the advantages of using relative risks see Breslow and Day[2]). The main situation where absolute excess risk may be a more suitable measure occurs when the cancer in question is rare in the absence of exposure.[7]

In this paper, we discuss how results from these types of epidemiological study can be used to assess risk, and in particular how these results might be used for extrapolation. Before considering individual situations, however, statistical limits on the degree of excess risk that epidemiology can detect will be discussed.

II. LIMITS OF RISK DETECTABLE BY EPIDEMIOLOGICAL MEANS, AND THE NEED FOR EXTRAPOLATION MODELS

The benefits to be expected from the control of environmental risks, either by individual or social action, can be more firmly evaluated if the risk at the exposure levels in question can be estimated directly. The question then arises, what degree of risk are epidemiological methods capable of detecting, or alternatively, what degree of risk is consistent with a substantial amount of negative epidemiological data?[2] An initial approach to setting limits

Table 1
COMPARISON OF RELATIVE RISK ESTIMATES FROM CASE-CONTROL AND COHORT STUDIES: CIGARETTE SMOKING AND LUNG CANCER

| Amount regularly smoked per day | Relative risk estimate (no. of cases in parentheses) | | | |
	Case-Control		Cohort	
Nonsmoker	1.0	(7)	1.0	(6)
< 5 cigarettes	3.7	(55)	5.8	(3)
5-14	7.5	(489)	7.0	(20)
15-24	9.6	(475)	18.5	(76)
25-40	16.6[a]	(293)	34.3	(96)
40+	27.6[a]	(38)	~ 30	(5)

[a] Smoking categories are 25 to 49 and 50+ respectively.

From Doll, R. and Hill, B., *Br. Med. J.*, 2, 1271, 1952 and Doll, R. and Peto, J., *J. Epidemiol. Commun. Health*, 32, 303, 1978. With permission.

Table 2
CONFIDENCE INTERVALS FOR AN ESTIMATED STANDARDIZED MORTALITY RATIO OF ONE, BASED ON A COHORT STUDY

| No. events in the exposed group | Comparison with an external standard | | Comparison with a nongroup of the same size as the exposed group | |
	Lower	Upper	Lower	Upper
100	0.801	1.216	0.750	1.334
200	0.866	1.149	0.816	1.225
1000	0.939	1.064	0.915	1.094
5000	0.972	1.028	0.959	1.042

of this type can be made by examining the width of the confidence interval associated with an estimated relative risk of unity, for different type and size of study. Some values are given in Tables 2 and 3, for cohort and case-control studies, respectively. Cohort studies of industrial exposures would rarely have more than a few hundred events of interest (e.g., deaths from the relevant cancer). Similarly, case-control studies with several thousand cases would be unusual. Thus, these two tables indicate that a negative study would not be inconsistent with a relative risk of 10%. Added to this 10% will be the additional uncertainty arising from uncontrolled bias and confounding, so that in the majority of situations excess relative risks of 15 to 20% will be compatible with an observed relative risk of one. That is, epidemiology data would usually be unable to distinguish between no excess risk, and an excess relative risk of 10%.

For the more common types of cancer in Western Europe or North America, lifetime risks are between 1% (approximately for leukemia in the U.S.) and 10% (lung cancer in males in the U.K.). The limits on observable excess lifetime risk are therefore of the order of 10^{-2}

Table 3

**95% CONFIDENCE INTERVAL FOR AN ESTIMATED ODDS
RATIO OF ONE, BASED ON A CASE-CONTROL STUDY
(UNMATCHED ANALYSIS, EQUAL NUMBER OF CASES
AND CONTROLS)**

Proportion of controls exposed	No. cases			
	100	**200**	**1000**	**5000**
5%	0.222—4.498	0.354—2.822	0.656—1.524	0.832—1.202
25%	0.502—1.992	0.620—1.612	0.812—1.231	0.913—1.096
45%	0.551—1.815	0.661—1.512	0.835—1.197	0.923—1.083

or 10^{-3}. For rare tumors these limits are lower. Mesothelioma and angiosarcoma of the liver are both exceedingly rare in the absence of the known risk factors, and one or two cases per ten thousand individuals would probably be noteworthy. Since for many exposures risks of 10^{-2} or 10^{-3} would seem unacceptably high, the determination of levels at which excess risk is lower by one or more orders of magnitude requires extrapolation beyond available data. This extrapolation requires some form of mathematical model relating risk to dose, and the question that arises is the extent to which the observed dose-response relationship suggests, or provides support for, the particular choice of model that is made.

This chapter examines the observed dose-response relationships for those situations for which the epidemiological information is most complete. The purpose is to demonstrate the type of risk estimates that epidemiology can provide, and the information it yields on the form of extrapolation to low risk levels that might be appropriate.

III. EPIDEMIOLOGICAL DATA ON DOSE-RESPONSE RELATIONSHIPS

Attention will be concentrated, in the main, on risk as related to levels of dose. Cancer risk is related to both level of exposure and duration of exposure, as well as other time variables such as time since first exposure, time since last exposure and, occasionally, age at exposure. For each of the exposures considered, the role of time will have to be considered, but the focus will be on the level of exposure.

A. Cigarette Smoking and Lung Cancer

More large-scale epidemiological studies have been undertaken to investigate this relationship than any other in the field of cancer, and the literature is vast (see for example Reference 8). There is an overall consistency in the results, so that the data from the British doctors study will be taken as representative. Duration of smoking is of major importance, and among those who have continued to smoke since they began smoking, the excess incidence of lung cancer increases with the fourth power of duration. On stopping smoking, the absolute excess incidence freezes and appears thereafter to remain roughly constant.[9] The absolute excess risk is approximately independent of the age at which smoking started.[10] To remove the effects that variation in these time and age variables will have an excess lung cancer risk, only continuous smokers will be considered who started smoking in early adult life (age 16 to 25), and whose smoking history was reported as changing little with time. This group is that described by Doll and Peto.[6]

The data are displayed in Figure 1, together with the best fitting linear dose-response curve, and a dose-response curve quadratic in a term of the form (dose + constant). This latter curve, given the arbitrariness of the constant, can be considered as a linear quadratic relationship. As pointed out by the authors, errors in the recording of smoking histories

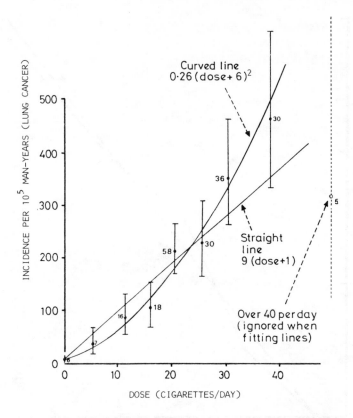

FIGURE 1. Dose-response relationship of lung cancer against daily cigarette consumption, standardized for age. The numbers of onsets in each group are given, and 90% confidence intervals are plotted.[6]

would tend to bias the shape of the dose-response curve away from a quadratic towards linearity. These data strongly suggest that some degree of upward curvilinearity is required to give a satisfactory fit, but with a nonzero coefficient for the linear term if a second degree curve is fitted. Nevertheless, with a dose-response curve for the relative risk of the form

$$RR = (Dose + 6)^2/36$$

as the dose becomes small, the curve becomes nearly linear, and at low doses an approximation of the form

$$RR \cong 1 + Dose/3$$

where dose is measured as cigarettes per day, would be nearly correct.

In these terms, the relative risk associated with passive smoking can be taken as corresponding[11,12] to smoking approximately two cigarettes per day. A study of serum cotinine levels among nonsmoking spouses of regular cigarette smokers would indicate whether such an estimate is plausible.

B. Alcohol and Tobacco Consumption and Esophageal Cancer

The data on which this discussion will be based come from the case-control study per-

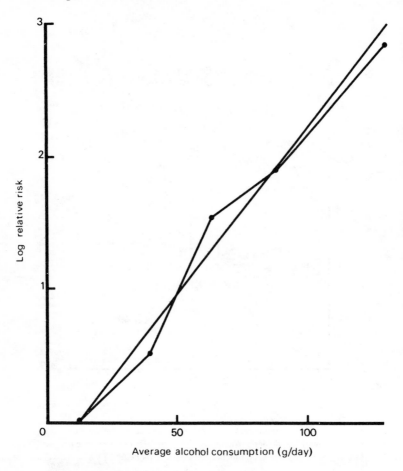

FIGURE 2.   The relative risk for esophageal cancer as a function of alcohol consumption.[2]

formed in Brittany.[13] Further analyses of these results are given in Breslow and Day.[2] This study, among those published, attempted the most precise quantification of alcohol consumption, in terms of grams of ethanol per day. As with the previous example, attention is confined to continuous smokers and drinkers who basically began exposure in early adult life. Neither the effect of age at exposure nor the effect of stopping exposure is known with much precision (see Day and Muñoz[14]). The increase in risk, expressed as a relative risk, with increasing consumption of alcohol is rapid and strongly curvilinear upwards. In fact, if the relative risk is put on a logarithmic scale, the relationship appears linear (see Figure 2). By contrast, for tobacco smoking, the relationship of relative risk to dose is curvilinear downwards (see Figure 3). There is no reason to suppose that smoking histories are less accurate in this study than in the majority of case-control studies of lung cancer, so that it appears as if the shape of the dose-response for tobacco smoking is different for esophageal cancer than it is for lung cancer. The difference may reside in the nature of the relevant exposure. Quantitatively, for esophageal cancer, the respective relationships with alcohol and tobacco are well fitted by the equation

Alcohol: Relative Risk $= \exp(\text{Constant} \times \text{Daily ethanol consumption})$

Tobacco: Relative Risk $= (1 + \text{Constant} \times \text{Daily tobacco consumption})^{0.5}$

The joint effect of alcohol and tobacco as seen both in this study and elsewhere[15] is close

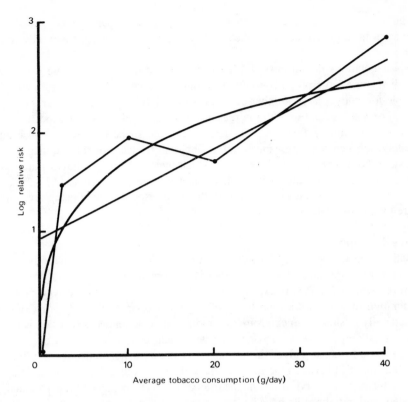

FIGURE 3.   The relative risk for esophageal cancer as a function of tobacco consumption.[2]

Table 4
LINEAR APPROXIMATION TO EXPONENTIAL (ALCOHOL AND ESOPHAGEAL CANCER) AND SQUARE ROOT (TOBACCO AND ESOPHAGEAL CANCER) DOSE-RESPONSE CURVES IN THE RANGE OF DOSES AT WHICH RELATIVE RISK IS LESS THAN TWOFOLD

	Relative risk predicted by various dose-response curves		
Dose	**Linear**	**Square root**	**Exponential**
0	1.0	1.0	1.0
D/100	1.01	1.015	1.007
D/10	1.1	1.14	1.07
D/2	1.5	1.58	1.41
D[a]	2.0	2.0	2.0

[a]   D is the dose at which risk is doubled, in the three cases.

to multiplicative, that is, the relative risk for the two factors multiply. A feature of the data, however, is the scarcity of individuals who drink but do not smoke. (There has been a suggestion[16] that among nonsmokers the effect of alcohol is inappreciable. This suggestion requires confirmation.)

It should be noted that for both these functions, as the dose becomes small the curve approaches the linear, and for dose levels at which the relative risk is less than twofold the linear approximation is fairly close (see Table 4). Thus even with marked curvilinearity, going in both directions, a linear low dose extrapolation, using parameters estimated from the observation in the range of dose where risk is appreciable, emerges naturally from the data.

The two associations just discussed refer to widespread habits responsible, in the populations studied, for a large proportion of the overall incidence of the cancer of interest. One has available for study a large number of cases with widely dispersed exposures, with high risks at high exposures, and in addition reasonably accurate measures of exposures. These conditions are particularly favorable for determining the shape of dose-response curves. Unfortunately, tobacco for lung cancer and alcohol and tobacco for the esophagus (and to a lesser extent the larynx and oral cavity for which there is similar behavior) are the only examples of common exposures with well defined dose-response curves. The two other exposures for which epidemiological data on the dose-response relationship are fairly extensive are ionizing radiation and asbestos.

C. Ionizing Radiation

Information is available to a limited extent on α radiation[17] and on continuous exposures lasting over several decades. The most extensive information, however, is for relatively short-term or even instantaneous exposure to γ radiation. Since both the type of radiation and the duration of exposure are probably of major importance, the following discussion refers specifically to short-term exposure to γ radiation. An extended discussion is provided in the third BEIR report, the value of which was reduced by a theoretical dispute on the shape of the dose-response curve: linear, linear-quadratic or quadratic, the data being inadequate to distinguish between them. Since then, the two studies that provided most of the information, the A-bomb survivors[18] and the irradiated ankylosing spondylitis patients[19] have been updated, and initial results have become available from a third large study, the follow-up of women given radiation treatment for cancer of the uterine cervix.[20] In addition, considerable further work has been done integrating the results of the studies on radiation and breast cancer risk. For illustrative purposes, attention will be confined in this discussion to cancers of the stomach and breast.

The main results for stomach cancer come from the three studies mentioned in the previous paragraph. The observed and expected numbers of cases, together with estimated exposure levels, are given in Table 5. The results from the three studies appear consistent. (The expected numbers from the A-bomb study were obtained internally, rather than from an external standard population as in the other two studies. Since, however, almost all the weight in this internal standardization comes from dose levels at which no differences in risk are seen, the results are essentially comparable.) Figure 4 combines the results, giving the observed to expected ratio as a function of dose. These data illustrate clearly the limitations of epidemiological evidence discussed in an earlier section (see Tables 1 and 2). Up to 200 rads, no excess risk is seen, with the observed to expected ratio remaining very close to one. After 200 rads, the risk begins to rise. The dose-response can clearly be equally well described as a threshold at about 200 rads, with a linear increase thereafter, or as a simple quadratic dose-response throughout the dose range. Surprisingly, however, even a linear dose-response over the entire range does not give an unacceptably poor fit. The results of fitting linear, quadratic and linear-quadratic dose-response curves are shown in Table 6. The predicted risks at low doses (less than 10 rads) are considerably different in the three models.

It is interesting to contrast these results with those for lung cancer and cigarette smoke in Figure 1. In the latter case, with many fewer cases, the curvilinearity of the dose-response could be demonstrated. In Figure 4, both the relatively low level of risk observed (a maximum relative risk of 1.65) and the concentration of cases at the low risk dose levels make the demonstration of curvilinearity more difficult.

For cancer of the breast, the main results have been summarized elsewhere[21] and are

Table 5
STOMACH CANCER AND IONIZING RADIATION: RELATIVE RISKS OBSERVED IN THREE DIFFERENT STUDIES

Dose (rads)	A-bomb survivors[a]	Ankylosing spondylitis series[b]	Irradiated cervical cancer patients[c]
0	0.99 (708)	—	—
1—9	0.98 (473)	—	
10—49	1.02 (340)	—	—
50—99	0.95 (91)	—	—
100—199	0.96 (64)	—	1.0 (86)
200—299	1.20 (32)	1.55 (31)	—
300—399	1.39 (17)	—	—
400 +	1.65 (29)	—	—

Note: Number of cases in parentheses.

[a] Kato and Schull.[18]

[b] Smith and Doll.[19] The dose estimate of 250 rads comes from Smith (personal communication).

[c] Day and Boice.[20]

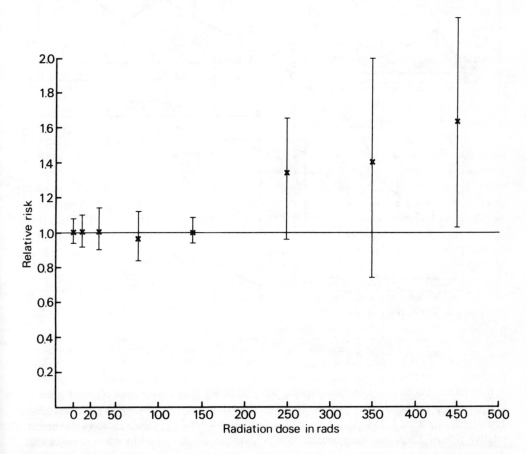

FIGURE 4. Relative risk for stomach cancer 10 years or more following radiation, or a function of radiation exposure (in rad).

Table 6
GOODNESS-OF-FIT OF LINEAR, QUADRATIC AND LINEAR-QUADRATIC DOSE-RESPONSE CURVES TO THE DATA OF TABLE 5 RELATING STOMACH CANCER RISK TO DOSE OF IONIZING RADIATION

Model	Goodness-of-fit[a]	Degrees of freedom
Linear	4.62	6
Quadratic	3.18	6
Linear-quadratic	3.14	5

[a]   Deviance — using GLIM.[24]

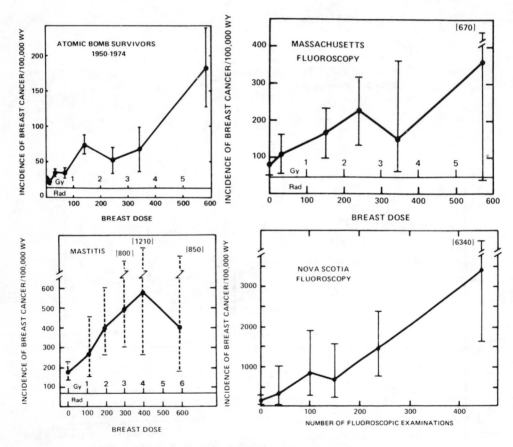

FIGURE 5.   Dose-response curves of breast cancer incidence[21] by radiation dose (or number of fluoroscopic examinations) for atomic bomb survivors,[25] Massachusetts TB-fluoroscopy patients,[26] Rochester mastitis radiotherapy patients[27] and Nova Scotia TB-fluoroscopy patients.[28]

shown in Figure 5. The contrast with Figure 4 is striking; the dose-response in Figure 5 presents little departure from linearity in each of the four studies. As with tobacco smoking and cancers of the lung and esophagus, the contrast between the radiation risks for breast and stomach cancer indicate that even for a single agent, the shape of the dose-response curve for cancer at different organs may be different. Risk estimation needs to be organ-specific.

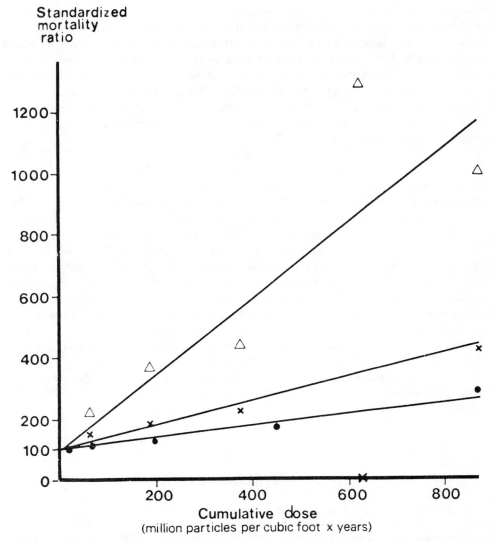

FIGURE 6.   Dose-response curves for lung cancer[23] related to asbestos exposure in New Jersey asbestos production and maintenance workers[29] and Quebec miners and millers.[30]

D. Asbestos Exposure

Asbestos exposure causes large increases in risk for both lung cancer and mesothelioma, with smaller increases in risk seen at other sites.[22] As with many industrial exposures, quantitative estimates of risk are made difficult by the scarcity of accurate measures of exposure. Many of the cohorts exposed to asbestos in which an excess has been demonstrated have no relevant exposure measures. Furthermore, methods of measuring fiber concentrations have changed over time. Acheson and Gardner[23] review the available quantitative data, those relating lung cancer to asbestos being summarized in Figure 6. Although the slopes of the dose-response curves are different in the three groups, each cohort yields an approximately linear dose-response with no evidence of a threshold. For mesothelioma, the quantitative data available are even fewer. Acheson and Gardner conclude "a 2% excess mortality from all asbestos related cancer may be associated with any dose within about a fivefold range".[23]

IV. DISCUSSION

The purpose of this chapter has been to display the type of results which epidemiology has produced. On many occasions, it is clear, no reliable information on exposure levels will be available, but the situations described earlier permit a few conclusions. First, for none of the exposures for which adequate information is available, is there any indication of a threshold dose. Second, although a wide range of dose-response curves has been observed, e.g., exponential, quadratic, linear-quadratic, linear, square root, most of them would be reasonably well approximated by a linear dose-response curve at low dose levels. That is, linear extrapolation from the lowest observed dose level to zero dose would give similar estimates of risk as those provided by the functional form used to fit the observation (square root, exponential, etc.). Thirdly, the assumption often made that epidemiology is an insensitive method of identifying risk needs qualification. In fact, for the situations discussed in this paper epidemiology would be more sensitive than, for example, long-term carcinogenicity bioassays. In most of these examples, the dose level at which an absolute excess risk of 1% occurs lies within the range of doses at which an excess is observed. That is, the 1% point can be determined by interpolation, without need for extrapolation. By contrast, very few experiments are designed to detect excess risks of this order of magnitude. The reason of course is evident; epidemiology at times is studying much larger numbers of individuals. For none of the four exposures considered, alcohol, tobacco, ionizing radiation, and asbestos, is it conceivable that animal carcinogenicity experiments would be performed capable of delineating so clearly the nature of the dose-response relationship.

In conclusion, for situations where large enough groups of individuals have been exposed to documented levels of the relevant agent, coherent dose-response curves can be estimated for the range at which exposure occurs, using either case-control or cohort study methods. Further development of dose-response estimation procedures as the necessary epidemiological information becomes available for a wider range of agents will provide a more solid empirical foundation on which to base models of human risk.

REFERENCES

1. **Breslow, N. E., Lubin, J. H., Marek, P., and Langholz, B.,** Multiplicative models and cohort analysis, *J. Am. Stat. Assoc.,* 78, 1, 1983.
2. **Breslow, N. E. and Day, N. E.,** Statistical Methods in Cancer Research, Vol. I., The Analysis of Case-Control Studies, International Agency for Research on Cancer, Lyon, *IARC Scientific Publications* No. 32, 1980.
3. **Cornfield, J.,** A method of estimating comparative rates from clinical data. Applications to cancer of the lung, breast and cervix, *J. Natl. Cancer. Inst.,* 11, 1269, 1951.
4. **Doll, R. and Hill, A. B.,** A study of the aetiology of carcinoma of the lung, *Br. Med. J.,* 2, 1271, 1952.
5. **Doll, R. and Peto, J.,** Mortality in relation to smoking: 20 years' observations on male British doctors, *Br. Med. J.,* 2, 1525, 1976.
6. **Doll, R. and Peto, J.,** Cigarette-smoking and bronchial carcinoma: dose and time relationships among regular smokers and life-long non-smokers, *J. Epidemiol. Commun. Health,* 32, 303, 1978.
7. **Day, N. E.,** Time as a determinant of risk in cancer epidemiology: the role of multi-stage models, *Cancer Surveys,* 2, 577, 1983.
8. Surgeon General, The Health Consequences of Smoking: The Changing Cigarette, U.S. Department of Health and Human Services, Washington, D.C., 1981.
9. **Doll, R.,** The age distribution of cancer: implications for models of carcinogenesis, *J. Roy. Stat. Soc. A.,* 64, 133, 1971.
10. **Kahn, H. A.,** The Dorn study of smoking and mortality among U.S. Veterans. Report on eight and one-half years of observation, *Natl. Cancer Inst. Monogr.,* 19, 1, 1966.
11. **Hirayama, T.,** Non-smoking wives of heavy smokers have a higher risk of lung cancer: a study from Japan, *Br. Med. J.,* 282, 183, 1981.
12. **Trichopoulos, D., Kalandidi, A., Sparros, L., and MacMahon, B.,** Lung cancer and passive smoking, *Int. J. Cancer,* 27, 1, 1981.

13. **Tuyns, A. J., Pequignot, G., and Jensen, O. M.,** Le cancer de l'oesophage en Ille-et-Vilaine en fonction des niveaux de consommation d'alcool et de tabac. Des risques qui se multiplient, *Bull. Cancer,* 64, 45, 1977.
14. **Day, N. E. and Muñoz, N.,** Cancer of the esophagus, in *Cancer Epidemiology and Prevention,* Schottenfeld, D. and Fraumeni, J. F., Jr., Eds., W. B. Saunders Company, Philadelphia, 1982, 596.
15. **Wynder, E. L. and Bross, I. J.,** A study of etiological factors in cancer of the esophagus, *Cancer,* 14, 389, 1961.
16. **Wynder, E.,** Epidemiological contributions to the prevention of cancer, in *Analytic and Experimental Epidemiology of Cancer,* Nakahara, W., Hirayama, T., Nishioka, K., and Sugano, H., Eds., University of Tokyo Press, Tokyo, 1973, 3.
17. **Thomas, D. C. and McNeill, K. G.,** Risk Estimates for the Health Effects of Alpha Radiation, Atomic Energy Control Board, Ottawa, Canada, Research Report, September 1982.
18. **Kato, H. and Schull, W. J.,** Studies of the mortality of A-bomb survivors. VII. Mortality, 1950-1978. I. Cancer mortality, *Radiation Res.,* 90, 395, 1982.
19. **Smith, P. and Doll, R.,** Mortality among patients with ankylosing spondylitis after a single treatment course with x-rays, *Br. Med. J.,* 284, 449, 1982.
20. **Day, N. E. and Boice, J. D., Jr., Eds.,** Second Cancer in Relation to Radiation Treatment for Cervical Cancer. Results of a Cancer Registry Collaboration, International Agency for Research on Cancer, Lyon, *IARC Scientific Publications* No. 52, 1983.
21. **Boice, J. D., Jr., Land, C. E., Shore, R. E., Norman, J. E., and Tokunaga, M.,** Risk of breast cancer following low-dose radiation exposure, *Radiology,* 131, 589, 1979.
22. **Hammond, E. C., Selikoff, I. J., and Seidman, H.,** Asbestos exposure, cigarette smoking and death rates, *Ann. N.Y. Acad. Sci.,* 330, 473, 1979.
23. **Acheson, E. D. and Gardner, M. J.,** Asbestos: scientific basis for environmental control of fibres, in Biological Effects of Mineral Fibres, Vol. 2, Wagner, J. C., Ed., International Agency for Research on Cancer, Lyon, *IARC Scientific Publications* No. 30, 1980, 737.
24. **Baker, R. J. and Nelder, J. A.** *The GLIM System, Release 3,* Numerical Algorithms Group, Oxford, 1978.
25. **Tokunaga, M., Norman, J. E., Jr., Asano, M., Tokuoka, S., Ezaki, H., Nishimori, I., and Tsuji, Y.,** Malignant breast tumors among atomic bomb survivors, Hiroshima and Nagasaki, 1950 to 1974, *J. Natl. Cancer Inst.,* 62, 1347, 1979.
26. **Boice, J. D., Jr., Rosenstein, M., and Trout, E. D.,** Estimation of breast doses and breast cancer risk associated with repeated fluoroscopic chest examinations of women with tuberculosis, *Radiat. Res.,* 73, 373, 1978.
27. **Shore, R. E., Hempelmann, L. H., Kowaluk, E., et al.,** Breast neoplasms in women treated with X-rays for acute postpartum mastitis, *J. Natl. Cancer Inst.,* 59, 813, 1977.
28. The Effects on Populations of Exposure to Low Levels of Ionizing Radiation, Report of the Advisory Committee on the Biological Effects of Ionizing Radiations, National Academy of Science, Washington, D.C., 1972.
29. **Enterline, P. E., DeCoufle, P., and Henderson, V.,** Mortality in relation to occupational exposure in the asbestos industry, *J. Occup. Med.,* 14, 897, 1972.
30. **McDonald, J. C., Liddell, F. D. K., Gibbs, G. W., Eyssen, G. E., and McDonald, A. D.,** Dust exposure and mortality in chrysotile mining, 1910-75, *Br. J. Ind. Med.,* 37, 11, 1980.

Chapter 2

ASSESSMENT OF HUMAN EXPOSURE TO ENVIRONMENTAL CONTAMINANTS WITH SPECIAL REFERENCE TO CANCER

H. F. Kraybill

TABLE OF CONTENTS

I. INTRODUCTION

Increasing awareness of the potential contribution of environmental factors and lifestyle in chronic disease causation has focused attention on the need to delineate the role of environmental etiologic agents. These would include atmospheric and aquatic pollutants which have additive or promotional effects on other causative agents such as contaminants in food, smoking, alcohol, and workplace exposures. In terms of cancer, estimates have been made on their relative contribution to the total cancer burden, to name one specific disease entity. State and national legislative bodies have mandated control of recognized and suspected carcinogens and, by association, mutagens, as potentially hazardous agents in air, water, and food; the elimination of drugs that have unanticipated long-term hazardous effects in current population or future progeny; the reduction or mitigation of workplace exposures that have proven carcinogenic; and other adverse effects in chronic disease induction. In this chapter, however, attention will be focused on those contaminants arising in our raw water supplies or those developed in water treatment or processing for drinking water in our municipalities.

The passage of the Safe Drinking Water Act of 1974 (PL93-523) in the U.S. was accelerated by an awareness of the fact that some of these contaminants — inorganic and organic chemicals plus radioactivity — in drinking water could pose a threat to human health. In late 1975 the U.S. Environmental Protection Agency (EPA) was directed under the powers of the Act to request the National Academy of Sciences to conduct a study of the adverse health effects attributable to these carcinogenic and/or noncarcinogenic contaminants in drinking water. The goal of this legislation was to provide a safe drinking water which would protect the health of the people, to the extent that it is feasible, taking costs and other factors into consideration. As an outgrowth of this legislation certain studies were required to identify and understand to a greater degree the level of human exposure to these contaminants or biorefractories (chemicals formed in water treatment or not removed by traditional technologies for processing), and the significance of human exposure to these by-products from the water chlorination process. The report issued by the Advisory Committee of the National Academy of Sciences (Safe Drinking Water Committee) should serve as a basis for revision of national interim drinking water regulations in accordance with the timetable presented in the Safe Drinking Water Act.[1]

The ultimate supply of drinking water depends to a great extent on the source or quality of water which is ultimately processed at the treatment plant. Many water sources are either ground water or surface water (rivers, lakes, etc.). The latter, which may be recycled many times and on which more dependence is placed as water resources, becomes critical. In surface water there is, in addition to industrial and municipal sewage effluents, agricultural runoff containing agricultural chemicals and, typically, a substantial amount of natural humic type organic chemicals, the latter giving rise to formation of new molecular species as developed in water chlorination. Cotruvo[2] estimates that over 200 million people in the U.S. drink chlorinated water since chlorination is required in most instances to achieve a drinking water free of pathogenic organisms which lead to many waterborne disease vectors in the past century.

In most instances, except for unusual pollution, the toxic contaminants (either noncarcinogenic or carcinogenic in nature) occur in low concentrations in the parts per billion (ppb) or trillion (ppt) range, but man is exposed continuously to such low doses over a long period of time. These agents can exert an effect either singly, as is the case of single inorganic cations in episodes of cardiovascular disease, or can act additively or synergistically as a multiple exposure. Modern chemical laboratory instrumentation using gas liquid chromatography (GLC), high pressure liquid chromatography (HPLC), and mass spectrometry (MS) facilitate the identification of an ever-increasing list of contaminants in the ($\mu g/\ell$) range.

According to the EPA there are 40,000 municipal water systems providing drinking water as well as 200,000 water supply systems for nonresidential locations.[3] Many of these systems are not equipped to treat water from contaminated sources but are designed to disinfect relatively pure water from selected sources. At the turn of the century most concern was centered upon processes which removed viruses, bacteria, and certain parasites. Sedimentation, filtration, and disinfection have virtually eliminated these potential disease vectors. However, as alluded to earlier, the human population exposure to inorganic and organic chemicals and, in some cases to radioactivity, has introduced the newer problems for public health concern. Consequently, many reports and comprehensive papers have appeared in recent years to provide some perspective on these various drinking water contaminants.[4-8]

Having identified the various contaminants, their frequency of distribution, geographic origin, and in some instances having quantitated the level of concentration with indications as to possible human exposure, the ultimate task is to assess the potential public health significance of such exposures in terms of lifetime risks. To achieve such an objective, the most obvious approach is to resort to epidemiological studies on population groups at risk and to buttress such studies with suggestive evidence on data from experimental in vivo and in vitro studies. These latter studies may indeed, along with leads from epizootics (toxicity and tumorigenicity in marine animals), provide the foundation for more constructive epidemiological pursuits. The main purpose then for such an assessment of the adverse health effects of these micropollutants is to provide a basis for intervention.

II. SOURCES OF CONTAMINANTS

Beyond the problem of bacteriological and parasitic contamination of water supplies, which are now easily destroyed and controlled, are the current concerns over contaminants of natural origin and those that are inadvertently added through man's lack of attention over the years to what was happening relevant to pollution of the oceans, rivers, lakes, streams, and underground aquifers that direct pollutants in the soil to wells which provide ground water supplies to communities. The view has been held that if a liter of water completely free of organic or inorganic contaminants were technologically attainable it would be prohibitively expensive. For example, natural waters, remote from industrial and population centers where pollution can be expected, contain molecular constituents or cations and anions of geologic origin. This of course, would include water supplies that course through deposits of uranium ores, and water supplies containing the leachings of radioactive chemicals from uranium mine tailings. Some water supplies are heavily loaded with inorganic cations which will be subsequently discussed in terms of their health significance.

A. Natural Origin
1. Asbestiform Materials
Asbestiform materials or asbestos particles have been found in many river systems at concentrations in excess of 10 μg/gal.[9,10] Concern exists about asbestos because its carcinogenic activity as inhaled particles is well documented in human studies, but the carcinogenic capacity of asbestos in its multiple mineralogical forms when ingested in foods, beverages, and water, remains to be assessed although studies have been undertaken on asbestos in water to ascertain whether it crosses the gastrointestinal mucosa and enters the bloodstream. These studies in which hamsters were exposed to asbestiform particles by ingestion have been completed and no significant treatment-related increase in the incidence of tumors in animals was demonstrated.[11] The Lake Superior problem resulted from taconite tailings containing mostly amphiboles (cummingtonite-grunerite) dumped into the Lake which supplied water to Duluth, Minnesota. Chrysotile was present in western Lake Superior at levels less than 1×10^6 fibers per liter.[12] While this was a man-made problem of pollution, there

Table 1
CONCENTRATIONS OF ASBESTIFORM MATERIALS IN U.S. WATERWAYS

Mineral	Location	Fiber count Fibers/ℓ	Fiber mass
Chrysotile	W. Lake Superior	$< 1 \times 10^6$	
Chrysotile	Thunder Bay, Ontario	0.8×10^6	2×10^{-4} µg/ℓ
Amphibole[a]	Duluth, Minn. water	1 to 30×10^6	30 µg/ℓ average
Amphibole	Grand Marais, Minn.	—	< 0.02 mg/ℓ
	Beaver Bay, Minn.	—	< 0.10 mg/ℓ
	Two harbors, Minn.	—	< 0.10 mg/ℓ
Amphibole	All locations (peak concentrations)	1×10^9	
Chrysotile	Rivers in Pennsylvania, Vermont, Massachusetts, and Connecticut	Range from 0.7 to 4 µg/ℓ	

[a]   Measurement of 1×10^8 fibers/mg < 2 µg of tailings.

Kraybill, H. F., *Progress in Experimental Tumors*, 20, 3, 1976.

are some locations near asbestos mines where the levels of asbestos fibers in water are quite high. For example, San Francisco, California (Marin County) had counts that approached 2×10^6 fibers per liter.[13] Andrew[14] reported that 31 streams in Minnesota and Wisconsin had no detectable concentration of amphiboles.

Some concentrations of asbestiform materials in some waterways in the U.S. are shown in Table 1.[8,9,14,15]

2. Radioactive Components (Radionuclides)

While everyone is exposed to some natural radiation from cosmic rays and terrestrial sources (the average background dose being 100 mrem/year), a small proportion comes from drinking water that contains radionuclides. The primary natural isotopes of interest include radium, uranium, and radon. The largest contribution to radioactivity in water is potassium-40. Fallout radionuclides such as ^{90}Sr and ^{137}Cs contribute some, plus ^{228}Ra, and a hypothetical typical water supply would have less than 10% of the annual background radiation from this source. It is estimated that 1 million people are exposed to radium levels greater than 3 pCi/ℓ and about 120,000 people are exposed to radium exposure levels greater than 9 pCi/ℓ.[1]

The radium content of fresh surface water can range from 0.01 to 0.1 pCi/ℓ with some ground waters escalating to as high as 100 pCi/ℓ.[1] Drinking water from surface supplies does not contain significant amounts of radium and for those that do, processing of water with flocculation and water softeners can remove the bulk of radium. Health effects of ^{228}Ra and radon are being studied since the information on ^{226}Ra seems adequate.[45] The average beta activity (^{3}H, ^{40}K, ^{90}Sr, ^{137}Cs, and ^{228}Ra), excluding tritium measured on interstate water supplies was 3.1 pCi/ℓ and the gross alpha activities for the same samples was less than 2 pCi/ℓ in all but one case.[16] The alpha activity comes from ^{228}Ra and ^{226}Ra and their daughter isotopes.

In a 1974 spectrometric analysis of Los Angeles drinking water composites, the probable doses over a 50-year span, at an intake of 2 ℓ/day, were constructed and are shown in Table 2. This study was done by Goldberg and reported on by the National Academy of Sciences.[1] While the biologic effects of radiation could produce teratogenic effects on the fetus, genetic disease, and somatic damage (primarily carcinogenic), most of these changes from back-

Table 2
POTENTIAL 50-YEAR CONSTANT INTAKE AND
DOSE RECEIVED FROM RADIONUCLIDES IN
DRINKING WATER

Radionuclide	Concentration (pCi/ℓ)	Potential doses — mrem/yr	
		Total body	Organ
^{40}K	5.20	0.09	
^{54}Mn	0.48	0.004	0.005 (GI tract)
^{90}Sr	0.09	0.0002	0.4 (Bone)
$^{95}Zn — ^{95}Nb$	0.61	0.000003	0.1 (GI tract)
^{137}Cs	0.05	0.003	
^{144}Ce	0.21	0.000004	0.03 (GI tract)
^{226}Ra	0.05	0.014	0.8 (Bone)

Note: Bone dose = 2 mrem/yr; all tissues = 0.12 mrem/yr.

Adapted from data of Goldberg as shown in Reference 1.

ground radiation (soils, building structures, foods, atmosphere, and water) are not generally increased enough in exposure for changes to be detected. The radioactive contamination in water would account for less than 1% of the incidence of cancers that could be attributed to natural background of radiation, the latter uncontrollable and a reality of life.[1] In some geographic regions where radioactivity levels are high such levels could cause an increased incidence of bone cancer from ^{226} Ra, ^{228} Ra and ^{90}Sr.[48] The standard or MCL for gross alpha particle activity is 15 pCi/ℓ .

3. Natural Marine Oil Seepage

A source of PAH (polycyclic aromatic hydrocarbon) pollution of water can occur through oil seepage from the ocean floor. Wilson et al.[16] have calculated that the probable range of seepage into the marine environment is 0.2×10^6 to 6×10^6 t/year. Only a few areas around the world are as seepage-prone as southern California. The Pacific Ocean appears to have more areas of high seepage potential, constituting about 40% of the world's total seepage. These PAH compounds can contaminate marine animals and thus the food chain is affected by such natural pollution. Of course, spillage of oil from tankers and other petroleum discards, and creosote applied to pilings in ship mooring places contaminate barnacles and shellfish.[17]

4. Inorganic Chemicals

Trace metals, or cations and anions have been found in ground water or surface water used for municipal water supplies. These inorganic solutes are associated with either natural processes or man's activities. A listing of these inorganic solutes is shown in Table 3. These chemicals have been identified through various studies or surveys.[18-20] As to the natural process, weathering of rocks, soil leaching, acid mine drainage, and other processes contribute to the contamination. All of this is dependent upon absorption characteristics, hydration, colloidal dispersion, coprecipitation, and hydration. As will be described later, the decaying of vegetation and breakdown of humus materials release inorganic solutes and organic compounds.

Ultimately, the processing of raw surface or ground water is undertaken to make water acceptable for consumption by removal of these trace metals or solutes. Of course, some treatment processes add chemicals such as alum lime, iron salts, chlorine, fluorine, and

Table 3
INORGANIC CONSTITUENTS DETECTED
FROM WATER SURVEYS

Cations

Aluminum	Indium	Scandium
Antimony	Iron	Selenium
Arsenic	Lanthanum	Silica
Barium	Lead	Sodium
Beryllium	Magnesium	Strontium
Bismuth	Manganese	Tantalum
Cadmium	Mercury	Tellurium
Calcium	Molybdenum	Thallium
Cesium	Nickel	Tin
Chromium	Palladium	Titanium
Copper	Platinum	Tungsten
Gallium	Potassium	Uranium
Germanium	Radium	Vanadium
Gold	Rubidium	Zinc

Anions and Gases

Bicarbonate	Fluorides	Phosphates
Bromides (bromine)	Iodides (iodine)	Sulfates
Chlorides (chlorine)	Nitrates	

From numerous surveys; Community Water Supply Survey, National Stream Quality Accounting Network and Rural Water Survey.[18-20]

others. The impurities in these chemicals may, of course, add some trace metals. Pipes which are made from iron, steel, copper, plastic, and asbestos cement, and used to transport water, add to the contamination. Also metals in pumps, valves, and small pipes add other types and amounts of trace contaminants. Finally, the corrosion process may account for small amounts of trace metals with which the water comes in contact. Kopp[21] has shown the frequency of detection and concentrations of trace metals in finished waters in the U.S. and these are shown in Table 4.

B. Agricultural Practices
   In the U.S. the agricultural industry has become quite dependent on the use of chemical fertilizers and pesticides. The literature is replete with many reports on the proliferation of these chemicals in the environment. Water resources reflect this great use by the identification of these inorganic and organic chemicals in the raw water and drinking water supply, national waterways, lakes, and oceans. No water resource escapes this contamination. Some of the contamination occurs through direct dumping of these agricultural chemicals into waterways, some from municipal sewage effluents and, in farming areas, by soil erosion and so-called agricultural runoff from heavy rains and flooded areas. Some contamination occurs from aerial drift and aerial spraying operations.
   It is impossible in this brief account to catalog all the inorganic and organic agricultural chemicals that contaminate the water resources on a worldwide basis since there are voluminous reports on identification and monitoring of these chemicals in the aquatic environment including the deposition of some of these chemicals in marine animals that are subsequently consumed by man. A listing of some of these organic chemicals in water that are carcinogenic, or potentially carcinogenic, will be presented later in this chapter.

Table 4
FREQUENCY OF DETECTION AND
CONCENTRATIONS OF SOME TRACE METALS
IN 380 FINISHED WATERS IN THE U.S.

Element	Frequency of detection	Concentration (μg/ℓ)		
		Minimum	Maximum	Mean
Zinc	77. 0	3.0	2010.0	79.2
Cadmium	0.2	12.0	12.0	12.0
Iron	83.4	2.0	1920.0	68.9
Manganese	58.7	0.5	450.0	25.5
Copper	65.2	1.0	1060.0	43.0
Silver	6.1	0.3	5.0	2.2
Lead	18.1	3.0	139.0	33.9
Chromium	15.2	1.0	29.0	7.5
Barium	99.7	1.0	172.0	28.6
Molybdenum	29.9	3.0	1024.0	85.9
Aluminum	47.8	3.0	1600.0	179.1
Beryllium	1.1	0.02	0.17	0.1
Nickel	4.6	1.0	490.0	34.2
Cobalt	0.5	22.0	29. 0	26.0
Vanadium	3.4	14.0	222.0	46.1

From Kopp, J. F., *Proc. 3rd Annu. Conf. Trace Substances Environmental Health,* Hemphill, D. D., Ed., University of Missouri, Columbia, 1970, 59.

C. Industrial Effluents

Industrial pollution of water resources has been one of the major problems of concern for those in environmental health programs. The Environmental Protection Agency has developed rigid standards under the Safe Drinking Water Act of 1974 to control reckless and unconcerned dumping. Episodes such as Minamata disease in Japan[22] from mercury dumping in Minamata Bay and formation of toxic methyl mercury taken up by foodfish, the numerous fish kills in rivers in the U.S.,[23] and the Love Canal problem[24] are a few of the grim results of irresponsible dumping of chemicals into waterways or effects of chemicals entering the water through percolation through the soil.

Petroleum spillage, as mentioned previously, results in major contamination. Walker and Colwell[25] have observed that in a heavy-metal-enriched environment, oil-soluble heavy metals, especially mercury, that concentrate in the oil through mercury-resistant microorganisms present in the oil layer are capable of degrading the petroleum. This is significant not only for transfer of mercury and oil in the food chain but because it becomes an important mechanism for degradation of PAH compounds, especially those that are carcinogenic. Not all hydrocarbons degrade readily, as for example benzene, which will not support the growth of microorganisms, whereas naphthalene is more susceptible to degradation.[26] In the Chesapeake Bay in Maryland, *Claudosporum penicillia*, *C. alternaria*, and *C. trichoderma* were found as oil degrading fungi. The ultimate seeding of waters with petroleum-degrading fungi or bacteria may offer one solution for decontamination.

If chemicals are discharged into waterways solubilized in solvents there may be some miscibility. The strongest possibility is that they may precipitate out of solution. Some examples here are some of the potent carcinogenic aromatic amines. Benzidine used in the dye industry has had a standard set by EPA regarding discharge of this chemical into waterways.[27]

Table 5
CARCINOGENIC CONTAMINANTS IN WATER THAT ARE RELATED TO ENERGY TECHNOLOGIES

Compound	Carcinogenic class	Compound	Carcinogenic class
Acrylonitrile	1	Ethane, hexachloro-	2
Benzene	1	Ethane, 1,2-dichloro-	1
Benzenamine, 2,4-dimethyl-	2	Ethylene, 1,1-dichloro-	1
Benzofluoranthene, 10,11-	2	Ethylene, 1,1,2,2-tetrachloro-	2
Benzofluoranthene, 11,12-	2	Chloroform	1
Benzopyrene 3,4	1	Carbon tetrachloride	1
Bis(2-chloroethyl) ether	2	Methane, iodo-	1
Bis(chloromethyl) ether	1	Indeno(1,2,3-c,d) pyrene	2
Bis(chloroethyl) ether	2	Vinyl chloride (chloroethylene)	1
Dioxane 1,4	1		

Note: Classification as to carcinogens based on evaluation in International Agency for Research on Cancer monographs and other sources: class 1 = carcinogenic in at least two animal species or in one animal species and confirmed in human epidemiology; class 2 = carcinogenic in one animal species.

Hushon, J., Clerman, R., Small, R., Sood, S., Taylor, A., and Thoman, D., Contract Report to Department of Energy/National Cancer Institute, 1980.

Another potent carcinogen, vinyl chloride, detected in the atmosphere miles away from a plant, has also been detected in water. This suggests that not only may marine animals or foodfish pick it up but it has been identified in drinking water as a carcinogenic contaminant which will be discussed later. The halo ethers, some of which are well characterized carcinogens such as BCME (*bis*-chloromethyl ether), have also been found in waterways and are obviously the end result of industrial pollution.[28,29]

D. Energy Technologies
In recent years, due to the shortages of petroleum resources in the U.S., an accelerated research and development program by the Department of Energy and industries was undertaken to provide for alternative resources. Solar technologies have been advanced and geothermal technologies considered and, as to the latter, radionuclides as pollutants in water might be considered. These have been discussed previously. Fossil fuel technologies such as petroleum recovered from shale oil and coal conversion projects, specifically the synfuels program, lead to potential resources for expanded energy supplies. These technologies are considered from the stage of mining or drilling for the resource through product conversion. These advanced coal combustion technologies and extraction processes, especially where effluents pass into the waterways and the atmosphere, are expected to yield pollutants in the raw water supplies where drinking water may be derived. These pollutants must be determined, and the extent of potential exposure assessed.

These pollutants could be associated with potential hazards to aquatic life and ultimately the potential hazard and toxicity for man including a carcinogenic exposure for the human population. The Department of Energy has given major attention to assessment of the biomedical effects. Table 5 lists some organic chemicals that are classified as carcinogenic which have been identified in water from energy technology activities.[30]

E. Effect of Water Processing or Treatment
Many of the organic compounds identified in drinking water are either formed as new molecular species through the chlorination process or are increased in concentration above

Table 6
CONTAMINANTS MONITORED IN DRINKING
WATER IN 113 U.S. CITIES

	Range in Concentration (μg/ℓ)		
Compound	Low	High	Average
Chloroform	0.002	470.0	83.5
Bromodichloromethane	0.02	180.0	17.5
Dibromochloromethane	0.05	290.0	11.9
Bromoform	0.15	280.0	4.2
1,1,2-Trichloroethylene	0.015	49.0	0.535
Carbontetrachloride	0.10	10.0	0.313
1,2,-Dichloroethane	0.02	1.8	0.346
Benzene	0.05	1.8	0.094
p-Dichlorobenzene	0.002	1.6	0.026
Vinyl chloride	0.05	0.18	0.052
1,2,4-Trichlorobenzene	0.002	0.58	0.008
Bis(2-chloroethyl)ether	0.005	0.36	0.016
2,4-Dichlorophenol	—	—	—
Polychlorinated biphenyls	0.05	0.20	0.07
Pentachlorophenol	0.007	0.7	0.008

Cotruvo, J., *Proc. Toxicol. Forum*, 1981, 247.

the value in raw water because of this treatment process. However, these organic contaminants constitute only a small percentage of total organic matter present in water. While many of the volatile organic compounds in drinking water have been identified and quantified, these represent no more than 10%, on a quantitative basis, of the total organic material. Of the nonvolatile fraction comprising the remaining 90% of the total organic matter in water, only 5 to 10% have been identified.

Worldwide, a 1976 listing of 2221 contaminants was compiled from survey data to represent an impressive group of contaminants in raw water. A 1980 listing of 1200 chemicals was prepared by EPA representing those present in drinking water[31,32] as organic contaminants. More recently the EPA has indicated that this list now comprises 1800 to 2000 contaminants.

For the last 80 years the chlorination process has been used to provide the elementary protection against pathogens occurring in drinking water. However, this essential function of disinfection has contributed another risk through formation of the by-products of the reaction of chlorine. Accordingly, the potentially negative side effect confronting public health authorities is whether these contaminants contribute an overall carcinogenic and/or mutagenic hazard through continuous exposure to these chemicals in drinking water.

F. Organic and Inorganic Distribution

The most commonly found and the one of the highest concentration in drinking water in the U.S. is chloroform and next are the other trihalomethanes — bromoform, dibromochloromethane, dichlorobromomethane. These, with the exception of chloroform, do not normally occur in drinking water unless the water was chlorinated at the treatment plant. Current water quality standards are oriented around the trihalomethanes and risk assessments are made as to relative safety of the drinking water in terms of concentration of chloroform. Table 6 lists the most significant of the contaminants identified and quantified in drinking water from a survey of 113 cities.

Chloroform, a trihalomethane, is ubiquitous in that it occurs in the atmosphere and food

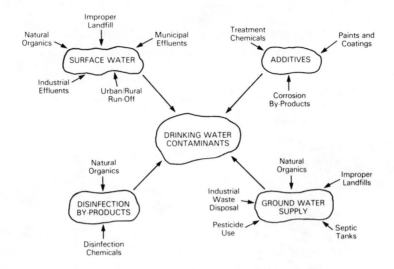

FIGURE 1.   Sources of contaminants in drinking water.

supply as well as water. For example, the uptake of chloroform in atmosphere for an adult is 204 mg/year (36%); for food supply it is 16 mg/year (3%); and for intake in water it is 343 mg/year, or 61% of total intake.[2]

The sources of organic contaminants in drinking water are shown graphically in Figure 1.

III. FREQUENCY AND GEOGRAPHIC VARIATIONS IN DISTRIBUTION OF CONTAMINANTS

The assessment of the potential health hazards from the presence of organic and inorganic pollutants in lakes, waterways, and estuaries, which ultimately distribute into coastal waters and the oceans, is complex. Contrary to common beliefs, the ocean is not a repository where dilution is expected to negate the effect of toxic agents on the marine environment and marine animals. Elias[33] has indicated that despite the normal expanse of the ocean area, only 10% is relevant for propagation and production of edible marine animals, and coastal and estuary waters are quite vulnerable to the discarded environmental pollutants. These contaminants build up in the food chain. Attempts at depuration of shellfish from contaminants are useful but cannot be predicted on a large scale.

A. Raw Water vs. Processed Water

Several types of surveys of raw and finished water supplies have been made by federal and state agencies in the U.S. This effort has been intensified in 1975 as described by Kraybill.[5-7] The concentration of contaminants in water sources varies and is dependent on the origin of the raw water to be processed to drinking water, which in turn is associated with the pathways of pollution. The extent of such contamination can be illustrated as shown in Table 7 by the relative concentrations of benzo(a)pyrene (BAP) in various types of water which could be the basic resource for production of municipal drinking water.[34] While some years ago pristine water supplies such as wells, natural springs, remote lakes, and mountain streams — relatively uncontaminated — could be the major sources of water for towns, villages, and cities, today with the great demand for water (industrial and residential) the water resources are limited, and more and more municipalities will have to depend on surface water and/or recycled surface water which would include types 2, 3, and 4 in Table 7.

Table 7
CARCINOGENIC POLYCYCLIC AROMATIC
HYDROCARBONS (BENZO(A)PYRENE) IN VARIOUS
WATERS

Type of water	Concentration BAP ($\mu g/m^3$)
Ground water	1 — 10
Clean or sparingly polluted river or lake water	10 — 50
Moderately polluted surface water	50 — 100
Highly polluted surface water	100 — 1000
Waste water	< 100,000 or > 100,000

From Borneff, J and Kunte, H., *Arch. Hyg. Bakteriol.*, 148, 585, 1964.

The trihalomethanes, especially chloroform, have been used as an indicator of contamination. Reference is made frequently to chloroform to illustrate the effect of chlorination on water to demonstrate that this process elevates the level of some contaminants and forms new molecular species. Values for levels of chloroform from 0.1 to 311 $\mu g/\ell$ (some escalation to 500 $\mu g/\ell$) with a median of 20 $\mu g/\ell$ have been recorded. Seldom detected in high quality raw water, the interaction of chlorine (used for disinfection) with the humic substances found in raw water escalates the transformation to the trihalomethanes.

While no one would question the necessity for disinfection with chlorine, and whereas treatment technologies remove many unwanted contaminants, organic and inorganic, the transformation mechanism as described has introduced a new problem. These formed organic contaminants can be subsequently removed by use of GAC (granulated activated charcoal) but this is not required by the EPA; however, it is being used in some locations. Costwise, according to Cotruvo,[2] the cost of chlorination with chlorine is 0.5 to 1.0 cents/1000 gal, for chlorine dioxide 1.5 to 2.0 cents/1000 gal, and for activated carbon, about 10 to 20 cents/1000 gal.

B. Organic and Inorganic Distribution

Woodwell et al.[35] reported that even if DDT production in the world was terminated in 1974, by the year 2000 there would still be 1 ppt in the water and the lower atmosphere. While DDT has been banned in the U.S., there is still a residue carried forth that prevails in foods and water worldwide and there are finite levels of DDT/DDE in mother's milk as shown in a survey in Michigan relevant to a study for assessment of PBB (polybrominated biphenyls) in the human population.[36] In that remote part of the world called Antartica, where one would assume there is no vegetation or plant source for DDT application or where water may be pure, penguins have measurable levels of DDT in the body fat. DDT is volatilized from lakes and rivers and carried through the stratosphere to be ultimately precipitated on the Antartica ice cap where the penguin obtains his ingestion of DDT in the water and consumption of other marine life. Peterle[37] has demonstrated DDT in Antartica snow and reported that about 2684 tons of DDT have collected in the snow of Antartica over 22 years. Measurements made in Barbados in front of the Lesser Antilles showed that northeasterly tradewinds could transport some of the pesticides from as far away as Europe or North America.[38]

The variation in concentration of pesticide contaminants in soil and water in different parts of the world is not only influenced by atmospheric migration but also is a reflection of the intensity of agricultural pursuits.

Data on inorganic carcinogenic contaminants have been reported by various investigators. Variations in some inorganic contaminants in water in various locations are illustrated in Table 8 as reported by Bouquiaux.[39]

Table 8
INORGANIC CONTAMINANTS AND THEIR VARIANT
CONCENTRATIONS IN CITY WATER SUPPLIES IN
THE NETHERLANDS AND FEDERAL REPUBLIC OF
GERMANY

Chemical	Location	Concentration ($\mu g/\ell$)
Arsenic[a]	Lindau	6.2
	Düsseldorf	2.4
	Mainz	8.1
	The Hague	1.0
Cadmium	Lindau	2.0
	Mainz	9.0
Chromium[a]	Lindau	10.0
	Düsseldorf	4.5
	Mainz	9.0
Selenium	Lindau	4.9
	Düsseldorf	6.0
	Mainz	3.1

[a]   Valence state unspecified. Some of these inorganic
      chemicals are suspect carcinogens (arsenic, cad-
      mium, and chromium VI). Selenium is a carcinostat.

Bouquiaux, J., Report Commission of the European Communities, 1974.

C. Identification and Classification of Carcinogenic Contaminants in Types of Water Including Drinking Water

Chemical contaminants appear in most all raw water supplies unless streams and rivers are in isolated regions not affected by industrial pollution. In the U.S. a national regional survey was mounted in 1975 to monitor organic contaminants in drinking water.[12] Hushon et al.[30] in a 1979 survey conducted for the National Cancer Institute and the Department of Energy, listed all the organic pollutants in surface water, ground water, drinking water, and those that appear in municipal effluents, industrial effluents, and from oil refining, oil shale, coal conversion processes, and coal mining. The total number of compounds was 4437 but of course, many of these are duplicated since many of the raw water contaminants may be carried through into ground water, surface water, and drinking water. The probable number of discrete chemical entities that are pollutants are estimated to be 2500. These pollutants are listed, by category, in Table 9. In their 1981 listing of water pollutants, the EPA showed 1200 (allowing for duplication in chemical isomers); however, latest estimates by EPA are in the range of 1800 to 2000.

In 1981,[40] in a collaborative study undertaken by the National Cancer Institute, the Department of Energy, Mitre Corporation, and SRI International, the following were listed: (1) 20 recognized carcinogens; (2) 23 suspected carcinogens; (3) 18 promoters or cocarcinogens, and (4) 45 mutagens. A listing of the carcinogens is given in Table 10, the promoters in Table 11 and the mutagens in Table 12. The basis for the classification as carcinogens, promoters, and mutagens (Ames Salmonella test system) is found in various reports, including the monographs of the International Agency for Research on Cancer,[41] the National Cancer Institute's Carcinogenesis Technical Report Series,[42] and other reference sources. The listing of carcinogens, mutagens, and promoters in this 1200-plus chemicals reflect, from a literature survey, only those on which carcinogenicity and mutagenicity studies have been conducted. Obviously, all chemicals have not been bioassayed.

This is an impressive array of recognized and suspect carcinogens, mutagens, promoters,

Table 9
AN INVENTORY OF ORGANIC POLLUTANTS
IDENTIFIED IN VARIOUS TYPES OF WATER

Pollutant category	No. of compounds
Oil refinery-related	370
Oil shale-related	150
Coal process-related	43
Coal conversion-related	488
Pollutants in surface water	781
Pollutants in ground water	154
Pollutants in municipal effluents	406
Pollutants in industrial effluents	1278
Pollutanta (biorefractories in drinking water)	767[a]

[a]   Estimate from EPA 1982 list = 1200.

Categorization by Hushon, J., Clerman, R., Small, R., Sood, S., Taylor, A., and Thoman, D., Contract Report to Department of Energy/National Cancer Institute, 1980.

Table 10
LIST OF RECOGNIZED AND SUSPECTED CARCINOGENS IN
DRINKING WATER

1. Acetamide (R)
2. Acrylonitrile (R)
3. Aldrin (S)
4. Alpha hexachlorocyclohexane (S)
5. Aniline (S)
6. Azobenzene (S)
7. Benzene (R)
8. Benzo(a)pyrene (R)
9. Benzo(b)fluoranthene (R)
10. Benzo(j)fluoranthene (S)
11. Benzo(k)fluoranthene (S)
12. Bis(2-Chloroethyl)ether (S)
13. Bis(chloromethyl) ether (R)
14. Butyl bromide[a] (S)
15. Bromoform (S)
16. Carbon tetrachloride (R)
17. Chlordane (S)
18. Chloroform (R)
19. p-Chloronitrobenzene (S)
20. 1,2-Dichloroethane (R)
21. Dieldrin (R)
22. DDT (R)
23. 4,4' DDE (R)
24. Dimethylaniline (S)
25. 1,4-Dioxane (R)
26. Heptachlor (S)
27. Heptachlor epoxide (S)
28. Hexachlorocyclohexane (S)
29. Hexachlorobenzene (R)
30. Hexachloroethane (S)
31. Hydrazobenzene (R)
32. Indeno (1,2,3-c,d)pyrene (R)
33. Isobutyl bromide[a] (S)
34. Lindane (S)
35. Methyl iodide (R)
36. n-Propyl iodide (S)
37. 1,1,2,2-Tetrachloroethane (S)
38. 1,1,2,2-Tetrachloroethylene (S)
39. 1,1,2-Trichloroethane (S)
40. 1,1,2-Trichloroethylene (S)
41. 2,4,6-Trichlorophenol (R)
42. Vinyl chloride (R)
43. Vinylidene chloride (R)

Note: R = Recognized; S = Suspected

[a]   Secondary and tertiary butyl bromide eliminated from list.

From Helmes, C. T., Sigman, C. C., Malko, S., Atkinson, D. L., Jafer, J., Sullivan, P. A., Thompson, K. L., Knowlton, E. M., Kraybill, H. F., Hushon, S., Thomas, D. P., Clerman, R. S., and Barr, N., U.S. Department of Health and Human Services, 1981.

Table 11
TUMOR PROMOTERS OR COCARCINOGENS

Benzo(g,h,i)perylene	Eicosane
Butylated hydroxytoluene	Fluoranthene
o-Chlorophenol	n-Octadecane
o-Cresol	Oleic acid
n-Decane	Phenol
1-Decanol	Pyrene
n-Dodecane	n-Tetradecane
2,3-Dimethylnaphthalene	n-Undecane
2,4-Dichlorophenol	2,4-Xylenol

From Helmes, C. T., Sigman, C. C., Malko, S., Atkinson, D. L., Jafer, J., Sullivan, P. A., Thompson, K. L., Knowlton, E. M., Kraybill, H. F., Hushon, S., Thomas, D. P., Clerman, R. S., and Barr, N., U.S. Department of Health and Human Services, 1981.

Table 12
LIST OF AMES TEST MUTAGENS IN DRINKING WATER

1. Acrylonitrile
2. Aniline
3. Azobenzene
4. Benzo(a)pyrene
5. Benzo(g,h,i)perylene
6. Bis(2-chloroethyl)ether
7. Bis(chloroisopropyl)ether
8. Bis(chloromethyl)ether
9. 1-Bromo-2-chloroethane
10. Bromomethane
11. Bromochloroethane
12. Bromodichloromethane
13. Bromoform
14. 2,3-Butane dione
15. Butyl bromide
16. Chloral
17. Chlordane
18. 1-Chloropropene
19. 2-Chloropropane
20. Dibromoethane
21. Dibromochloromethane
22. Dichloroacetonitrile
23. 1,2-Dichloroethane
24. 1,2-Dichloropropane
25. 1,3-Dichloropropene
26. 4,6-Dinitro-2-aminophenol
27. 2,6-Dinitrotoluene
28. Fluoranthene
29. Furfural
30. Hexachloro-1,3-butadiene
31. Indeno(1,2-3-c,d)pyrene
32. Methylchloride
33. Methylenechloride
34. Methyl iodide
35. Perylene
36. m-Phenylenediamine
37. Pyrene
38. Styrene
39. 1,1,2,2-Tetrachloroethane
40. 1,1,2,2-Tetrachloroethylene
41. 1,1,2-Trichloroethylene
42. Tris(chloroethyl)phosphate
43. 1,1,1-Trichloroethane
44. Vinyl chloride
45. Vinylidene chloride

From Helmes, C. T., Sigman, C. C., Malko, S., Atkinson, D. L., Jafer, J., Sullivan, P. A., Thompson, K. L., Knowlton, E. M., Kraybill, H. F., Hushon, S., Thomas, D. P., Clerman, R. S., and Barr, N., U.S. Department of Health and Human Services, 1981.

and/or cocarcinogens as contaminants in drinking water. With more and more surveys and with the sophistication of analytical methodology, more and more of these carcinogenic/ mutagenic contaminants will be identified.

IV. EPIDEMIOLOGICAL CONSIDERATIONS AND ASSESSMENT OF POTENTIAL ADVERSE HEALTH EFFECTS

In more recent years, the major emphasis has been on experimental and epidemiological studies on the organic chemicals or contaminants in drinking water. These recent studies have been focused on the possible development of malignant disease or cancer. In an earlier era more emphasis was placed on the qualitative and quantitative aspects of inorganic chemicals and cations in drinking water and the probable association with cardiovascular disease. Since some of the inorganic chemicals are identified as carcinogenic it would seem that great emphasis needs to be placed on studies on the inorganics in water and cancer.

A. Cancer Epidemiology — Organic Contaminants

Since epidemiological studies could provide the clues toward establishment of a positive association between cancer incidence and pollutants in drinking water it was inevitable that during 1976 to 1980 a series of epidemiological programs on drinking water would emerge.

Experimental studies on which the current knowledge of toxicity tests are based are translated to effects on man. However, doses and dose rates are much larger than those that correspond to the usual concentrations of toxic materials occurring in drinking water. Thus, there is a problem here of uncertainty in estimating the magnitude of risk to health from ingestion of these contaminants in water. Additionally, one can always anticipate that in terms of health hazard the effects of two or more contaminants or a multiple exposure should be envisioned, i.e., an additive or synergistic effect or perhaps, inhibition, might prevail.

Depending on what literature source is searched, there have been roughly 20 epidemiologic studies. To date, four hypotheses have emerged from these studies; namely that the risk of cancer may be greater in the following situation: (1) surface water compared with ground water; (2) chlorinated water compared with nonchlorinated water; (3) high levels of trihalomethanes (THM) compared with low levels of THM; and (4) water from highly polluted surface water such as the Mississippi compared with surface water from less polluted waters. The THM studies were given greater weight or significance than the other studies.[43] The THM studies suggest association between this "index of contamination" and increased frequency of bladder cancer. The evidence that long-term exposure to organic carcinogens in water results in cancer is highly variable and a statistically significant correlation has not yet been demonstrated.[44] The credibility of the studies is generally questioned for the following reasons: (1) lack of standardized data collection procedures; (2) lack of standardized statistical techniques to draw conclusions from collected data; (3) lack of clear definition of independent variables that must be considered; (4) lack of consideration of confounding variables (studies on smoking and alcoholism); and (5) lack of sophisticated techniques to track migration patterns (population movements are important). An overview or summary on the epidemiology projects addressing drinking water pollutants and cancer is set forth in Table 13.[30]

While experimental and epidemiologic studies provide some suggestive evidence on the potential hazard of these carcinogenic biorefractories, one cannot describe with assurance any causality for cancer. However, the probability remains that these ingested water contaminants may contribute to the total cancer burden.

B. Inorganics/Radioactivity in Water

Previous reference has been made to the radionuclides in water. Relevant to the background radiation to the U.S. population which is 100 mrem/year, part of this background comes from drinking water. It is estimated that 1/1000 rem/year (0.244 mrem) to total body would be accumulated which is less than 1% of the background. The radiation to the bone from strontium and radium radioisotopes would be about 10% of the total background.[44,45]

Table 13
OVERVIEW AND SUMMARY OF SOME EPIDEMIOLOGY STUDIES REFERENCED TO DRINKING WATER CONTAMINANTS AND CANCER

Geographic location and date		Investigators	Focus of study chemicals/water type
Colorado	1979	F. P. Savage, B. J. McCarthy	Nitrosamines
Miami, Fla. area	1979	H. Enos, et al.	Trihalomethanes
Ohio	1979	D. C. Greathouse	THMs and organics
Washington City, Md.	1979	C. Krause, J. McCabe	Chloroform
Multistate study	1978	K. P. Cantor et al.	Trihalomethanes
Multistate study	1979	T. Mason et al.	Trihalomethanes
7 Counties, N.Y.	1978	M. Alavanja et al.	Chlorinated organics
Cincinnati, Ohio	1979	K. C. Kopfler	PAHs
Multistate study	1979	N. A. Reiches	Halogenated organics
New Orleans, La.	1979	E. S. Hyman	PAHs and halogenated organics
New Orleans, La.	1979	M. S. Gottlieb/F. Mather	Surface vs. ground water
New Orleans, La.	1976	T. Page et al.	Mississippi river water
Missouri	1979	C. J. Marienfeld	Surface vs. ground water
North Carolina	1979	C. M. Shy/R. S. Shrube	Surface vs. ground water
Cincinnati, Ohio	1976	R. A. Harris/N. A. Reiches	Surface vs. ground water
88 Counties, Ohio	1977	R. J. Kuzma et al.	Surface vs. ground water
North Carolina	1979	R. F. Christman/M. Ibrahim	Not mentioned
Pennsylvania	1979	J. B. Andelman/D. C. Greathouse	Not mentioned

From Hushon, J., Clerman, R., Small, R., Sood, S., Taylor, A., and Thoman, D., Contract Report Department of Energy/National Cancer Institute, 1980.

The radiation risks would be of three types, namely, teratogenic, genetic, and somatic.[44] It is suggested that the radiation levels are so small that the measurable effects during gestation from water radionuclides would not be found. Similar evaluations have been made as to the genetic and somatic effects. Insofar as somatic and carcinogenic effects, where total background radiation may be 4.5 to 45 cases of cancer per million depending on risk model used, the contribution of radiation from water other than exposure to the bone would cause a negligible increase in the total cancer burden.[45,46]

While the above is but a capsule version on data available, and these calculations of dose and risk are presented elsewhere, the actual radiation exposure from most water supplies is so small a proportion of the normal background to which all humans are exposed, that it appears almost impossible to measure any of the adverse health effects cited above.[47] This is not to say that where a few water supplies may show higher radium levels (on a case-by-case basis) they should not be evaluated since they could pose a higher risk of bone cancer for those specific population groups.[48]

As to future directions, surveys on water supplies should ascertain the complete distributions of beta and alpha radiation in the counting measurements. The ratio between ^{228}Ra and ^{226}Ra should be measured extensively in ground and surface water where ^{226}Ra is known. The MCL for ^{226}Ra and ^{228}Ra is 5 pCi/ℓ and the MCL for gross alpha particle activity is 15 pCi/ℓ according to EPA. Monitoring for radium and other radionuclides is required every 3 years. Where some water supplies have high radon content, water vapor such as showers, and humidifiers might yield an inhalation exposure that constitutes a hazard, and thus this measurement should be made and the hazard estimated.[49]

C. Epidemiology of Inorganic Components and Cardiovascular Disease
Most of the emphasis in drinking water quality has been focused on the organic contaminants since, as previously described, many of these microcontaminants have been labeled

carcinogens and mutagens. However, other investigations have shown that there is a negative correlation between the hardness of local drinking water supplies and local cardiovascular mortality rates.[50,51] Perhaps these constituents that determine hardness and softness are not viewed the same as organic contaminants but since geographic variation is relevant to hardness or softness, these trace element contaminants must be viewed as such even though most people consider hardness itself as not a constituent of drinking water but a property.

In consideration of the hardness of drinking water in areas of the world, some biologically important trace metals such as chromium, copper, zinc, cadmium, and lead have been studied epidemiologically as to association with low cardiovascular activity.[52,53] In general, two cations, calcium and magnesium in finished drinking water supplies have been associated with hardness and in turn with cardiovascular mortality rates. For example, Elwood et al.[54] have shown in their epidemiological studies that calcium was associated with ischemic heart disease mortality at most ages. They consider that another element, iron, may warrant further consideration. They also assert that the association between water hardness and mortality is not explained away in terms of associations between mortality and any of the other 12 elements they studied which included Li, Na, Mg, Al, K, Mn, Fe, Cu, Zn, Sr, Ba, and Pb.

Apparently, because of certain discrepancies in the correlation of hardness in water to cardiovascular diseases, some of the trace elements have been looked at in conjunction with calcium in drinking water to ascertain whether these trace elements in some regions were playing a role in the cardiovascular disease associations. Masironi,[55] for example, studied the role of radioactivity in conjunction with hardness of local water in relation to cardiovascular disease mortality.

Schroeder and others have speculated that soft water contributes to cardiovascular mortality in causing hypertension resulting from cadmium corroded from plumbing materials.[52] In some areas soft water left standing overnight in lead plumbing picks up considerable quantities of lead.[56] This lead contamination was postulated as a probable associated agent but plumbosolvency was recorded in hard waters as well.

The negative correlation between water hardness and cardiovascular disease mortality can be explained according to Bull and McCauley[57] on the basis of three hypotheses such as (1) presence of trace metal(s) in hard water protects against cardiovascular disease; (2) presence of toxic trace metals in soft water that may be causally related to cardiovascular disease; and (3) the relationship between water hardness and cardiovascular disease is spurious and merely covaries with nonmetallic contaminants in drinking water.

Calabrese and Tuthill[58] have reported that in studies on high school students in two closely matched communities with sodium levels in drinking water of 8 and 10 mg/ℓ, respectively, there was a statistically significant elevated systolic and diastolic blood pressure in both males and females with the high sodium level in one community.

Despite all of the above, cardiovascular disease is a complex entity affected by age, diet, stress, exercise, genetics, and other factors. Consequently, some caution must be exercised in the development of a rational interpretation of the extent and magnitude of the drinking water variable in the etiology of this disease. Therefore, the relationship of hard and soft water to cardiovascular disease remains to be more fully assessed.

In studies by Southwick et al.[59] where arsenic levels in water were four times the EPA standard and ranged to levels as high as 0.18 to 0.21 mg/ℓ, no adverse health effects were observed in populations. The effects looked for were cancer, dermatological signs, anemia, nerve conduction, and other symptoms of arsenic intoxication.

D. Risk Assessment

While demonstrating some interesting statistical associations, most of the epidemiological studies do not establish causality since the studies may be confounded by factors such as cigarette smoking of the population and/or failure to identify a stable population in areas where water quality was studied. Large populations may have to be studied to demonstrate any strong statistical associations because in the instance of the trihalomethane concentrations, they are not large enough to establish any magnitude of differences. Furthermore, most attention is given to the volatile organics in drinking water, at least predominantly in the experimental studies. Yet the major risk could reside in the nonvolatile products and/or the products of disinfectant reactions with naturally occurring chemicals (i.e., humic acids). Additionally, the most intriguing factor in the assessment of risk is the potential for additive effects of carcinogens and mutagens present in water which could contribute to the total cancer burden, Moreover, some carcinogens, of course, may persist in body tissues making quantification difficult.

The quantification of risk is limited by certain factors such as (1) diverse mathematical models which yield extreme variations and the estimates of risk; (2) studies with different species sensitivities producing a wide range in risk estimates; and (3) failure to consider either additive or synergistic, or even inhibitory or antagonistic effects which may underestimate the true risk. If positive relationships do exist between drinking water contaminants and certain cancers, it is reasonable to assume they will be discerned through the utilization of more precise exposure and control data, and perhaps by other analytic designs which should permit applicable risk assessment.

Obviously, large populations are repeatedly exposed to potentially toxic contaminants in drinking water in minute amounts (ppb or ppt level) over many years or a whole lifetime. Toxicity data obtained from laboratory animals will generally have to be relied upon for estimating human risk. Epidemiological studies can buttress, supplement, or contradict laboratory data. However, in this instance, where epidemiological studies on drinking water have their inadequacies, the reverse may be true in that new directions may come from innovative experimental studies.

In the absence of dose-response data from experimental studies in the area of low level exposure, extrapolations are made to the human population relevant to risk assessment. These estimations on predicting cancer incidence for a prescribed number of the population are usually based on 1–, 10–, or 100 million members of the population. Certain assumptions are inherent in the application of mathematical models to this risk estimation process. These assumptions are that the dose-response curve is linear through both high and low dose regions of the dose-response curve and also that the biochemical and pharmacokinetic mechanisms that may prevail in a mammalian response to such an environmental insult are similar in the high dose and low dose regions. In many instances this latter situation has been shown to be incorrect. By this procedure the estimation of lifetime cancer risk factors are calculated and are shown in Table 14[60] for six organic carcinogenic contaminants in drinking water. The estimation of lifetime cancer risk is based on an assumed concentration of 1 µg of that chemical per liter of water.

More recently, as information and data are revealed on carcinogenesis mechanisms, it is revealed that some of these organic chemicals may evoke a carcinogenic response through a nongenotoxic or epigenetic mechanism. In this case it implies that such responses are indeed dose-dependent. This may mean that some of these chlorinated aliphatics such as chloroform may not evoke a response at low levels or at least the dose or level as represented in drinking water. This suggests essentially that the risk factor would be reduced by many orders of magnitude or, at levels of occurrence, present an insignificant risk.

V. OTHER APPROACHES TO ASSESSMENT OF ADVERSE EFFECTS

As implied in the preceding discussion on epidemiological studies, since there are limitations on their utility as to demonstration of causal relationships, it follows then that one

Table 14
ESTIMATION OF SOME HUMAN LIFETIME CANCER RISK FACTORS FROM MATHEMATICAL EXTRAPOLATIONS OF ANIMAL DATA ON KNOWN LEVELS OF CARCINOGENS IN DRINKING WATER

Name of chemical	Observed concentration	Estimate of human lifetime cancer risk (per $\mu g/\ell$)
Carbon tetrachloride	5.0	1.1×10^{-7}
Chloroform	366.0	1.7×10^{-6}
Vinyl chloride	10.0	4.7×10^{-7}
Chlordane	0.1	1.8×10^{-5}
Dieldrin	8.0	2.6×10^{-4}
PCB (Aroclor 1264)	3.0	3.1×10^{-6}

From Kraybill, H. F., *J. Am. Water Works Assoc.*, July, 370, 1981.

Table 15
ALTERNATIVE PROCEDURES FOR ASSESSMENT OF SAFETY AND/OR HAZARD OF DRINKING WATER

Approach #1   Traditional carcinogenesis rodent bioassay of samples of water that are concentrated both before and after GAC treatment.

Approach #2   Synthetic water samples bioassayed with rodents; i.e., highly purified or distilled H_2O to which is added back organic contaminants at level of occurrence of multiples of such levels in drinking water.

Approach #3   Using drinking water concentrates, samples are bioassayed by short-term tests.

Approach #4   Large animal bioassay resources having many control groups of rodents which are at various locations on variant water supplies. Study focused on tumor incidence difference of control animals and association with chemical profiles of these water supplies.

Approach #5   Use of fish as animal model to bioassay a wide spectrum of drinking H_2O supplies.

From Kraybill, H. F., *Water Chlorination: Environmental Impact and Health Effects*, Vol. 3, Jolley, R. L., Brungs, W. A., Cumming, R. B., and Jacobs, V. A., Ann Arbor Science Publ., Ann Arbor, 1980, 973.

must resort to experimental approaches as a surrogate for direct evidence on man or really to serve as a guide for future directions on epidemiological studies. The frustration and futility of using human population studies to provide the requisite supporting evidence on hazard really fails to provide a strong argument on behalf of needs for certain water filtration and processing technologies.[61]

A. Alternative Experimental Approaches

In these considerations, the author has proposed five alternative procedures for assessment of safety and/or hazard of drinking water samples based on experimental approaches. These alternatives are delineated in Table 15.

It is feasible by use of low temperatures and selective precipitation of dissolved inorganic solutes to obtain drinking water samples that are concentrated many times to yield a sample of water that could be bioassayed at different dose levels, the latter being dependent on the degree of concentration. Some of these samples could be processed through GAC for removal of the organic contaminants. Such concentrates could be bioassayed through traditional procedures using the rodent as the test animal.

In another operation, such as that delineated in approach 2 in Table 15, a control drinking water sample for testing would be triple distilled or passed through an ion exchange column and GAC filters for removal of all organics. The test group of animals would be maintained on a synthetic water sample viz the highly purified water as described above to which variant concentrations of the various chlorinated aliphatics are added at levels as quantified in drinking water and then evaluated in dose-response studies.

Another approach considered is, in effect, an "animal epidemiological" study. This procedure is predicated on the assumption that drinking water samples are variable, qualitatively and quantitatively, as to carcinogenic organic contaminants. On this assumption a national bioassay resource could be exploited where tumor incidence data on thousands of control animals (rats and mice) could be used. Variation in tumor response data may be associated then with the variation in the qualitative and quantitative descriptions or profiles on the drinking water from multiple locations. The dose-response data would be predicated on the assumption that variants such as diet composition and air pollutants would be controlled and therefore the sole variant would be drinking water. Another attractive and less costly study than the use of a large rodent study is the use of the fish as the animal model to bioassay a wide spectrum of water samples, again with variable qualitative and quantitative characteristics. These samples can be evaluated in dose response studies to reflect a tumorigenic response conditioned on these qualitative and quantitative features in selected water samples. The use of the fish bioassay may offer many advantages in that large numbers of test animals can be used, the economics and logistics of using fish are favorable, the limitation on test dose to minimize the aspect of intoxication is possible, and finally, the assay or test dose would be within the tolerable range of exposure for the sensitive fish which in turn may be comparable to the actual concentration of these organic contaminants in drinking water. Therefore, the marine animal bioassay holds considerable promise as one of the best alternative procedures in the assessment of carcinogenicity of contaminants.

B. Other Evidence on Adverse Effects of Contaminated Drinking Water

Heartlein et al.[62] have demonstrated that some municipal waters in an agricultural area were quite mutagenic, dependent on time of year. They found that in the Lake Bloomington area of Illinois tap water derived from this lake water resource, when tested for mutagenicity in the Ames system and in the waxy-90 reverse mutation tests in Zea mays was more mutagenic in May water concentrates than in any other concentrates of tap water taken in succeeding months, up to October. This would imply that agricultural and industrial chemicals may be contaminating the drinking water to a greater amount in that region at that time of year since a number of organic contaminants were identified by chemical analysis. Control concentrates prepared from deionized and distilled treated well water may not have these mutagenic contaminants which may pose a potential health hazard.

Another unusual situation which may be termed aquagenic pruritis has been reported by British workers. Some patients at the Institute of Dermatology at St. John's Hospital for Diseases of the Skin in London complained of intense itching following brief contact with water at any temperature. Some patients could have the pruritis subside if they took antihistamines prior to bathing or after bathing. The pathogenesis of this disorder is not quite known even though, according to the paper by Greaves et al.[63] other patients have been observed with this condition. Here again, one might speculate that the contaminants might be sufficient to induce this pruritis but the authors of the paper do not even suggest this possibility. It would be interesting to explore this.

Another interesting piece of work was conducted by Canadian workers on the toxicology of compounds resulting from the use of chlorine in food processing. This work was conducted by Cunningham and co-workers.[64-66] A series of papers they published in 1976 and succeeding years on such products as cake flour, milk, meats, and poultry demonstrated that chlorine gas added to cake flour, for example, results in chlorinated lipids and proteins, both of which are toxic to rats, resulting mainly in reduced body weight gains and increases in the size of liver, kidney, and heart. Chlorination apparently reduces the digestibility of lipids and had a greater effect on more unsaturated lipids. These chlorinated lipids can be distributed throughout the body and ultimately dechlorinated and deposited in most tissues especially the muscles. If they escape dechlorination and are deposited in adipose tissue, then half-life could be equivalent to that of unchlorinated lipids. In rats, placental and mammary transfer of chlorinated fatty acids was found to occur with the potential for toxicological hazard. This instance with chlorination in various foods is mentioned since chlorination of drinking water has been shown to also produce a spectrum of transformation products that have adverse effects as previously cited.

VI. DISCUSSION AND SUMMARY ON ASSESSMENT OF POTENTIAL HAZARDS FROM EXPOSURE TO ENVIRONMENTAL CONTAMINANTS IN WATER: KNOWNS AND UNKNOWNS

Since man may receive exposure to many environmental insults, the drinking water he ingests represents just one source of his total exposure. Gone are the days when primitive man may have drank relatively contaminant-free water. The sources of drinking water impounded in reservoirs is essentially derived from lakes, springs, wells, and rivers, all of which may have a contaminant burden of naturally occurring materials, additives, from industrial effluents, municipal sewage effluents, or runoff from agricultural chemicals. The contaminants in water may be categorized as microorganisms, particulate matter, inorganic solutes, organic solutes, and radionuclides. Around the turn of the century, prior to so-phistication in water treatment technologies, gastroenteric diseases resulting from pathogenic microorganisms were a more common occurrence. For example, as late as 1933 in Chicago, Illinois, an outbreak of amebiasis occurred. However, this was due to what is called a "cross-connection" where, through back siphonage, hotel drinking water became contaminated with wastewater. In earlier periods gastroenteric diseases occurred more from drinking water from questionable sources. For the traveler today, in visiting other countries of the world, pathogenic contaminants still present a major problem. Modern technologies in the U.S. relevant to water treatment include a series of steps in processing (coagulation, sedimentation, filtration, and disinfection). The latter process, disinfection based on chlorination or ozonation, provides the ultimate in assurance of safety. However in some instances, if there is a breakdown in, for instance, a chlorinator, then contaminants may build up in such non-chlorinated surface or ground water resulting in acute diseases for those dependent on public water supplies. It should be emphasized that even in this country gastroenteric disease episodes may occur in isolated cases. The application of chlorination, while guaranteeing safe water insofar as bacterial contamination, has a trade-off in that this process introduces, through chemical conversion, a whole series of molecular species as microcontaminants which have been identified in newly chlorinated raw water supplies which become drinking water for a municipality. Of course, some of the organic contaminants in the original water supply may be carried through if not fully processed. The micropollutant organic contaminants in the drinking water are believed to be formed by action of chlorination on the humic acids in the original water supply.

Worldwide raw water supplies, predominantly surface water (rivers) from which drinking water is derived, contain thousands of inorganic and organic contaminants, limited in their identification only by the sensitivity of analytical methodology. It is conceded that insofar as organic contaminants in drinking water are concerned, one can only identify and quantify about 10% of the contaminants, primarily the volatiles. The other 90%, or nonvolatiles, are essentially unknowns. The EPA in their water quality programs and projects has advanced our knowledge about U.S. water supplies through their extensive monitoring and research programs. Whereas a few hundred organic contaminants (contaminants) were identified in drinking water only 5 years ago, in 1980 this list had escalated to about 1200 unique chemicals and by 1982 it may have increased to well over 1200, maybe reaching 1800. Of course, there may be some repetitive compounds such as isomers or qualitative identifications because of chromatography and mass spectrometry that are not quantitatively established. One does not want to leave the impression that the pollution picture has worsened. The increase in chemicals identified is really a reflection of the intensive methodology and monitoring. In a search of the literature on 767 biorefractories in drinking water, relating only to those chemicals which have been tested for carcinogenicity, mutagenicity, and promotion activity, the National Cancer Institute, with the aid of a contractor, was able to identify 20 recognized carcinogens and 23 suspect carcinogens, 45 mutagens, 18 promoters or cocarcinogens. In the experimental area, using in vitro as well as in vivo procedures, contaminated water (raw water), when bioassayed, showed some mutagenic activity.

In recent years there has been a noticeable acceleration in epidemiological studies. With the number of studies increasing, it is possible at the time of publication of this chapter that 25 to 30 studies will have been performed. Most of the studies have been concentrated in the Ohio and Mississippi River basins. These studies have been directed toward various sources such as ground water vs. surface water, chlorinated vs. nonchlorinated water, and studies oriented around THM (trihalomethane) levels as a contaminant indicator. All of these variables or factors are studied with reference to any association with increased cancer incidence. Some investigators feel contented with these approaches and not dismayed with the findings which are at best some statistical associations which fall short of causality. These studies are limited insofar as uncontrolled variants. It may be expecting too much to prove any causal relationship relevant to these micropollutants and cancer incidence, since one is titrating here to extremely low levels (ppb or ppt) as a human exposure in a "sea of other environmental insults" which are much, much greater and perhaps difficult to exclude or control in the conduct of a study. Thus far, the few experimental studies and the spectrum of epidemiological studies have provided some presumptive evidence but it remains mostly for further experimental studies to elucidate whether these ingested organic or even inorganic micropollutants, which are carcinogenic and mutagenic, are capable of contributing a small fraction to the total cancer burden.

Directing attention to the inorganic contaminants in drinking water, about 10 years ago another problem developed when the community of Duluth, Minnesota, which receives its water supply from Lake Superior, became aware that their drinking water was contaminated with asbestiform particulates or taconite tailings from nearby mines. These tailings were mostly amphiboles. Earlier, there had been concern about the development of lung cancer and lung diseases among workers who inhaled or were exposed to asbestos. Therefore, it logically followed that some doubts would be raised relevant to the effect of ingested asbestos in drinking water in terms of potential health hazard over a long-term exposure and whether some of these various forms of asbestos (chrysotile, crocidolite, amosite, etc.) could indeed be absorbed through the gastrointestinal tract and migrate in the blood supply eventually to reside or build up in target organs such as liver, lung, and kidney. To resolve this problem, one obviously could not wait for population findings, so experimental studies, rather large in scale, were started after these observations in Duluth. Feeding experiments with rats and hamsters, using a specifically designed asbestos (particle size), have been completed. Thus far, there appears to be no carcinogenic hazard from ingested asbestos since the animals receiving the asbestos exposure in their diet showed no statistically significant increase in tumor incidence over controls. These experimental findings appear to confirm the speculations made by other investigators that asbestos particles are rapidly transported through the gut and not absorbed through the microvilli in the intestinal mucosa.

While much attention has been focused on the organic contaminants in water, interest has more recently been revived in the inorganic contaminants. There is a wide array of inorganic solutes in water but only 22 have been reviewed extensively as to their potential health effects. It should be emphasized that 13 of the 22 inorganic cations are trace elements or essential nutrients. Nutrients, per so, do not introduce a serious health problem, unless, like most any chemical, they are taken in excess when toxicity could prevail. The cation lead, which is usually considered in the category of a toxic element, may be of concern. Lead is an element that builds up in the body and the body burden for this element increases with age. The maximum contaminant level (MCL) for lead in drinking water is 50 ppb. By the same token, the current standard for arsenic may be reexamined. Presumably, the origin of concern about arsenic comes from earlier studies on arsenic in drinking water and the population exposures and skin cancers in Taiwan, Cordoba, and South America. Arsenic has been incriminated as a carcinogen in occupational exposures[68] but more recently scientists at the U.S. Department of Agriculture, Grand Forks, North Dakota, Nutrition Laboratory, have shown that low levels of arsenic are essential in the functioning of the hematopoietic system or, more specifically, for red blood cell integrity.[69]

There is apparently a continuing debate on hard water vs. soft water. Given some uncertainties, the prevailing view from a preponderance of evidence is that there is an inverse relationship between cardiovascular disease incidence and hardness of water. Further studies may, however, be indicated as to the role of the metal ions in water associated with hardness and softness and cardiovascular disease.

Radioactivity in water occurs because of natural sources and man-made activities or technologies. The primary natural isotopes of interest are radium, uranium, and radon. Of the two isotopes of radium only the occurrence and health effects of ^{228}Ra are being investigated since the existing information and data on ^{226}Ra seem adequate. As to the man-made additions of radioactivity, one, of course, considers radioactive isotopes or radiopharmaceuticals inadvertently disposed of in the water supplies or the radioactive material from fallout. However, in communities monitoring alpha and beta activity, these ground water supplies (wells) originate in water passing through geologic deposits of radioactive material such as uranium and radium. The EPA has studies underway to develop adequate procedures for uranium removal. The occurrence and potential health risks of radon are being studied. The EPA indicates that the state of Maine has a feasibility study underway on a possible epidemiologic association relating to the effects of radon in drinking water. Radioactivity in water contributes only a small fraction to the total normal background radiation to which all people are exposed. To detect with any degree of certainty what the adverse health effect may be could be quite difficult. Since radium is a bone seeker and occurs in some water supplies wherein it could reach a concentration exceeding the standard, in this instance a risk of bone cancer should be considered. Thus, there are systematic surveys of municipal water supplies to determine the levels of occurrence of radioactivity.

Realizing that the epidemiological studies have fallen short of establishing causality there may be a need for reinforcement through other approaches and such approaches should envision assays with in vitro systems to delineate where the mutagenic activity resides in water samples and fractions. Using an in vivo approach, the National Cancer Institute has implemented studies where fish of various strains are used as an animal model to bioassay a mixture and subfractions of water samples made up to simulate the concentration and multiples of that concentration of chemicals that are major contaminants in drinking water. The objective, of course, is to ascertain whether this combination of organic contaminants will evoke a tumorigenic response in fish.

Attempts have been made to assess the risk of these organic carcinogenic biorefractories in water. Estimates of potential risk are predicated on quite a few assumptions. Furthermore, as our orientation on the possible carcinogenic potency of each of these contaminant chemicals is enhanced, it becomes clear that not all such contaminants should be appraised on an equal basis as to their potential carcinogenic insult. This is particularly significant in that at the present time we do not know what chemicals are acting through a genotoxic or epigenetic mechanism. With time, as we obtain more information on mechanisms, perhaps, this problem may be clarified.

Indeed, further explorative efforts, predominantly in the area of experimental research, need to be fostered in order to achieve a better appraisal as to the significance of all these micropollutants in our drinking water and how they may contribute singly or additively, if at all, to the total cancer burden. However, in the interim it would seem prudent to recommend and initiate control technologies for reduction and elimination of these contaminants in drinking water.

REFERENCES

1. National Academy of Sciences, Advisory Center on Toxicology, Assembly of Life Sciences, National Research Council Safe Drinking Water Committee, *Drinking Water and Health,* NAS Sciences Publishers, Washington, D.C., 1977.
2. **Cotruvo, J.,** Current practices in chlorination of drinking water, in *Proc. of the Toxicol. Forum,* Arlington, VA, 1981, 247.
3. **Pojasek, R. B.,** How to protect drinking water sources, *Environ. Sci. Technol.,* 11, 342, 1977.
4. **Tardiff, R. G., Craun. C. F., McCabe, L. J., and Bertozzi, P. E.,** Suspect carcinogens in water supplies, Interim Report to Congress, Appendix B - Health Effects Caused by Exposure to Contaminants, EPA Progress Report, Cincinnati, Ohio, U.S. EPA, 1975.
5. **Kraybill, H. F.,** Origin, classification and distribution of chemicals in drinking water with an assessment of their carcinogenic potential, in *Water Chlorination: Environmental Impact and Health Effects,* Vol. 1, Ann Arbor Science Publ., Ann Arbor, MI, 1975, 11.
6. **Kraybill, H. F.,** Distribution of chemical carcinogens in aquatic environments, in *Progress in Experimental Tumors,* Vol. 20, S. Karger, Basel, Switzerland, 1976, 3.
7. **Kraybill, H. F.,** Global distribution of carcinogenic pollutants in water, *Ann. N.Y. Acad. Sci.,* 1977, 298, 80.
8. New York Academy of Sciences, Annals of Conference on Aquatic Pollutants and Biological Effects with Emphasis on Neoplasia, H. F. Kraybill, R. G. Tardiff, C. J. Dawe and J. C. Harshbarger, Eds., NYAS Publishers, New York, 1977, 198.
9. **Cook, P. M., Glass, G. E., and Tucker, J. H.,** Asbestiform amphibile minerals: detection and measurement of high concentration in municipal water supplies, *Science,* 185, 853, 1974.
10. **Nicholson, W. J. and Farger, A. M.,** Environmental contamination sources. Meeting on Biological Effects of Asbestos, National Institutes of Health Seminar, Bethesda, Maryland, February, 1973 (unpublished).
11. National Toxicology Program, NTP Technical Report, Carcinogenesis bioassay of Syrian golden hamsters, 1982, in press.
12. Environmental Protection Agency, Office of Toxic Substances, Report to Congress on preliminary assessment of suspected carcinogens in drinking water, Table 11.2-7, Washington, D.C., 1975.
13. **Robeck, G.,** Personal communication to author from Water Quality Laboratory, Environmental Protection Agency, Cincinnati, Ohio, 1974.
14. **Andrew, R. W.,** Mineralogical and suspended sounds measurements of water sediment and substrate samples for 1972 Lake Superior Study II, Stream Sediments Data Report, April 1973.
15. **Cook, P. M.,** Distribution of taconite tailings in Lake Superior water and public water supplies, Environmental Protection Agency, National Water Quality Laboratory, Second Progress Report, Duluth, Minn., July, 1973.
16. **Wilson, R. D., Monoghan, P. H., Osanik, A., Price, L. C., and Rogers, M. A.,** Natural marine oil seepage, *Science,* 184, 857, 1974.
17. **Shimkin, M. B., Koe, B. K., and Zichmeister, L.,** An instance of the occurrence of carcinogenic substances in certain barnacles, *Science,* 113, 650, 1951.
18. Community Water Supply Surveys, Environmental Protection Agency, Office of Drinking Water, Health Effects Research Laboratory, Cincinnati, Ohio, 1969 and 1978.

19. Report on Rural Drinking Water Supplies, Environmental Protection Agency, Office of Drinking Water, State Programs Division, Washington, D.C., 1978.

20. Report of National Stream Quality Accounting Network, U. S. Geological Survey, Survey Circular #719 and Open File Report 78-200, USGS Water Resources Division, Reston, VA., 1978.

21. **Kopp, J. F.,** The occurrence of trace elements in water, in *Proceedings of the Third Annual Conference on Trace Substances in Environmental Health,* D. D. Hemphill, Ed., University of Missouri, Columbia, 1970, 59.

22. **Harada, Y., Araki, Y., Sudon, H., and Mujamoto, Y.,** Minamata disease in children, *Kumamoto Domon Kaishi,* 6, 268, 1965.

23. **Mount, D. I. and Putnicki, G. J.,** Summary Report of 1963 Mississippi River Fish Kills, Trans 31st North American Wildlife and Natural Resources Conference, 1966, 177.

24. Report of Hazardous Waste Enforcement Task Force, List of Chemicals Found in the Love Canal, Environmental Protection Agency, Office of Enforcement, Washington, D.C., 1979.

25. **Walker, J. D. and Colwell, R. R.,** Mercury resistant bacteria and petroleum degradation, *Appl. Microbiol.,* 27, 285, 1974.

26. **Walker, J. D. and Colwell, R. R.,** Microbial ecology of petroleum, Proc. of American Petroleum Institute, EPA/US Coast Guard Conference on Prevention and Control of Oil Spills, American Petroleum Institute, Washington, D.C., 1973, 685.

27. Anonymous, Fed. Register 39, 3756, 1974.

28. **Kloepfer, R. D.,** Identification of specific organic chemicals in the Ohio and Wabash Rivers, U.S. Environmental Protection Agency Report, Evalsville, Indiana, 1971-1972, 1972.

29. **Piet, G. J., Zoeemen, B. C. T. J., Nettenbreiger, A. H., and Ruijgrok, G. T. H,** Bis(2-chloroisopropyl)ether in surface and drinking water in the Netherlands, Orientation Report of Rijksinstituiut voor Drinkwater, Voorziening Parkweg 13 s'Gravenhage, Netherlands, 1973.

30. **Hushon, J., Clerman, R., Small, R., Sood, S., Taylor, A., and Thoman, D.,** An assessment of potentially carcinogenic, energy related contaminants in water, Contract Report to Dept. of Energy/National Cancer Institute, McLean, VA, May, 1980.

31. **Shackleford, W. S. and Keith, L. H.,** Frequency of organic compounds identified in water. Report of Environmental Protection Agency, Research Laboratory, Athens, Georgia, 1976.

32. Commission of the European Communities, Analysis of Organic Micropollutants in Water, Cost Project 64b, EUCO/MCU/73/76, XII/476/76, 1976.

33. **Elias, P.,** The medical significance of marine pollution by organic chemicals, *Proc. Royal Soc. London Ser.,* B, 189, 443, 1975.

34. **Borneff, J. and Kunte, H.,** Kanzerogene substanzen in wasser and boden. XVI. Nachweis von polyzyklischen aromaten in Wasserproben durch direkte Extraktion, *Arch. Hyg. Bakteriol.,* 148, 585, 1964.

35. **Woodwell, G. M., Craig, P. P., and Johnson, H. A.,** DDT and the biosphere where does it go? *Science,* 174, 1101, 1971.

36. Program data on eleven state survey of DDE, PCB, and PBB levels in human milk, Environmental Protection Agency and Michigan State Department of Health (Michigan Data), 1976.

37. **Peterle, T. J.,** DDT in Antartica snow, *Nature,* 224, 620, 1969.

38. **Risebrough, R. W., Huggett, R. J., Griffin, J. S., and Goldberg, E. D.,** Pesticides: Transatlantic movements in the northeast trades, *Science,* 159, 1233, 1968.

39. **Bouquiaux, J.,** Nonorganic micropollutants of the environment, Report prepared for the Commission of the European Communities V/F/1966/74E, Luxenbourg, 1974.

40. **Helmes, C. T., Sigman, C. C., Malko, S., Atkinson, D. L., Jafer, J., Sullivan, P. A., Thompson, K. L., Knowlton, E. M., Kraybill, H. F., Hushon, S., Thomas, D. P., Clerman, R. S., and Barr, N.,** Evaluation and classification of the potential carcinogenicity and mutagenicity of chemical biorefractories identified in drinking water, U.S. Department of Health and Human Services, Public Health Service, National Institutes of Health, February 1981.

41. International Agency for Research on Cancer, Monograph Series on Evaluation of Carcinogenic Risk of Chemicals to Man, Volumes 1, 3, 4, 5, 7, 9, 11, 15, 19, and 28, 1972, 1973, 1974, 1975, 1976 and 1979.

42. National Cancer Institute, Bioassay of Chemicals for Possible Carcinogenicity, Carcinogenesis Technical Report Series Numbers 2, 8, 9, 13, 14, 21, 27, 34, 55, 68, 74, 80, 92, 130, 154 and 155, U.S. Department of Health Education and Welfare, Public Health Service, Bethesda, Maryland.

43. National Academy of Sciences, National Research Council Safe Drinking Water, Volume III, Epidemiological Studies — Cancer Frequency and Certain Organic Constituents of Drinking Water: a review of recent literature published and unpublished, A Report for the Environmental Protection Agency, Washington, D.C., 5, 1978, Chap. 2.

44. BEIER Committee, The effects of populations of exposure to low levels of ionizisang radiation, Report of Advisory Committee on the Biological Effects of Ionizing Radiation, National Academy of Sciences, National Research Council Publication, Washington, D.C., 1972.

45. **Hickey, J. L. S. and Campbell, S. D.,** High radium-226 concentrations in public water supplies, *Pub. Health Rep.,* 83, 551-1968.

46. **Lucas, H. F.,** Correlation of the natural radioactivity of the human body to that of the environment: Uptake and retention of [226]Ra from food and water, in *Radiological Physics Division Semiannual Report,* Argonne National Laboratories, Report July-December, *ANL* 6297, 55, 1971.

47. **Lucas, H. F. and Krause, D. P.,** Preliminary survey of radium-226 and radium-228 contents of drinking water, *Radiology,* 74, 114, 1960.

48. **Rowland, R. E., Lucas, H. F., and Stehney, A. F.,** High radium levels in the water supplies of Illinois and Iowa, in *Proc. Int. Symp. on Areas of High Natural Radioactivity,* T. L. Cullen and E. P. Franca, Eds., Academia Brasileira de Ciencias, Rio de Janeiro, Brazil, 1976.

49. **Wrenn, M. E.,** Internal dose estimates, in *Proc. Int. Symp. on Areas of High Natural Radioactivity,* T. L. Cullen and E. P. Franca, Eds., Academia Brasileira de Ciencias, Rio de Janeiro, Brazil, 1976.

50. **Comstock, C. W.,** Water hardness and cardiovascular diseases, *Am. J. Epidemiol.,* 110, 375, 1979.

51. **Sharrett, R. A.,** The role of chemical constituents of drinking water in cardiovascular diseases, *Am J. Epidemiol.,* 110, 401, 1979.

52. **Schroeder, H. A.,** Relation between mortality from cardiovascular disease and treated water supplies, *J.A.M.A.,* 172, 1902, 1960.

53. **Neri, L. C., Hewitt, D., Shreiber, G. B. et al.,** Health aspects of hard and soft waters, *J. Am. Water Works Assoc.,* 67, 403, 1975.

54. **Elwood, P. C., St. Leger, A. S., and Morton, M.,** Mortality and the concentration of elements in tap water in county boroughs in England and Wales, *Brit. J. Prevent. and Soc. Med.,* 31, 178, 1977.

55. **Masironi, R.,** Cardiovascular mortality in relation to radioactivity and hardness of local water supplies in the USA, *Bull. WHO,* 43, 687, 1970.

56. **Beattie, A. D., Moore, M. R., Devenay, W. T., Millar, A. R., and Goldberg, A.,** Environmental lead pollution in an urban soft water area, *Brit. Med. J.,* 1, 491, 1972.

57. **McCauley, P. T. and Bull, R. J.,** Experimental approaches to evaluating the role of environmental factors in the development of cardiovascular disease, in *Drinking Water and Cardiovascular Disease,* Pathotox Publ., Park Forest, Ill., 1980, 27.

58. **Calabreese, E. J. and Tuthill, W.,** The influence of elevated levels of sodium in drinking water on elementary and high school students in Massachusetts, in *Drinking Water and Cardiovascular Disease,* Pathotox Publ., Park Forest, Ill., 1980, 151.

59. **Southwick, J. W., Western A. E., Beck, M. M., Whitley, J., Isaacs, R., Petajan, J., and Hansen, C. D.,** U.S. EPA Project Summary Report from Health Effects Research Laboratory, Cincinnati, Ohio, February, 1982.

60. **Kraybill, H. F.,** Carcinogenesis of synthetic organic chemicals in drinking water, *Jour. Am. Water Works Assoc.,* July 1981, 370.

61. **Kraybill, H. F.,** Animal models and systems for risk evaluation of low level carcinogenic contaminants in water, in *Water Chlorination: Environmental Impact and Health Effects,* Vol 3, R. L. Jolley, W. A. Brungs, R. B. Cumming and V. A. Jacobs, Ann Arbor Science Publ., Ann Arbor, Michigan, 1980, 973.

62. **Heartlein, M. W., DeMarini, D. M., Means, Katz, A. J., J. C., Plewa, M. J., and Brockman, H. E.,** Mutagenicity of municipal water obtained from an agricultural area, *Environ. Mutag.,* 3, 519, 1981.

63. **Greaves, M. W., Black, A. K., Eddy, R. A. J., and Coutts, A.,** Aquagenic pruritus, *Brit. Med. J.,* 282, 2008, 1981.

64. **Cunningham, H. M. and Lawrence, G. A.,** Effect of exposure of meat and poultry to chlorinated water on the retention of chlorinated compounds and water, *J. Food Sci.,* 42, 1504, 1977.

65. **Cunningham, H. M. and Lawrence, G. A.,** Placental and mammary transfer of chlorinated fatty acids in rats, *Food Cosmet Toxicol.,* 15, 183, 1977.

66. **Cunningham, H. M., Lawrence, G. A., and Tryphonas, L.,** Toxic effects of chlorinated cake flour in rats, *J. Tox. Environ. Health,* 2, 1161, 1977.

67. **Tseng, W. F.,** Prevalence of skin cancer in an endemic area of chronic arsenicism in Taiwan, *J. Natl. Cancer Inst.,* 40, 453, 1968.

68. **Lee, A. M. and Fraumeni, J. F.,** Arsenic and respiratory cancer in man: An occupational study, *J. Natl. Cancer Inst.,* 42, 1045, 1969.

69. **Nielsen, F. H., Givand, S., and Hyron, D.,** Evidence of a possible requirement for arsenic by the rat, *Fed. Proc.,* 34, Abstr. 3987, 923, 1975.

Chapter 3

THE INFLUENCE OF NUTRITION, IMMUNOLOGIC STATUS, AND OTHER FACTORS ON THE DEVELOPMENT OF CANCER

Paul M. Newberne

TABLE OF CONTENTS

I. INTRODUCTION

Two out of every ten people that die in the United States each year are victims of cancer of one type or another. Beginning with the publication of the Surgeon General's Report on smoking and health in 1979,[1] the public appears to have accepted the fact that cigarettes are a prime factor in the occurrence of lung cancer in human populations. While the amount of tobacco consumed remains relatively constant and the number of people who smoke remains about the same, the population using cigarettes continues to shift, generally to a younger age group. The numbers of cancer cases is increasing steadily as the population grows but the age adjusted total incidence in mortality rates for sites other than the respiratory tract, the pancreas and stomach have remained rather stable over the past 30 to 40 years. Cancer of the lung and of the pancreas have increased; stomach cancer has steadily decreased.

There has been much discussion and debate in the scientific literature and in the public press during the past 10 years relative to factors which influence cancer incidence. Some of the discussions have been based on epidemiologic studies, some on laboratory animal studies, and much on personal opinions. The public is now asking about the causes of cancer which are not associated with smoking; they wish to know what these causes are and how they can be avoided. This is of particular concern because of pronouncements from governmental agencies (NCI, NTP) which end up in the lay press about putative carcinogenic agents. These are often nutrients or food additives, and the basis for the allegations depend solely on studies in one or two species of animals, under sometimes bizzare conditions, which suggest an effect but without scientific scrutiny and interpretation.

In a general way the causes of cancer may be categorized into two major areas, namely, endogenous factors and exogenous factors (Table 1). These classifications do not identify any single factor or condition that may be causally related to human cancer but they do provide an overall framework for approaching an assessment of cancer cause and prevention.

As shown in Table 1 lifestyle stands out as one of the primary areas for potential growth of knowledge regarding cancer cause and prevention.[1-3] Considerable information has accumulated from epidemiologic studies of migrants moving from one area of the world to another.[4,5] Japanese migrants have provided probably the most interesting and useful information because they are an easily identified population; in moving from Japan to either Hawaii or to California they provide a subset of the population with specific types and frequencies of tumors which can be compared to that of the area to which they migrate. Observations of cancer in these individuals and known types and frequencies in Japan and in the United States have given us considerable information relative to gastric, colon, and breast cancer.[5,6] In these three organ sites Japanese migrants tend to assume the frequency[7] and incidence of the population living in the area to which they take up residence (Table 2). Clearly this implies an environmental influence with diet and nutrition important potential factors.

Two subpopulations within the United States have provided much information on tumor incidence and lifestyle; these are the Mormons and the Seventh Day Adventists (SDA).[8-10] These two groups of citizens have personal habits which are quite different from those of non-Mormons and non-SDAs. Diet, alcohol use, and smoking habits are important differences.

Genetics and familial influences on certain types of tumors have been known for a long period of time and these will not be considered here.[11] Rather, exogenous and endogenous factors which may be modulators of cancer risk will be considered in as much detail as space permits. In particular, those factors which are associated with substantial negative or positive risk will be given priority.

Many factors in our environment are real or potential sources of cancer causing agents including the air we breathe, the water we drink, the regions in which we live, the type of work we do, and, of course, the quality and quantity of foods that we eat. Exposure to

Table 1
ENDOGENOUS AND EXOGENOUS FACTORS CONTRIBUTING TO CHRONIC DISEASE IN HUMAN POPULATIONS

A. Endogenous Factors
1. Internal chemicals
2. Hormonal Factors
3. Metabolic deficiencies
4. Dietary deficiencies

B. Exogenous Factors
1. Air pollution
2. Industrial exposures
3. Food and water contaminants
4. Lifestyle
   (a) Diet
   (b) Tobacco
   (c) Alcohol
   (d) Sexual behavior
   (e) Infection
   (f) Medication
   (g) UV light
   (h) Radiation

Table 2
MORTALITY FROM GASTROINTESTINAL CANCER IN JAPANESE AND IN CAUCASIANS[a]

| Site of cancer | Japanese in Japan | | California | | | | | |
| | | | Foreign born Japanese | | U.S. Born Japanese | | Caucasian | |
	M	F	M	F	M	F	M	F
Colon	1.9	2.1	6.1	7.0	6.3	10.4	7.9	8.3
Stomach	58.4	30.9	29.9	13.0	11.7	13.3	8.0	4.0

[a] Mean/per 10^5 population.

Abridged from Ackerman, L. V., *Nutr. Today*, 2, 1972. With permission.

many of these factors by human populations varies in ways that cannot be measured precisely. For example, it is a difficult task to assess diets that are consumed by different human populations because records are not maintained in any accurate form. Recall of diets that have been consumed in the past week or the past year is a very imprecise way of determining food intake. Furthermore, the types of diets that are consumed, even though they fall into the same categories of food ingredients, vary considerably in their nutrient and nonnutrient contents (including contaminants), depending in part on the area of the country where the food is produced.[12] This is in striking contrast to the relative ease of determining the population that smokes.

Another area of very little concern until recently is that of the influence of immunologic status on cancer. Observations in transplant patients who are immunosuppressed to prevent rejection of the transplant,[13] and the therapeutic use of immunosuppressive drugs in man,[14,15] have established a role for immunocompetence in cancer causation. In addition, in a significant number of cases second cancers appear following radiotherapy of certain types of primary tumors.[16] Whether or not this phenomenon operates via the immune system is, of course, not known at this time but it is established that radiation has a profound effect on the immune system.

Probably the most advanced area of knowledge about immunocompetence and its relation to cancer is that of hepatocellular carcinoma. This tumor develops in patients with chronic inflammatory liver disease and the risk of developing it is greatest (1) in patients with liver disease from hepatitis B virus (HBV), (2) or those consuming excess alcoholic beverages, and (3) those with hemachromatosis. In these high risk groups the associated agents themselves are suspected by some to be directly carcinogenic, although, in the case of alcohol, contaminants may be more important. HBV-DNA is found in an integrated form in most HBVs antigen-positive tumors.[17,18] Dietary contamination with aflatoxins and other environmental toxicants, and exposure to factors which enhance or inhibit mixed function oxidases must also be considered as other risk factors along with diminished immunocompetence and hepatitis viral infection.[19,20]

Diseases of the liver are associated with a high risk of primary liver cancer; this type of neoplasm most commonly affects men and has usually reached a stage of macronodular cirrhosis by the time the tumor is apparent. These observations in human subjects have been repeated in animal studies helping to further confirm the suspicions.[21]

This chapter will introduce observations from human studies suggesting interrelations between lifestyle factors and other conditions associated with cancer in human populations and provide the accessible experimental data which tend to either support or deny such associations.

II. NUTRITION AND CANCER

There are a number of ways by which the diet and the nutritional status of the host may influence the induction of cancer in human or animal subjects. For example, contaminated food may result in exposure to carcinogens; there may be interactions between nutrients and chemicals in the diet which enhance or inhibit a carcinogenic effect. Nutritional status influences microsomal enzyme activity and in this manner may promote the activation or deactivation of procarcinogens or carcinogens. In addition, nutrients have a profound effect on the immune system, influencing the body's surveillance mechanisms. There can be little doubt that environmental factors play a role in both human and animal cancer.

It is generally accepted that some natural substances (i.e., aflatoxin) are human carcinogens; in concert with hepatitis B virus, aflatoxin and, perhaps, other toxins is probably a major cause of primary liver cancer in some areas of the world. Some food additives may contribute to human cancer in some small way but the overall contributions from this source are probably not significant.

It is well established from studies in animals that nutrients can affect the biological response of animals and, presumably, man, to some additives. Furthermore, it is clear that some of the macronutrients (i.e., fats) act as promoters in animal carcinogenesis. It is likely that nutrients such as vitamins E and C influence the formation of nitrosamines in human diets or in vivo and in this way influence the incidence of some types of human cancer. One of the most common food additives, table salt (sodium chloride), is associated with gastric cancer in Japan, perhaps by producing trauma to the gastric mucosa and thus permitting increased exposure to nitrosamines or other carcinogens or by increasing sensitivity to carcinogenic agents.

The evidence is circumstantial but it would appear that the contribution of the diet toward

Table 3
DIETARY AND
LIFESTYLE HABITS AND
RISK FOR CANCER

Factor	Relative Risk
Total Calories	0.70
Animal Protein	0.70
Total Protein	0.61
Total Fat	0.89
Cigarettes	0.08
Coffee	0.55
Tea	0.06

Abridged from Armstrong, B. and Doll,
R., *Int. J. Cancer*, 15, 617, 1975. With
permission.

the etiology of cancer is most likely to be an effect of the overall composition of the diet and not to any one single nutrient. Although the National Academy of Sciences recent report on Diet, Nutrition and Cancer,[22] has made some specific recommendations to the general public in regard to nutrition and its potential effects on risk for cancer, there is as yet no dietary regimen that can be prescribed with confidence toward alleviating or preventing cancer. Much of the evidence has been produced in animal systems where models have been designed to test hypotheses, based on observations from epidemiologic studies. Examples of some of the real or potential influences of some of the major dietary nutrients on susceptibility to cancer will be presented in the following pages.

A. Calories
Most of the epidemiologic data relating calories to cancer incidence do not permit quantitation of total dietary intake, and thus, to total calories. Many of the dietary studies in human populations have been based on preselected food lists. The international distribution of hormone dependent cancers has generated suspicion that these types of tumors may be related to affluence and in this way related to an increased overall total caloric intake. Berg[23] has suggested that diets consumed by affluent populations, if ingested from early life on through old age, may overstimulate the endocrine system and in this manner lead to metabolic aberrations resulting in increased cancer induction. Doll, Armstrong and colleagues[3,35,26] have shown that relative risk is high in regard to calories (Table 3).

Results of a study in Hong Kong populations by Hill et al.[24] indicated that the affluent group had more than twice the mortality from colo-rectal cancer, compared to the poorest group. While the relative proportions of nutrients in their diets were similar the estimated daily caloric intake was 2700 in the lowest socioeconomic group and 3900 in the highest. Gregor et al.[25] concluded, on the basis of data analyzed as to caloric intake and the incidence of gastrointestinal cancers, that the mortality rates for gastric cancers fall as the per capita food intake increases but for intestinal cancer the rates rise. This may reflect a longer lifespan in the populations where increased per capita food intake is associated with decreased gastric cancer, allowing more time for large bowel cancer to develop.

Doll and associates[3,26] have reported on the relation of cancer incidence to total caloric intake. In both reports significant correlations have been found for total calories and rectal cancer and leukemia in males, and for breast cancer in females. Gaskill et al.[27] also observed a correlation between breast cancer and total caloric intake in women of the United States but when these data were controlled for age at first marriage, the correlation lost significance.

Some of the data in the literature present conflicting results on caloric effects; for example, Miller et al.[28] studied a number of dietary variables, including total caloric intake, and found no association between the total caloric intake and breast cancer. In addition Jain et al.[29] examined data regarding cancer of the colon and rectum and reported a direct association with caloric intake for both colon and rectal cancer. In both of these studies it is unclear as to whether or not the total caloric intake was associated with a specific class of nutrients or if calories were derived from balanced sources.

Total body mass has has been considered one of the positive correlates to breast cancer in women; deWaard et al.[30] found such an association in post-menopausal women in The Netherlands. Similar observations have been noted in studies from Brazil,[31] and Taiwan.[32] The suggestions from these investigations have indicated that total body mass might be related to nutritional status and thus to the occurrence of breast cancer, but this has been questioned by others.[33] Despite these questions further studies by deWaard et al.[34] have demonstrated that the heavier and taller post-menopausal women had the highest incidence of breast cancer. If indeed height and weight are important in the incidence of breast cancer this may relate in some important way to total caloric intake as appears to be the case in comparing breast cancer incidence between women in Holland and women in Japan.

The American Cancer Society has carried out long-term prospective studies examining the relationship between mortality of cancer and other diseases and how these data may relate to variations in weight.[35] These studies have shown that cancer mortality was significantly elevated in both sexes only among those that were 40% or more overweight and that males under similar weight excess had increased mortality from cancer of the colon and rectum. Women, in addition to greater risk for breast cancer, with increased overweight, also had an increased risk for endometrial cancer. Despite the less than clear evidence of these epidemiologic studies it appears reasonable that a high total caloric intake may be a risk factor for cancer of some sites in subsets of the human population.

Some of the epidemiologic studies described above have been extended to animal models. It was the fascination of the possible relationship between nutrition and cancer for Tannenbaum that provided much of the early evidence for an influence of diet and nutrition on cancer.[36] This was long before the subject became fashionable in scientific circles or of significant interest to the public.

Tannenbaum showed that tumors induced by benzo(a)pyrene in mice were inhibited by caloric restriction. The level of dietary fat had a profound effect on the growth of skin tumors or of spontaneous and chemically induced breast tumors.[36-40] While these studies were not as well controlled as would be the case with today's protocols, they clearly pointed to an effect of caloric restriction and a reduction in both chemically induced and spontaneous tumors of several types by reduced caloric intake. Table 4 illustrates data typical for caloric studies. Mice, with a daily dietary intake of 11.7 calories had 25% more spontaneous mammary tumors than mice with a daily caloric intake of 9.6 calories. Furthermore, it appeared that dietary fat had more of an enhancing effect if the calories from fat were 18% compared to 2%. This seemed to indicate to Tannenbaum that dietary fat exerted a specific influence beyond its caloric contribution. This investigator must be given credit for introducing the idea of caloric rather than nutrient restriction as a factor in reducing risk of developing cancer, a concept which still has been largely ignored by the scientific community.

Contemporaries of Tannenbaum, Lavik and Bauman,[41] studied the effect of dietary fat on 3-methylcholanthrene-induced skin tumors in mice. Using lard as a source of fat these investigators found that a low fat, low calorie diet was associated with fewer tumors and that a highly saturated fat with a low melting point resulted in fewer skin tumors.

The above brief references to the influence of caloric intake on cancer suggest that the evidence from epidemiologic studies is weak and for the most part indirect. Much of the data associates body weight and obesity with cancer and with the caloric source more often from fat than from other macronutrient energy sources.

Table 4
INFLUENCE OF CALORIES ON SKIN
TUMORIGENESIS FROM DMBA

Diet During Skin Painting (10 Weeks)	Diet Following Skin Painting (20 Weeks)	Tumors At End of Period (%)
Ad libitum	Ad libitum	69
Ad libitum	Restricted	34
Restricted	Ad libitum	55
Restricted	Restricted	24

Abridged from Tannenbaum, A., *Cancer Res.*, 4, 673, 1944. With permission.

Unfortunately, the experimental evidence in animals is clouded by the lack of control of experimental diets. In most of the animal studies the intake of all nutrients was simultaneously depressed; thus the effects may have been related to other nutrients as well as fat and total caloric intake. These studies are difficult to interpret; none permit an adequate assessment of the overall influence of total calories on cancer incidence. However, based on nutritional concepts and the recognized interactions of nutrition and disease it seems reasonable to assume that total caloric intake may interrelate with factors that influence the risk for cancer in human and animal populations.

B. Protein and Amino Acids

The literature records a number of reports which associate dietary protein with cancer of the kidney, pancreas, colorectum, prostate, endometrium, and breast. A problem with interpreting the data is that the major sources of dietary protein in most studies also contain other nutrients, and sometimes, nonnutritive components. Thus if protein is in fact associated with cancers of various sites the influence may be more indirect than direct.

There are a few epidemiological studies purporting to show an effect of protein on cancer in human populations. Kolonel et al.[42] analyzed the diet histories of more than 4000 subjects and observed a correlation of 0.7 between total protein and total fat consumption. In a larger study the incidence rates for 27 types of cancer in 23 countries and the mortality rates for 14 types of cancer in 32 countries were examined by Armstrong and Doll.[26] They reported relationships between many variables including the correlations of total protein and total fat; these correlations turned out to be 0.7 and 0.93 respectively. On the basis of these studies Armstrong and Doll reported that per capita intakes of total protein and animal protein were significantly correlated with the incidence of and mortality from breast cancer.

Knox observed that data from 20 different countries[43] indicated a strong correlation between per capita intake of animal protein and mortality from breast cancer. This compared to the findings of Armstrong and Doll that there was a stronger association for animal protein than for total protein. In both studies however the correlations of breast cancer with per capita total fat intake were generally stronger than those for animal protein. The mortality rates for breast cancer in 41 countries was correlated with per capita food intake for 1964 to 1966 by Hems;[44] there was a direct correlation with intake of protein, calories and total fat from animal products, independent of other components of the diet. A later study by Hems,[4,5] in England and Wales supported an association of breast cancer with fat more strongly than with proteins. Gray et al.[46] found a direct correlation for breast cancer and per capita intake of animal protein after controlling for height, weight and age at menarche. Gaskill et al.[27] observed a direct correlation between breast cancer mortality and per capita

Table 5
**STANDARDIZED MORTALITY RATES (MALE) FOR
STOMACH AND INTESTINAL CANCER AND THE
DAILY INTAKE OF ANIMAL PROTEIN IN VARIOUS
COUNTRIES**

Country	Standard Mortality Rate		Daily Intake of Animal Protein	
	Stomach	Intestines	1947 to 48	1962 to 63
Austria	43.04	9.59	21.1	47.5
Belgium-Luxembourg	29.46	12.99	33.1	45.9
Denmark	23.93	13.91	59.1	57.9
Finland	44.80	5.20	34.6	54.7
France	23.03	11.34	29.7	43.0
Federal Republic of Germany	38.41	9.15	21.1	49.2
Italy	34.22	7.97	15.2	29.8
Norway	27.67	7.89	47.9	48.8
Portugal	32.60	6.65	18.0	27.3
Sweden	24.83	9.78	62.6	54.3
Switzerland	27.81	10.53	43.8	51.3
Yugoslavia	21.72	2.16	11.2	25.5
Poland	39.79	2.95	17.3	40.0
Czechoslovakia	43.59	6.49	28.3	43.0
Hungary	46.40	6.26	15.9	37.0
Canada	18.69	13.88	59.2	63.8
USA	14.87	12.25	61.4	64.2
Australia	17.06	12.68	59.9	59.6
Chile	64.63	2.73	24.7	29.2
Colombia	21.20	2.24	22.5	20.0
Japan	67.96	2.97	9.7	16.9

Abridged from Gregor, O., Toman, R., and Prusova, F., *Gut*, 10, 1031, 1969.

protein intake but the finding was not statistically significant, after controlling for age at first pregnancy. Based on diet histories Kolonel et al.[42] reported a direct correlation between the consumption of animal protein and breast cancer in Hawaii in 5 different ethnic groups. Gregor et al.[25] have compared two periods (1947 to 48 and 1962 to 63) for gastrointestinal cancer and protein intake (Table 5). Their data indicate a direct correlation between protein intake and mortality from intestinal cancer in 28 countries.

Miller et al.[28] and Phillips,[8] in two of three case control studies of diet and breast cancer, found a significant direct association with dietary fat only in two of them but a direct association was found in the third between animal fat, protein and breast cancer. Thus, in epidemiologic studies it appears that there is some evidence for a correlation between dietary protein and breast cancer but the evidence is somewhat weakened because of the effects of other nutrients or factors in the diet which may be difficult to separate from protein.

A number of investigators have associated breast and colon or colorectal cancer with the consumption of protein.[25,47] Armstrong and Doll[26] found a strong correlation between per capita intake of total protein, animal protein and total fat on the incidence of and mortality from colon and colorectal cancer. On the other hand, Bingham[48] found no association for intakes of animal protein when attempting to correlate the average intakes of foods, nutrients and fiber in different regions of Great Britain. In a case control study of large bowel cancer Jain et al.[29] reported a direct association between the consumption of high levels of protein and risk for both colon and rectal cancer; there was, however, a stronger association for saturated fats and colorectal cancer.

One of the problems in attempting to identify clearly protein effects in illustrated by some of the studies which have examined the effects of meat intake (as opposed to protein per se) on colorectal cancer. Data from such studies are difficult to interpret; meat contains fat, some small amount of carbohydrates, and variable amounts of vitamins, minerals and non-nutrients. In some studies there have been strong correlations between international mortality rates for colon cancer and meat consumption, particularly beef.[49,50] Knox[43] has recorded the strongest correlations between cancer of the large intestine with the consumption of eggs; this was followed by beef, sugar, and pork. On the other hand time trend data for per capita beef intake and colorectal cancer incidence showed no clear association in studies in the United States.[9]

In studies related to large bowel cancer and selected types of foods, Haenzel et al.,[51] examining Japanese cancer cases in Hawaii, compared to hospital controls found an association between cancer of the colon and consumption of legumes, starches, and meats with the strongest association for beef. Further investigations in indigenous Japanese and in other populations and geographic areas did not confirm these findings.[52-55]

A strong direct correlation has been observed between the intake of animal protein and pancreatic cancer in studies from 33 countries examining up to 22 different types of neoplasms.[56] In an additional study from Japan,[57] Ishii et al. found an association of pancreatic cancer with the consumption of high meat diets by men. These data however were based on responses to a questionnaire completed by relatives of the deceased and some margin of error should be expected. A relative risk of 2.5 for daily meat intake in pancreatic cancer incidence in Japan in a cohort of 265,000 subjects followed prospectively has been reported by Hirayama.[58] While these observations do not provide strong evidence for a role for protein in pancreatic cancer they do suggest that meat may in some way be associated with a higher risk. This of course requires confirmation.

Other types of tumors have been associated with the intake of meat and/or protein. Armstrong and Doll[26] found a strong correlation between animal protein consumption and the incidence of renal tumors. Prostate cancer was significantly correlated with the consumption of total protein and animal protein in the study by Kolonel.[42] Hirayama[58] reported a close correlation between the increase in meat consumption in Japan and the increase of prostatic cancer, indicating a parallelism between the two. Others have indicated an increased incidence of endometrial cancer correlated with the intake of total protein; however, this may relate to the reported occurrence of breast cancer which parallels the intake of protein.

Experimental evidence for a correlation between cancer and protein intake has been derived from better controlled studies than the epidemiologic evidence. Despite better experimental design much of the data are not wholly convincing. An inhibitory effect of selected amino acid deficiencies on tumor responses in laboratory animals was reported as early as 1936.[59,60] Following the report of Voegtlin and Maver,[59] a number of reports appeared in the literature relative to the effect of protein intake in experimental animal models. Generally, animals that have been fed lower dietary concentrations of protein have developed fewer tumors. Since several factors in addition to protein were being varied at the same time it is difficult to interpret the results of some of these studies.[61-64] It should be noted also that in some studies the levels of carcinogen varied between the high and the low dietary protein groups of animals.[64] Furthermore, the total food intake was less for animals fed very high levels of protein.[62] Variations in conditions may have had an influence on tumor induction, as noted previously.

Another factor involved in protein intake and cancer is the balance of amino acids in the diets. The growth limiting sulfur amino acids must be present in sufficient quantity and balance to provide an adequate diet for optimal growth; this in turn can relate to tumor

Table 6
EFFECT OF DIETARY PROTEIN
ON AFLATOXIN CARCINOGENESIS

Treatment Aflatoxin, Plus	No. Rats With Tumors
High Protein Diet	11/30
Low Protein Diet	0/12

Abridged from Madhavan, T. V. and Gopalan, C.,
Arch. Pathol., 80, 133, 1968.

incidence since imbalanced protein in effect results in lower caloric intake and diminished utilization of the available protein.

In regard to protein sources and quality some studies have shown no significant difference in tumor incidence when animals were fed diets containing either animal protein or plant protein, compared on an equal basis.[65-67]

Amino acid balance in proteins as noted above is highly important. Larsen and Heston,[67] observed that cysteine added to a low casein diet fed to strain A mice increased the incidence of spontaneous pulmonary tumors. On the other hand, in other observations, a cysteine deficient casein diet had no effect on mammary gland tumors in female C3H mice. However, when the diets were supplemented with cysteine most all of the animals developed mammary tumors in a very short period of time. These two observations are somewhat similar to studies done with lipotrope deficient diets and effects on liver tumors.[20,21] In numerous studies in our laboratory using this animal model, it has been shown that cysteine added to the diet above the normal level actually enhances liver tumor induction.

From these and other observations it would appear that tumor enhancement by dietary protein in experimental animal models occurs only when the amino acid profile of the protein is imbalanced and that the effect is probably not due to any specific amino acid. In addition to the question of protein quality and quantity on cancer in experimental animals, even less is known about potential effects of protein on initiation or promotion stages.

The work of Ross and co-workers[63] has served clearly to point out the effects of protein on the development of spontaneous tumors in experimental animals. These workers have shown that the total incidence of various types of tumors was related directly to caloric intake and tumors appeared sooner if the caloric intake was high.[68] Since the rats used in the studies of Ross and Bras developed so many different types of tumors, they could not easily compare the effect of diet on specific types. The highest number of any tumor type for any one diet occurred as fibrosarcomas in animals fed a 30% casein diet. In some of the groups however the authors noted that when the caloric intake was identical there were more tumors in the group with the higher percentage of calories from protein. This was in particular the case with bladder tumors.

There are a number of significant reports relative to the relationship of dietary protein to chemically induced tumors in animal models. If aflatoxin is fed in diets with varying levels of protein, the incidence of liver tumors is generally lower in the animals fed the diets with lower concentrations of protein.[69] Young rats intubated with a carcinogenic dose of aflatoxin and then fed either 5% or 20% casein diets for one year clearly showed the effects of protein (Table 6). The rats fed 5% protein had no tumors whereas 50% of those fed the 20% casein diet had tumors. The studies of Wells et al.[70] and Temcharoen et al,[71] essentially confirmed the work of Madhavan and Gopalan,[69] by showing that higher protein diets enhanced the induction of liver tumors in animals exposed to aflatoxin.

In general the effects of the quality or quantity of dietary protein on liver carcinogenicity in rats is almost always opposite to the influence of protein level on hepatotoxicity; this has been shown by many investigators. The effects appear to be related to the metabolism of the carcinogen such that a lower protein diet results in less activating enzyme in the liver, less toxic intermediate and thus, less hepatotoxicity; the parent compound is not converted to the reactive intermediate as rapidly with low protein as it is with a high protein diet. In the case of aflatoxin B_1 this phenomenon is related to the rate of activation of the parent compound by the mixed function oxidase enzyme system to the 2,3 epoxide which binds to DNA, and, while less well understood, in some degree to rate of repair of DNA.

Engel and Copeland[72] reported that dietary protein had no influence on tumors induced by 2-AAf in rats, whether the diets contained 9% or 27% casein. On the other hand, Morris et al.[73] reported more and a greater variety of tumors in rats treated with 2-AAF when they were fed diets containing 18%, or 24% casein, compared to those fed 12% casein diets. Some of these reports are difficult to interpret, as noted above, because of the complexity of the diets and failure in some cases to fully explain how the diets were constructed and how the animals were exposed to the carcinogen.

The work of Walters and Roe,[74] using the infant mouse as the test animal and DMBA as a carcinogen, found that mice fed 25% casein developed significantly more lung tumors than those fed diets containing either 10% or 15% casein. On the other hand there are reports with an opposite effect where a reduction in the protein content of the diet enhanced the formation of dimethylaminoazobenzene (DAB) liver tumors.[75] Further to the effects of protein in an area that has been somewhat neglected was the work of Clinton et al.[76] These investigators showed that the effect of dietary protein on the incidence of DMBA induced mammary tumors in rats was dependent on whether the diet was fed before or after the administration of the carcinogen. This has been further investigated by Topping and Visek,[77] who studied the effect of dietary protein on adenocarcinomas of the small and large intestines of rats induced by dimethylhydrazine. These investigators observed larger and more numerous tumors in rats fed diets containing 15% or 22.5% protein compared to those given 7.5% protein diets. Furthermore, the lag time was less for appearance of tumors in the group given 22.5% protein.

A number of studies have been conducted in attempts to define the mechanism of how protein may influence carcinogenesis. It has been shown, as noted above, that a low protein intake depresses the mixed function oxidase enzyme systems responsible for metabolism of carcinogens such as aflatoxin B_1.[78,79] This is related to the formation of aflatoxin B_1-DNA covalent adducts.[80]

While the evidence for an effect of dietary protein on metabolism of aflatoxin B_1 is strong and no doubt has an important influence on metabolic activation and initiation processes, there is additional evidence from more recent studies which indicates that protein may have a profound effect following the initiation of the carcinogenic process and thus, metabolism is not involved.[81,82] One of the currently used markers for initiated liver cells, gamma glutamyltranspeptidase (GGT), has been shown to be more often present in the liver of rats fed a 20% casein diet compared to 5% casein diet, when both were given after the administration of aflatoxin B_1. Furthermore, a post initiation effect of a low protein diet appeared to be capable of overcoming the potential carcinogenic effects of a higher AFB_1-DNA adduct level established by feeding high levels of protein during the administration of AFB_1.

A number of studies have been reported relative to the effects of protein on tumor transplants. The majority of the evidence indicates that low protein diets tend to inhibit the growth of transplanted tumors.[83-88]

In attempting to evaluate the evidence for an effect of protein on carcinogenesis it appears that whatever influence this important class of nutrient has is probably not straightforward or direct. Low protein does appear to suppress the carcinogenic process and the progress of the transformed cells to tumor formation but there are some important exceptions to this general statement. For example, the DMBA induced tumor yield in animals fed low protein diets may be increased in some cases. In addition, nominal amounts of protein (20 to 25%) would appear to enhance the induction of tumors whereas levels in excess of 25% appear to have no further enhancing effect or in fact sometimes inhibit tumorigenesis.[70,77,82] In epidemiologic studies there are suggestions for a possible association between high levels of dietary protein and risks of cancers of certain organs and tissues but the literature on protein is much more limited in this respect than that concerning fat and cancer. Taken together the evidence from both epidemiological and laboratory studies only suggest that protein may have an effect on risk of cancer of certain sites but the lack of adequate data on this important dietary ingredient precludes firm conclusions.

C. Carbohydrates

When fiber is excluded and considered separately from other carbohydrates there is very little in the literature relative to this class of dietary ingredients and its effect on cancer. Armstrong and Doll[26] found a direct correlation between sugar intake and pancreatic cancer mortality in women; they also reported a weak association between liver cancer incidence and the intake of potatoes. Hems[44] reported that a high intake of refined sugar was associated with an increased incidence of breast cancer. This report however is in direct opposition to the finding of these same investigators,[89] where an inverse relationship was found between breast cancer and starch. In a study involving data from 37 countries, Drasar and Irving[90] found a direct correlation between the intake of several sugars and breast cancer. In other studies, starchy foods including bread and potatoes, have been reported to correlate with both esophageal and gastric cancer.[91,92]

In addition to the evidence from epidemiologic studies there are a few reports from experimental investigations which suggest an association between carbohydrates and certain forms of cancer.

Neither Roe et al.[93] nor Friedman et al.[94] found an effect of sucrose on tumor induction in mice and rats. This was the case even when sucrose was fed at a level as high as 77% of the diet equivalent to 40 g/kg body weight. Sucrose, as 20% of the diet fed to rats and to dogs, has yielded conflicting results in the laboratories of some investigators.[95] After 2 years of consuming 20% sucrose in the diet female rats had a higher incidence of hepato-cellular tumors. Later studies however did not confirm this nor was there any evidence in female beagle dogs for such an effect.

The limited studies to date leave open the question of sucrose per se and its influence on tumorigenesis.

The report of Gershoff and McGandy[96] is interesting; these investigators examined the interaction of dietary lactose fed at a level of 49%, or sucrose varying from 43 to 55% of the diet, along with vitamin A deficiency. Urinary bladder tumors in Charles River rats was used as the model. One aspect of this study of particular interest, but not a new observation, was the induction of bladder stones by the high lactose diet. The stones appeared to be related to the induction of tumors because of the irritant effect of the calculi. Gershoff and McGandy reported that about 60% of the rats fed lactose in a vitamin A deficient diet developed bladder calculi. About 30% of those with bladder calculi also developed transitional cell carcinomas. Groups of rats fed sucrose without vitamin A deficiency did not develop calculi nor did they develop histological changes which suggest preneoplastic changes in the bladder epithelium. This is an interesting observation since it represents a rare case of experimental tumor induction in the absence of a known carcinogen.

D. Fiber

The real or potential effects of fiber as a type of carbohydrate, as opposed to refined carbohydrates such as sucrose and other sugars, is sufficiently different from conventional carbohydrates to direct it to be considered separately. There has been a considerable amount of attention in the last few years directed to the physiological significance of dietary fiber. The tendency in the beginning was to consider all indigestible fiber the same, but it is now recognized that fibers differ remarkably in digestibility and in the chemical constituency such as cellulose, lignin, hemacellulose, pectins, and gums. A major characteristic common to all of these is the capacity to form bulk in the gastrointestinal tract.

Dietary fiber has a complex composition that varies from one type of plant to another. Earlier studies were directed toward "crude fiber" and the measurements generally underestimated the fiber content because the analytical techniques for determining crude fiber only measured lignin and cellulose.

It has been established that populations in the Western world during the past several decades have decreased their consumption of fiber.[97] On the basis of this knowledge and in considering other populations, particularly Africans, where the food fiber intake is high, Burkitt and Trowell[98] suggested that chronic diseases of the bowel, including colon cancer, are associated with a low intake of dietary fiber; this hypothesis has been examined by a number of investigators in epidemiology as well as in controlled laboratory studies.

As noted above, most of the data derived from epidemiologic studies have been directed toward the possible role of dietary fiber in protection against large bowel cancer. The proposed mechanisms have ranged broadly from the capacity of fiber to dilute carcinogens that are present in the large bowel, to decreasing intestinal transit time of ingesta; an additional consideration is the influence of fiber on the composition and metabolic activity of the microflora of the gut, the latter involved in the production of putative carcinogens by modification of fecal bile acids. At best, the epidemiological studies have yielded conflicting results. It has been suggested[99] that the differences in colon cancer incidence among East Indian populations in the North of India and those in the South of India might be explained by the high levels of roughage in the North Indian diet, compared to the very low levels in the diet of Indians living in the south of the country. The Punjabis in the north, who consume a high level of vegetable fiber have a very low incidence of large bowel cancer. This compares to a higher incidence in the south Indians where vegetable fibers are rare or absent in the diets.

In a study involving adult males from Denmark who are at a high risk for colon cancer and comparing them to a group from Finland, a low-risk population, it was found that the Danes consumed less fiber and stool weights were much less than those of the Finns,[100] a finding similar to that in the Indian study. Bingham et al.[48] in examining United Kingdom subjects, found no significant correlation between total fiber intake and mortality rates from colorectal cancer. However, this latter study pointed out that the mean intake of the Pentosan fraction of dietary fiber in vegetables was inversely correlated with mortality from colon cancer suggesting a different effect by pentosan-containing fiber compared to the other types of fiber. This finding and others have led investigators to examine in a more detailed manner specific components of fiber rather than just the total crude fiber.

There have been reports of studies which have yielded results in opposition to those recorded above. For example, Liu et al.[101] examined the mortality from colon cancer in 20 industrialized countries and compared these rates to per capita food intake. The examinations were made between 1967 to 1973 and compared to those from 1954 to 1965. The relationship of fiber to colorectal cancer in these studies was not significant. The authors concluded that cholesterol was more closely associated to colorectal cancer, as an important risk factor, than fiber. Additional studies by Drasar and Irving,[90] from 37 countries indicated no relation of fiber consumption to colon cancer. A low risk group in Utah was compared to the United States national population as a whole and there was little if any difference observed in fiber intake between the 2 populations.[102]

Table 7
FIBER, DIMETHYLHYDRAZINE AND
COLON TUMORS IN RATS

Treatment	%Colon Tumors	Tumors per Tumor-Bearing Rat
4 DosesDMH	68	1.9
Beef Fat, No Bran		
Beef Fat, Plus Bran	38	1.1
Corn Oil, No Bran	65	1.5
Corn Oil, Plus Bran	43	1.6
8 Doses DMH	70	3.0
Beef Fat, No Bran		
Beef Fat, Plus Bran	63	3.1
Corn Oil, No Bran	90	2.8
Corn Oil, Plus Bran	66	3.4

Abridged from Wilson, R. B., Hutcheson, D. P., and Wideman, L., *Am. J. Clin. Nutr.*, 30, 176, 1977. Copyright American Society for Clinical Nutrition. With permission.

A number of case control studies have also reported inconsistent results. These have included studies in hospital and neighborhood controls,[103] of populations in Minnesota and in Norway;[104] populations with colorectal cancer and those without it in the United States;[105] and a study in Canada.[29]

Hill et al.[24] reported that the highest socioeconomic group that was studied in Hong Kong had the highest incidence of colon cancer; this correlated with a high intake of fiber and calories. The lowest socioeconomic groups had correspondingly lower incidence rates and lower fiber intake. This was somewhat in line with the observations of Martinez et al.,[106] who found higher consumption of fiber and total residue in Puerto Rican cancer cases than in controls.

Since the Japanese have a relatively low colon cancer incidence, studies were conducted by Glober et al.,[107,108] comparing the bowel transit times in men from three different populations. They examined the Japanese in Japan at low risk and compared them to Japanese and caucasians in Hawaii who have a high risk of colon cancer. Bowel transit times were similar in both Japanese populations but were shorter than that of caucasians.

In reviewing these various studies it seems clear from the epidemiologic evidence that it is difficult to support the hypothesis that dietary fiber per se protects against colon cancer by whatever mechanisms have been proposed.

There have been a relatively large number of studies done in animal models in attempts to correlate fiber and colon cancer. Tumors have been chemically induced by dimethylhydrazine (DMH), azoxymethane (AOM), methylazoxymethanol acetate (MAM), dimethylaminobiphenyl (DMB), and nitrosomethylurea (NMU). These various chemicals have permitted studies relative to the effects of various types of fiber on induced cancer. Some of these will be reviewed here.

There are reports which indicate that bran has some protective effect in rats against DMH induced colon cancer.[109-111] Table 7 illustrates typical results of animal studies. Freeman et al.[112,113] have also observed a protective effect by cellulose when fed to rats also exposed to DMH. Other studies have been conducted, the results of which are equivocal in many cases and conflicting in others.[114-116] It is of particular interest however that some investigators have demonstrated that some types of fiber result in a denuding of the epithelium and an increase in cell turnover in the colon which exposes more of the cells to the effects of carcinogens.[117]

Table 8
NUTRIENT INTAKE IN RELATION TO CANCER INCIDENCE IN HAWAII: INFLUENCE OF FAT. TEN ETHNIC GROUPS. INCIDENCE PER 100,000 POPULATION, ANNUAL RATE.

Target Organ	Correlation Fat		Coefficient Protein	
Lung	Cholesterol	0.30	—	
Breast	Total	1.49	—	
	Animal	2.50	Animal	2.34
	Saturated	4.17	—	
	Unsaturated	2.59	—	
Uterus	Total	0.60	—	
	Animal	1.06	Animal	0.94
	Saturated	1.79	—	
	Unsaturated	1.05	—	
Prostate	Animal	0.87	Animal	0.76
	Saturated	1.27	—	
Stomach	Fish	10.21	Fish	3.18

Abridged from Kolonel, L. N., Hankin, J. H., Lee, J., Chu, S. Y., Nomura, A. M. Y., and Hinds, M. W., *Br. J. Cancer*, 44, 332, 1981. With permission.

The question of whether or not fiber, in its various forms has any influence on colorectal cancer in human populations or on induced colon cancer in animals is still unanswered and the evidence equivocal. There is no conclusive evidence to suggest that dietary fiber is protective; however, the animal data suggests that some forms of fiber may offer protection from cancer at some sites. It seems reasonable then that an increase in fiber consumption, particularly from fruits, vegetables, grains, and cereals might be encouraged in human populations.

E. Dietary Lipids

The influence of quality and quantity of dietary fats have been studied from both epidemiological and experimental standpoints, in regard to specific tumor types, probably more than any other single dietary ingredient. Moreover, there is more convincing evidence regarding fat and its relation to cancer than there are for any of the other nutrients. Despite the fact that there have been large numbers of studies, it is sometimes difficult to interpret the findings because protein is almost always an accompanying nutrient and it is difficult to say that the observed effects are in fact a result of fat alone.

There have been a large number of epidemiologic studies relative to dietary fat and breast cancer. A number of these have shown direct associations between the per capita intake of fat and breast cancer incidence or mortality.[26,27,42-44,46] Table 8 shows data suggesting such associations. In most of these studies the correlations have been higher for total fat than for other dietary factors considered at the time, including animal protein, meat, or specific fat components in oils.

Both intracountry and intercountry studies have been used to compare dietary fat intake with breast cancer. A large study by Gaskill et al.,[27] compared the per capita intake of various foods, by states within the United States, with the breast cancer mortality rates and found that there was a significant correlation of fat intake when all states were combined. Other factors however, including age at first marriage and geographic location, seemed to nullify some of these correlations. In England and Wales it was noted that time trends for breast cancer mortality from 1911 to 1975 correlated best with corresponding intake patterns of a per capita basis for fat, sugar and animal protein 10 years earlier.[45] Moreover, Kolonel et al.[42] correlated the consumption of fat with breast cancer incidence in Hawaii and found that there were significant associations between breast cancer and total fat consumption, with animal fat, and with both saturated and unsaturated fats. In other studies, Phillips,[8] observed a direct association between the frequency of consumption of high fat foods in breast cancer in case control investigation among Seventh-Day Adventists in California. Luben et al.,[118] Miller et al,[119] and Nomura et al.,[120] using case control studies, have also associated the consumption of fat with breast cancer in a positive way.

In addition to breast cancer there is also epidemiological evidence for an association between dietary fat and prostate cancer. This has been supported by the observations of Kolonel et al.,[42] Armstrong and Doll,[26] Hirayama,[58] Blair and by Schuman et al.[121] These epidemiologic observations support the suggestion of many investigators that dietary fat is associated in some as yet unclear manner with increased risk for prostate cancer in men.

Perhaps the most studied aspect of fat in cancer interrelationships, aside from the breast, is that of the gastrointestinal tract. In these investigations dietary fat has been associated with cancer at a number of sites along the gastrointestinal tract, including the stomach and large bowel. Major emphasis however has been placed on large bowel cancer and the relation of dietary fat to neoplasms at this site.

Higginson was one of the first to report an association between gastric cancer and frequent consumption of fried foods and animal fats.[122] As early as 1966 this investigator reported a correlation between gastric cancer and frequent consumption of fried foods along with increased use of animal fats for cooking. A study by Graham et al. however,[123] could not confirm the suggestion of Higginson, based on a study of 168 gastric cancer cases and matched hospital controls. Armstrong and Doll,[26] have reported correlations between colon and rectal cancer incidence in mortality and per capita intake of total fat. These data were based on an international study and represented many countries with differing use rates per capita for fat.

A strong correlation between mortality and cancer of the large intestine and per capita total fat intake was reported by Knox.[43] Others have reported varying results from epidemiologic studies in relation to cancer of the large bowel and fat intake but the correlations have been largely based on international studies. Studies from individual countries have been less convincing and, generally, have not shown a significant correlation between total fat intake and large bowel cancer.[48,102,124]

It is interesting to observe the striking difference between intracountry associations and international correlations. A number of factors may be involved including the variation in fat intake amongst countries whereas within a given country the fat intake may be more uniform. In addition, some implications have been based on fat disappearance rather than on human consumption

It was referred to earlier that the diets of Danish residents are associated with a high risk and that of the Finns with a low risk, based on food fat content. Consumption of total fat is similar for both groups but the differences in fiber may account for part of the variation of large bowel cancer. Phillips,[8] has also reported a direct association between colon cancer and the frequent consumption of high fat food by Seventh-Day Adventists. Dales et al.,[105] observed a direct association between colon cancer and frequent consumption of foods high in saturated fat, the association stronger for those who not only consumed the diets high in saturated fat but low in fiber content. Numerous investigations clearly point to a suggestion for fat and colon cancer interactions but many are conflicting and provide data that are very difficult to interpret.[50-54] The data of Enstrom[9] is probably as correct as any in pointing out that trends in beef intake (and thus fat intake) in the United States do not correlate with trends in the incidence of colorectal cancer, either from incidence or from mortality.

Results from many epidemiological studies only suggest an association between fat intake and gastrointestinal tract tumors, particularly the large bowel, breast, and prostate cancer, and definitive answers await further investigations.

The experimental evidence for an effect of dietary fat on mammary carcinogenesis was reported by Tannenbaum.[37] This investigator reported that dietary fat enhanced the development of mammary tumors in mice regardless of whether they were spontaneous or induced by chemicals. It was interesting in this study to observe that the incidence of spontaneous tumors was greater when the high fat diets were started at 24 weeks of age compared to waiting until they were 38 weeks of age, nearing 1/2 of their lifespan. These findings point to increased susceptibility of the developing or young mammary tissue to the effects of fat.

A question of considerable significance is whether the effects are mediated during early exposure or late exposure to the nutrient or to the carcinogen.[126] It has been shown that chemically induced tumors can be affected by caloric restriction but the incidence of tumors can be modified by increasing the dose of the carcinogen.[127,128] Tannenbaum also showed that fat, rather than calories per se, was responsible for enhancing tumorigenesis by feeding mice isocaloric high and low fat diets.[37]

The quality as well as the quantity of fat was shown by Carroll and Khor to be as important a factor in the induction of breast cancer as was the carcinogen and its dose.[129] These investigators used the mammary gland as the target organ to test this hypothesis and found that the incidence of tumors at this site was uniformly high with all dietary fats, when they were fed at a level of 20% in the diet, but the number of tumors per group was greater in the rats fed unsaturated fats. These investigators have further concluded, on the basis of more recent studies, that dietary fat exerts its effect during the promotional stage of carcinogenesis.[130] Recent studies,[131] in our own laboratory suggest that it is a function of the type of fat that determines whether or not it influences carcinogenesis during exposure to the chemical or afterwards (initiation or promotion stages of tumorigenesis). In most of the studies reported to date it appears that the amount of unsaturated fatty acids in the fat under consideration has a significant effect on response of animals to chemical carcinogenesis.

It would appear that the polyunsaturated fat used in a particular study must provide sufficient amounts of essential fatty acids (about 3% in the case of the rodent) to demonstrate an effect. Furthermore, the diet must contain relatively high levels of total fat, in addition, in order to increase the incidence of breast tumors.[132]

Chan et al.[133] reported a reduction in the latent period for the development of NMU induced breast tumors in Fischer rats fed high fat diets, compared to rats fed low fat diets. The data from this study however are difficult to interpret since the NMU model produced an increase in the incidence only and not in the multiplicity of breast tumors.

Studies by McCay et al.,[134] have shown that both the quality of the fat and its concentration in the diet influences the growth rate of DMBA induced breast tumors. A diet containing a high level of polyunsaturated fat resulted in an average tumor growth rate considerably greater than that of tumors in rats fed a diet containing high levels of saturated fat. These studies have indicated that tumor growth rate may be determined in part by total dietary fat and in part by the polyunsaturated fat content of the diet, probably reflecting the importance of essential fatty acids.

Transplantable tumors have been used to study the effects of dietary fat. In general, transplanted mammary tumors appear to develop more readily in mice fed a high level of polyunsaturated fat compared to mice fed an equivalent level of saturated fat.[135] A number of other investigators have reported similar evidence but there is conflicting interpretations of effects of various types and amounts of dietary fat on transplantable tumors.[136-138]

Additional studies in animal models have shown that the effects of a high fat diet on breast carcinogenesis can be modified, particularly by diets that are marginal in the lipotropic factors, choline and methionine.[131,139,140] The tumor incidence induced by AAF was lower and the death rate from tumors occurred later in DMBA treated rats if the diet was marginal in lipotropic factors, compared to those fully supplemented.

An interesting further development in the effects of dietary fat on cancer of various sites are the observations that Fischer rats, generally resistant to breast cancer induced by AAF, exhibit an increased incidence of hepatic carcinomas if the diet is high in fat and marginal in the lipotropic factors. This clearly points to dietary effects on target organs where either the susceptibility of the target organ is altered or an additional target organ is added, in part by the quality and/or quantity of dietary fat.[140]

Dietary fat can also modify heptocellular carcinomas induced in rats by aflatoxin B_1. When beef fat was fed to rats the number of tumors was the same irrespective of when the beef fat was fed, either during the period of exposure to the carcinogen or before and after exposure to the carcinogen.[141] On the other hand, feeding polyunsaturated fat before and after carcinogen exposure resulted in 100% tumor yield compared to only 66% tumors when the oil was fed only after tumor induction. It was concluded in this study that unsaturated fats increased tumor yield more effectively than saturated fats and that this effect may occur during the initiation or early promotional phases of hepatocarcinogenesis.

The incidence of pancreatic adenocarcinomas induced by azaserine is also amendable to modification by dietary fat. Longnecker et al.[142,143] have clearly demonstrated that unsaturated fat, in the form of either corn oil or safflower oil, was much more effective in inducing tumors than was saturated fat at comparable levels. Furthermore, if the rats were fed a diet that was deficient in the lipotropic agents there was a marked increase in hepatocellular carcinomas induced by azaserine.

There has been an increased interest in the effects of dietary fat on experimental carcinogenesis of the large bowel in animals. One of the earlier demonstrations that dietary fat might have an effect on colon carcinogenesis was a report by Nigro et al.[144] These investigators demonstrated that rats treated with AOM developed more intestinal tumors when fed a diet containing 35% beef fat, compared to those fed a regular chow type control diet that contained considerably less fat. These data are difficult to interpret because the laboratory chow is an ill-defined diet and one cannot effectively sort out the effect of calories from that of fat on tumorigenesis.

Reddy et al.[145,146] have produced a number of reports regarding large bowel carcinogenesis induced by DMH or AOM and the effects of quality and quantity of fat on incidence and multiplicity of tumors. These investigators have reported that low fat corn oil diets resulted in more tumors than low fat lard diets; however, rats fed 20% fat, whether it was saturated or unsaturated, have about the same tumor incidence. It appeared from these studies that the animals ate the same quantity of diet but since the caloric density of low and high fat diet differed those eating the high fat diets received more calories. This has been associated with an increased tumor risk. Other investigators have also reported on the effects of dietary fat on carcinogenesis in rodents and the data are somewhat equivocal when examined in detail as to the effect of fats on colon carcinogenesis.[147-149]

Experiments in our laboratory using both DMH (Table 9) (which requires metabolic activation) and NMU (a direct acting carcinogen) to induce colon tumors in rats have failed to support the hypothesis that quality and quantity of fat in our experimental systems do influence carcinogenesis.[150] In using a direct acting carcinogen which requires no metabolic modification (NMU) or one that requires activation (DMH), we have observed no effect on tumor incidence by either the quality (corn oil, beef fat, or Crisco®) or the quantity (5% or 20% of fat in the diet) of fat on tumor induction.

There are a number of reports in the literature relative to the influence of cholesterol on human and animal cancer.[101,147] Pearce and Dayton[151] clearly indicated, in a study using groups over 400 men in each trial treatment, that those fed high levels of polyunsaturated fats as a means of lowering serum cholesterol had a much higher incidence of cancer deaths

Table 9
TUMOR INCIDENCE IN DMH-TREATED RATS
FED LOW AND HIGH FAT DIETS

		5% Mixed Fat	24% Beef Fat	24% Corn Oil	24% Crisco®
Number of Rats[a]		39	40	40	40
Animals with Intestinal Carcinoma					
Colon	No.(%)	30(77)	27(68)	25(63)	22(55)
Small Intestine	No.(%)	2(5)	3(8)	5(13)	3(8)
Colon Tumors					
Adenocarcinoma (Total No.)		56	40	35	34
Benign Polyps		1	1	0	0
Tumors/Tumor Bearing Rat		1.60	1.48	1.40	1.50
Mean Tumor Size (mm ± SD)		9.0±5.8	9.2±4.3	9.4±6.5	10.7±9.8
Tumor Location (No.)					
Proximal Colon		20	17	18	24
Mid Colon		21	13	15	9
Distal Colon		15	10	2	1
Frequency of Tumor Size (%)					
<5 mm		13	15	23	15
5 to 10 mm		66	50	55	59
11 to 20 mm		17	34	16	10
20 mm		6	3	9	12
Tumor Bearing Rats With Multiple Tumors (%)					
1 Tumor		47	67	60	59
2 Tumors		33	22	40	32
3 Tumors		10	7	0	5
4 Tumors		8	4	0	5
5 Tumors		3	0	0	0

[a] N = effective number of starting animals.

than those fed a conventional diet. Other studies have suggested a similar effect.[152] However, Ederer et al.[153] in examining 5 control trials of cholesterol-lowering diets, failed to find any significant difference in relative risks between those fed the diets affecting serum cholesterol concentrations and those that did not receive cholesterol-lowering diets. There are numerous other epidemiologic reports in the literature to which the reader may turn for further information.[53,154-160] Results of studies referred to above are conflicting and inconsistent. Nevertheless cholesterol may have an influence on certain types of carcinogenesis and certain types of cancer, in part by way of its influence on cell membranes.

A brief word regarding the relationship of fecal steroid excretion and bowel carcinogenesis is in order at this point. It has been pointed out by a number of investigators that the metabolites of fats in the colon of individuals at risk might provide evidence to confirm or deny an influence of fat on carcinogenesis. It appears that the amount of neutral and acidic fecal steroids correspond to the level of dietary fat intake. Studies of the ratios of primary to secondary bile acids or the ratio of cholesterol to its metabolic products reveal no differences among populations at risk or those that are not at risk for large bowel cancer.[153,161-163] A number of others have alluded to evidence for and against the influence of dietary fat on potential carcinogens produced from bile acids but the evidence is at best equivocal.

It would appear at this point that the evidence for dietary fat in both experimental and in epidemiological studies about cancer causation is greatest for breast cancer. There is also evidence for dietary fat having an influence on prostate cancer in man and perhaps uterine cancer in women. The data on fat and large bowel cancer however are equivocal. On the other hand, there is excellent evidence for an effect of fat on the experimentally induced liver and pancreatic cancer. Thus, it would appear that fat quality and quantity may be more closely associated with cancer of some sites in humans and in experimental animal model systems than any of the other nutrients. These suggestions however require further confirmation.

F. Vitamins

A major interest in diet, nutrition and cancer is centered on some of the vitamins. In recent years, vitamin A and the retinoids, ascorbic acid or vitamin C, and vitamin E, have come under intense study. Vitamin K, as one of the fat soluble vitamins has received essentially no attention and the B complex vitamins have been much less addressed. However, as we shall point out, some areas have been examined in depth but mainly in animal models and without much interpretation as to how vitamins may impact on cancer in human populations.

Vitamin A is a generic term for a family of substances possessing vitamin A activity. To these naturally occurring materials have been added a number of synthetic analogues, all of which are now categorized under the generic term, "retinoids." While vitamin A was the first vitamin isolated and identified as one of the nutrients of significance to human health, most knowledge about its use and biological effects centers only on its relation to vision. Despite this, there has been an accumulating data base indicating that vitamins A (the retinoids) are very much involved with regulation of cell proliferation and in this way may relate to the neoplastic process.

A number of epidemiological investigations have indicated an inverse relationship between "vitamin A" and a variety of cancers of various organ sites, primarily those of the epithelial surfaces which require vitamin A for their integrity. In the epidemiologic studies the estimates of vitamin A intake have been based primarily on the frequency of ingestion of the group of foods known to be rich in vitamin A, in beta carotene, and in certain other members of this family of nutrients. The green and yellow vegetables are those most often associated with vitamins A in the diet. For this reason the studies have usually measured an indirect effect of vitamin A or of beta carotene.

One of the first investigators to report the correlation between vitamin A and lung cancer was Bjelke.[164] This investigator studied Norwegian men and observed lower values for lung cancer cases in nonsmokers. Additional studies by Gregor et al.[165] and MacLennan et al.[166] while varying somewhat in the assessment, have indicated that those with lung cancer had consumed less vitamin A. Others have also pointed out that lung cancer incidence has varied inversely with carotene intake and with foods that contain higher concentrations of vitamin A.[167-169]

Mettlin[168] has pointed to a relationship between vitamin A consumption and urinary bladder cancer. This has also been indicated in cancer of the larynx,[170] the esophagus,[171,172] and the stomach.[58,122,173] More recent cohort studies in the United Kingdom and in the United States have suggested an inverse relationship between serum levels of vitamin A and cancer risk in general.[174,175] The latter studies are difficult to interpret because serum levels are usually buffered by liver stores of vitamin A and these reports have concerned populations which are generally not deficient in vitamin A.

There are a number of lines of evidence based on experimental animal models indicating that vitamin A and other members of the retinoid family have a marked influence on the induction of cancer. These reports include a large number of studies done with the synthetic retinoids including 13-cis retinoic acid.

Some of the earlier studies in regard to vitamin A and its influence on neoplasia or

Table 10
VITAMIN A, 13-CIS-RETINOIC ACID AND LUNG
CANCER IN THE SYRIAN GOLDEN HAMSTER

Treatment (Diet)	Malignant Tumors of Respiratory Tract	
	Number	%
Control, 2μg/g retinyl acetate, BP	46/89	51.7
Low Vitamin A 0.3 μg/g retinyl acetate, BP	102/127	80.3
High Vitamin A 30 μg/g retinyl acetate, BP	40/88	45.4
Control, 2 μg/g retinyl acetate + 13-cis- retinoic acid during dosing BP	38/83	45.8
Control, 2 μg/g retinyl acetate + 13-cis- retinoic acid during and after dosing BP	4/91	4.4
Control, 2 μg/g retinyl acetate + 13-cis- retinoic acid after dosing BP	11/84	13.1

From Newberne, P. M. and McConnell, R. G., *J. Environ. Pathol. Toxicol.*, 3, 323, 1980. With permission.

neoplastic-like lesions were actually done in cultures.[176] There are a number of features in common between early neoplasic lesions in the respiratory tract and those associated with uncomplicated vitamin A deficiency. Squamous metaplasia of the tracheal or bronchial epithelium is associated with both carcinogen-induced changes and with vitamin A deficiency. Crocker and Sander[176] demonstrated that vitamin A inhibited the induction of squamous cell metaplasia and proliferation of the epithelium of the respiratory tract in vitro, caused by benzo(a)pyrene. Nettesheim and Williams[177] found that vitamin A deficiency enhanced the induction of respiratory lesions by 3-methylcholanthrene (MCA), however the interpretation of these data is difficult; the early lesions have as yet not been established as definitive precursors of cancer although they are associated with it. It should be re-emphasized here that the lesions referred to as ''preneoplastic'' cannot be taken as a priori evidence for exposure to a cancer inducing agent or condition.

There are conflicting reports regarding vitamin A effects on cancer induction, some of which derives from studies in our own laboratory. For example, Smith et al.[178] reported that an excess of vitamin A had no inhibitory effect on lung tumors induced by benzo(a)pyrene in hamsters. This is in contrast to the report of Saffiotti et al.[179] who observed that a high intake of vitamin A appeared to protect against squamous metaplasia and respiratory neoplasms in hamsters induced by benzo(a)pyrene. These data also are difficult to interpret because it is not clear how the results from the different animals and experiments were compiled.

Some studies have indicated that certain types of experimental tumors are enhanced by vitamin A. For example, retinyl acetate has been shown to enhance hormone induced mammary tumorigenesis in mice,[180] Rogers et al,[181] reported that a high intake of vitamin A increased the frequency and incidence of large bowel tumors in rats exposed to DMH.

A number of studies regarding vitamin A and, in particular the synthetic retinoids, have been reported by Sporn et al.[182-183] This area of research has a great deal of promise because many of the synthetic retinoids have unquestioned inhibitory effects on experimentally induced tumors of many sites in animal models. We have observed a protective effect on lung tumors in hamsters by 13-cis retinoic acid[184] (Table 10). It is still too early to know whether or not this can be extrapolated to human populations; some very interesting clinical studies are now in progress which should provide some answers within a few years, relative to the inhibitory effects of retinoids on cancer in subsets of human populations at high risk.

In addition to vitamin A and the synthetic retinoids a major interest has focused recently on the carotenoids, particularly beta carotene. These are the precursors of vitamin A in mammalian systems; carotenoids can be converted into vitamin A in vivo. In addition, in humans and in some animal species, they are absorbed unchanged from the gastrointestinal tract and are stored in tissues throughout the body, particularly in the lipid storage depots. Peto et al.[185] have put forth the hypothesis that beta carotene, rather than vitamin A, may be the beneficial factor in tumor inhibition when carotenoids are consumed.

Matthews-Roth[186] has reported that beta carotene and phytoene exert a significant protective effect against the development of ultraviolet induced skin tumors in hairless mice. Phytoene does not have vitamin A activity and any protective effect that is exerted, must reside in the carotenoid structure per se rather than in its being converted to vitamin A and acting through this mechanism. In addition Shamberger,[187] in some earlier studies, reported a relationship between vitamin A and carcinogenesis. This investigator conducted studies in which beta carotene was applied to the skin of mice along with croton oil; it was observed that this treatment increased the induction of epidermal tumors initiated by DMBA. Thus, more research is required into the actions of vitamin A and carotenoids in relation to carcinogenesis before it can be said with certainty that vitamin A and the retinoids, including carotenes and carotenoids, are effective in inhibiting or reducing certain types of cancer in animals and in human populations.

1. Vitamin C (Ascorbic Acid)

It was early in the 1960s when reports appeared in the literature noting that consuming foods rich in ascorbic acid was inversely related to the appearance of certain types of cancer.[122,188] Shortly thereafter there were other reports that suggested that vitamin C protected against gastric cancer, perhaps by blocking the reaction of secondary amines with nitrite to form N-nitroso compounds, some of which are gastric carcinogens.[42,173]

Some of the epidemiologic studies conducted early in the 1970s were concerned with human populations along the Caspian littoral of Iran. In these studies there was an inverse association between esophageal cancer and the consumption of fresh fruits which contain ascorbic acid.[172,189,190] More recently another report has appeared,[191] which indicates an inverse relationship between vitamin C consumption and uterine cervical dysplasia in women in New York. On the other hand, Jain et al.[29] failed to find an association between ascorbic acid consumption and colon cancer in a case control study.

There is additional evidence suggesting a role for ascorbic acid from animal studies. It has been shown conclusively that ascorbic acid can prevent nitrosation of amines which in turn prevents the formation of nitrosamines, some of which are gastric carcinogens.[192-194] It is well established that ascorbic acid can act in this capacity to prevent nitrosation reactions which produce carcinogenic nitrosamines; however, this class of compounds is only one of several which are associated with gastric cancer.

There have been a number of reports relative to the effects of ascorbic acid on cells in culture. These have indicated that ascorbic acid can inhibit the effects of carcinogens on cells in tissue culture.[195,196] Currently however the mechanisms for such inhibition is not clear. Based on evidence available to date one can only suggest that vitamin C or ascorbic acid may have an influence on human cancer; in animal systems it has been established only that it inhibits carcinogenesis through its effect on nitrosamine formation. Other considerations require further investigation.

2. Vitamin E

To date there is no epidemiologic evidence to suggest that vitamin E has an influence on tumor induction in humans. The only remotely possible correlation has been through studies related to the consumption of certain foods that contain vitamin E. These however are still speculative and will not be covered in this report.

There are data in the literature indicating that vitamin E or alpha tocopherol can have an influence on tumors induced in experimental animals. There is some indication that vitamin E acts similarly to ascorbic acid in blocking the formation of nitrosamines.[196,197] Vitamin E is fat soluble and nitric oxide nitrosation takes place most easily in lipophilic solvents. Thus, inhibition of nitrosation through this mechanism can occur in a lipid milieu; blocking of nitrosation by ascorbate occurs in the gastrointestinal tract, an aqueous environment.

Several investigators have studied the effects of alpha tocopherol on DMBA induced mammary tumors in animals.[198,199] It has been reported that the ingestion of high levels of alpha tocopherol during the initiation stage of mammary tumors had no effect. However other reports,[199] indicate that a large vitamin E supplement fed prior to DMBA exposure decreased tumor incidence by as much as 50%; on the other hand there are reports that alpha tocopherol at a reduced level in the diet resulted in an increased tumor incidence, in keeping with the above reported results. There was no significant difference in the tumor incidence between mice fed the high or the low doses of vitamin E, however, the average number of tumors per animal was less if the mice were fed the high vitamin E diet.[200]

Shamberger and colleagues,[201] reported an effect of vitamin E on skin tumors in mice promoted by croton oil. If vitamin E were applied topically a number of days after the carcinogen was applied there was a significant decrease in the number of tumors. Mice with benzo(a)pyrene-induced sarcomas survived longer if they were given vitamin E (alpha tocopherol). Haber and Wissler,[202] noted that there was a marked decrease in the carcinogenicity of methycholanthrene in mice if the animals were given diets containing supplemental vitamin E. A number of other investigators have reported some effects of vitamin E on tumorigenesis. From these studies it is clear that the results of vitamin E and its effects on tumorigenesis is variable. The results appear to depend in part on the carcinogen that is used and in part on the animal species that is used as a model.

The well known effect of vitamin E as a blocking agent in nitrosation confirms a role as a protective agent in the environment, however. Vitamin E competes for available nitrite and in this way, as with ascorbic acid, it blocks the nitrosation of amines and amides, reducing the amount of nitrosamines in foods and in the environment. Segregating the effects of ascorbic acid from alpha tocopherol on blocking nitrosation is yet to be done but as noted above one (vitamin C) is hydrophilic while the other (vitamin E) is lipophilic.

3. Choline and other B Vitamins

A number of the B vitamins have been investigated in relation to their effects on experimental cancer. Much of the earlier work was complicated by failure to control for the intake of other dietary nutrients, particularly protein and total calories. Some of the data are in fact conflicting or negative in terms of the protective effects of B vitamins on cancer.[65,203] In other cases some of the B vitamins have been shown to have a protective effect, primarily through the influence on metabolic pathways. Kensler et al.[204] demonstrated that riboflavin offered some protection against liver cancer caused by azo dyes in rats because it enhanced the detoxification of the carcinogen by a flavin dependent enzyme system.[205] On the other hand, instead of a protective effect one can envision that riboflavin might also be an enhancer under some conditions, particularly if it is required for activation of the parent compound to the ultimate carcinogen.

The complex interrelationships between the B vitamins choline, methionine, folic acid, and B_{12} have been examined in detail with reference to effects on tumor induction by N-nitroso compounds, mycotoxins, and polycyclic aromatic hydrocarbons. The most consistent results obtained with animals fed diets low in the lipotropic factors choline and methionine are an enhancement of cancer of the liver, the pancreas and, to a lesser extent, the colon. The diets used by Rogers and Newberne,[206] do not have a consistent effect on tumor induction in target organs other than the liver and the colon; the effects of lipotropic agents on carcinogenesis appear to be mediated through effects on metabolism but other possible mechanisms are under investigation.

A concerted effort has been applied in recent years to determine whether or not the metabolism and transport of vitamins or the binding of the appropriate coenzyme forms to apoenzymes are altered in tumor cells.[207-209] In some cases the transport and binding of the coenzyme forms have been reported but it is not known whether these alterations have any influence on the induction of or protection from carcinogenesis.

Thus the B vitamins require considerably more investigation to determine whether or not they are related to human cancer, and to further elucidate their real or potential role in experimental animal models. At this point there are no valid conclusions on B-vitamins except in the case of experimental models consuming lipotropic factors where it is clearly shown that they do play a role in susceptibility to chemical carcinogenesis. The lipotropes appear to be associated with methylation or hypomethylation of nucleic acids and by this mechanism may influence control of cell replication. This issue is receiving significant attention in a number of laboratories in regard to control mechanisms for cell proliferation.

G. Minerals

There are very few studies reported in the literature relative to the effects of oral ingestion of minerals on carcinogenesis. Only in recent years has there emerged an interest in regard to some of the trace elements; the macroelements (Ca, P, Mg) have been largely ignored. Many of the studies reported in experimental systems have been associated with parenteral administration of essential and of toxic agents. This is not a realistic approach for assessing human risk although it provides information of academic interest. These studies have been reviewed in some detail and will not be covered further in this review.[210,211] While the results of these studies are interesting they have done little to elucidate the risk or protective effects of minerals, particularly trace elements, in the amounts that occur naturally in the diet or in the environment of humans.

The earlier work of Schroeder and colleagues,[212-218] raised questions about the influence of environmental heavy metals and trace elements on carcinogenesis. These investigators studied the influence of at least 20 trace elements, some of which were heavy metals, on mice and rats using biological assay systems. In most cases, when administered at levels of as much as 3 mg/liter of water, there was little effect on growth and survival. These observations as well as the difficulty in attempting to interpret results from studies conducted under the less than adequate methods of trace element analysis available at that time leave many conclusions in question. Nevertheless, the studies of Schroeder and colleagues pointed the way toward investigations which have yielded some very interesting data carried out under more controlled conditions and where elemental analysis is much more sophisticated and accurate.

Of the essential elements only selenium and zinc have data, sufficient in quantity and quality, to be of significance. The heavy metals, arsenic, cadmium and lead will not be considered in this chapter but references are provided for entry into the literature for those interested.

1. Selenium

Selenium deficiency and excess have been economic problems in livestock for more than 100 years; it was only in the 1930s when as association was clearly identified between selenium deficiency or excess and important livestock diseases. There was little progress toward a better understanding of selenium and its significance to animal and possibly human disease until the 1950s.[219] The discovery that selenosis (an excess) and selenium deficiency occurred in some animal species was responsible for mapping the distribution of selenium in the soils and in plant foods in animal and human tissues in the United States and other countries. It is important to know what is in the soil because of all of the elements that are dependent upon the soil concentration, for transport into plants and animal tissues, selenium is the most prominent member of the trace elements directly related to soil content. When there is an excess, some plants tend to concentrate selenium and when the deficiency is moderate to severe the plants contain very little of it. Because of this food products grown on such soils are directly affected by the selenium content of the soil.

The National Academy of Sciences issued a report in 1971, on epidemiological correlations of diseases in humans and in animals and pointed the way for investigators to further develop knowledge regarding selenium.

Shamberger and colleagues,[220-222] have reported in numerous publications on the relationship between selenium and cancer. They have correlated selenium levels in forage crops with mortality by state in the United States and reported an inverse relationship for cancer of the gastrointestinal and genitourinary tracts with the selenium content of staple foods. Shamberger and colleagues,[223] examined the selenium levels in the blood of 100 cancer patients and compared these with the selenium content in the blood of 48 normal subjects. The analyses revealed that patients with gastrointestinal cancer and Hodgkin's disease had significantly lower levels of selenium than normal subjects but there were no differences between normal individuals and patients with cancer at other sites.

Another group of workers, Schrauzer and colleagues, have correlated per capita intake of selenium with cancer mortality rates in more than 20 countries.[224,225,226] A weakness of these reports however is that the correlations were based on food disappearance rate of major staple foods rather than actual estimates or measurements of food intake. This does not detract from the fact that these investigators found an inverse relationship between selenium intake and leukemia as well as cancers of other sites including the colon, breast, ovary, prostate, bladder, and lung. The blood levels of selenium and corresponding cancer mortality rates in 22 different countries were compared; these revealed an inverse relationship between selenium levels and for most of the cancer at various sites.

Shamberger et al,[224-226] examined cancer mortality rates in the United States by county, comparing the rates in the Northeastern part of the United States with corresponding levels of selenium in the water supply. They reported an inverse correlation between selenium levels in the drinking water and mortality from colorectal cancer.

Results of experimental animal studies with selenium have been controversial and conflicting as to the evidence for a carcinogenic or a protective role for this important element. Studies that were conducted at the FDA during the 1940s indicated that high levels of selenium induced or enhanced tumor formation.[227] In addition, results of studies from Russia,[228] reported two decades later, suggested that animals given high selenium supplements in the diet developed liver cirrhosis as well as heptocellular carcinoma.

There are a number of major defects in both of these studies; the Food and Drug Administration investigators used diets that were low in protein and, to compound the stress, they used a protein of low quality. The difficulty in interpretation of Russian studies was related to the fact that a selenium-free control group was not used. The results from three separate trials were all different; the final, third experiment failed to produce any tumors in a trial using 100 animals.

Results of both of these studies appear to have been influenced by extraneous dietary factors or conditions unrelated to selenium. Selenium, therefore, is no longer considered a carcinogen by the scientific community although it is a highly potent hepatotoxin when administered at levels only moderately above those considered essential for animals.

There is an accumulation of data which clearly indicates a protective role, in animal models, for selenium against certain types of tumors. For example, Schrauzer et al.[229] reported that breast tumors in female C3H mice were reduced from 82% in controls to 10% in selenium supplemented mice. Harr et al.[230] and Ip and Sinha,[231] demonstrated beneficial effects of supplementation of selenium at dietary concentrations very little above those exhibiting physiological effects. An interesting aspect of the studies by Ip and Sinha related to their finding that the incidence of mammary tumors, induced by DMBA, was enhanced by diets high in polyunsaturated fats and by a dietary deficiency of selenium. Supplementation at a level of 0.1 μg/gm, barely above the nutritional requirements, resulted in protection against tumor formation.

Numerous other investigators have reported an effect of selenium on tumor incidence in animal models; generally it has reduced the incidence or frequency of tumors of various sites.[232-237]

Studies conducted in our own laboratories relative to the acute and chronic effects of selenium on aflatoxin induced tissue damage have clearly indicated an effect of selenium within the range of safety concentrations for animals.[238] For example, the acute effects of aflatoxin are enhanced by either a deficiency or an excess of selenium and these observations carry over into the chronic effects,[239] where it is clear that a protective effect is exerted by levels somewhat higher than nutritional needs. When the toxic level and above is reached the carcinogenic effect is enhanced.

Further studies in our laboratory,[240] have shown that selenium, interacting with various fats, does indeed have an effect on biological response but that this does not appear to be through its influence on modulating tissue peroxidation.[241] These data point to a potentially very important aspect of the biological activity of selenium. It appears that the in vivo and in vitro effects of selenium may be to decrease the activity of hydroxylating enzymes that activate parent chemicals, such as aflatoxin B_1, to their reactive carcinogenic metabolites. This appears to be one of the mechanisms of action of selenium, in this case acting through glucuronyltransferase (GSH-TX).[234,242] However, additional effects may be through the selenium dependent enzyme glutathione peroxidase (GSH-PX). The report of the National Academy of Sciences,[219] alluded to the effects of selenium on certain biological systems. It was reported that while the mechanism of action is not clear, selenium does protect against acute and chronic toxicity of some of the heavy metals and this protection does not appear to be through an increased elimination of the toxic elements. Rather, it appears, that there is an increased accumulation of nontoxic forms of the heavy metal.

The animal data regarding selenium and cancer strongly suggest an effect of this trace element on the induction of cancer in animal models, but the limited evidence does not permit firm conclusions. There are suggestions for an effect of selenium on the rate of DNA repair and on its influence on activating enzymes. In addition it appears to influence cell cycle by extending the time for the G_2/M phase. It is clear that increasing selenium to a point just below the toxic level protects against certain types of tumors in animals; in other cases the results are equivocal or negative. This area of nutrition and cancer is open for some fruitful further investigations.

2. Zinc

The question of whether or not zinc is involved in cancer is relatively recent. An hypothesis has been based on assumptions made from a few studies in human populations,[244,245] and a number of animal experiments. Zinc is essential to more than 100 enzyme systems; furthermore, it has been demonstrated to be effective in reversing certain types of diseases in human populations. Its role in animal nutrition has been established for many decades. It is also accepted that severe zinc deficiency does occur in humans and that more moderate forms of deficiency are not uncommon.[246]

Zinc deficiency in humans and in animals results in depressed immune function in both

Table 11
ZINC AND COPPER LEVELS IN SERUM AND HAIR OF NORMAL HUMANS AND THOSE WITH DISORDERS

Specimen	Zn	Cu
Serum (µg%)		
Normal subjects	102.7 + 18.5[a](15)[+]	133.4 ± 30.5[a](15)
Patients with esophageal cancer	78.0 ± 14.9[b](20)	159.4 ± 44.4[a](20)
Patients with other types of cancer	114.4 ± 31.8[a](6)	148.6 ± 10.2[a](6)
Patients with other disorders	96.2 ± 15.0[a](9)	146.7 ± 39.4[a](9)
Hair (ppm)		
Normal subjects	195.0 ± 29.0[a](16)	
Patients with esophageal cancer	162.0 ± 33.0[b](19)	
Patients with other types of cancer	169.0 ± 37.0[b](4)	
Patients with other disorders	212.0 ± 48.0[a](8)	

Note: Values in each group with different superscripts (a or b) are significantly different. Mean ± S.D. Numbers in parentheses are the number of subjects.

From Lin, H. J., Chan, W. C., Fong, Y. Y., and Newberne, P. M., *Nutr. Rep. Int.*, 15, 635, 1977. With permission.

the cell mediated and humoral arms of the system.[247-250] The levels of zinc in blood and other body issues have been examined in cancer patients and compared to controls by a number of investigators. Schrauzer et al,[225,226] observed that the mean zinc concentration in food and in blood correlated inversely with mortality rates from cancer of the large bowel, breast, ovary, lung, bladder, and oral cavity. Strain,[250] in examining patients with bronchogenic carcinoma, observed that zinc levels differed between the cancer patient and age-matched controls. Furthermore, copper levels were lower in controls resulting in high ratios of zinc to copper in cancer patients. Davies et al.[251] had reported earlier that the level of zinc in the plasma of bronchogenic carcinoma patients were lower than those of other patients with other types of cancer and were also lower than normal laboratory values.

In studies from our own laboratory we have reported that the levels of zinc in serum, hair and from the diseased but noncancerous esophageal tissue from esophageal cancer patients were significantly lower than levels from patients with other types of cancer, with patients with other types of disease, or in normal subjects.[252] The concentration of zinc in the hair was lower in all cancer patients than in normal subjects (Table 11). In agreement with the observation of Strain et al.[250] the work of Lin et al. found significantly elevated copper levels in esophageal cancer patients, affecting the zinc to copper ratios. Furthermore, the serum of such patients had much lower iron levels than normal subjects.

The studies of Petering et al.[253] working with transplanted Walker 256 carcinoma in rats, showed that zinc deficiency inhibited the growth of transplanted tumors in animals and also prolonged the survival time. These data were confirmed by DeWys and Poires.[254] The effects of zinc were extended to leukemia, lung carcinoma and plasmacytoma TEPC-183.[255]

While these observations are not surprising, in view of the fact that rapidly growing tissues require zinc for cell proliferation, they are in contrast to animal studies conducted in our own laboratory using the esophagus as the tumor target organ.[265,257] We have consistently observed an enhancing effect of zinc deficiency on nitrosamine induced esophageal cancer and an inhibitory effect if supplements were administered at levels of nutritional requirements (Table 12). In our model the enhancing effect by zinc deficiency is further elevated by administration of ethanol.[257]

In contrast to our deficiency studies and protective effects by modest zinc supplementation, we have not found that an excess of zinc, over and above the nutritional requirement, has

Table 12
ANIMAL GROUPS, TREATMENT, AND TUMOR INCIDENCE

Group No.	Total No. of rats	Dietary zinc, ppm	MBN[a]	4% ethyl alcohol in drinking water	13-cis-RA[b]	No. of rats with tumors/No. of survivors %	Total No. of tumors	Average No. of tumors per rat
1	12	60	−	−	−	0/12(0)	0	0.00
2	40	60	+	−	−	14/35(40.0)	63	4.5
3	40	7	+	−	−	25/33(75.7)	76	3.04
4[c]	40	7(60)	+	−	−	18/35(51.4)	87	4.83
5	40	7	+	+	−	29/34(85.3)	99	3.41
6	40	7	+	+	+	33/35(94.3)	76	2.30

Note: The + and − signs mean presence or absence of diet component.

[a] 2 mg/kg body wt, twice weekly, 8 doses.
[b] 13-cis-RA (67 µg/g) was added to the diet at the end of dosing.
[c] Returned to control diet, 60 ppm zinc, at end of MBN administration. Groups 3, 5, and 6 were significantly different from group 2, P <0.05. Groups 3 and 4 were significantly different, P <0.05.

From Gabrial, G., Schrager, T., and Newberne, P. M., *J. Natl. Cancer Inst.*, 68, 785, 1982.

any further appreciable effect on tumor inhibition. Poswillo and Cohen,[258] on the other hand have reported that a high level of dietary zinc, greatly exceeding the nutritional requirements, suppressed the carcinogenic effect of DMBA in hamsters. Additional evidence for an effect of excess zinc was reported also by Duncan and Dreosti.[259]

The seeming conflict in reports from our laboratory and other laboratories in regards to zinc deficiency can be explained largely on the basis of the organ site being examined. Zinc deficiency inhibits the growth of all tissues in mammals so far examined, with the exception of esophagus. In this case a zinc deficiency actually enhances the proliferative activity of esophageal epithelium, permitting more cells to be exposed to the carcinogenic process. This very likely provides the explanation for the conflicting results reported from several laboratories.

From the foregoing conclusions it seems clear that the epidemiologic evidence while suggestive, is not yet available to make specific comments on the effects of zinc in human cancer but in experimental animal models it does have a profound effect, the results of the studies depending on the tissue examined and the animal model used in the investigation.

III. IMMUNOCOMPETENCE, NUTRITIONAL STATUS, AND CANCER

Investigators working in developing countries around the world recognized decades ago that malnourished populations were more susceptible to infectious disease than were those consuming adequate amounts of nutrients.[248,249] Only recently has significant efforts been made to design field and laboratory studies which would help to better understand the complex interrelationships between nutrition and the immune system. The reports by Cicely Williams in 1931 and 1932, in which kwashiorkor was so adequately described, served as a catalyst for research ultimately relating malnutrition to deranged immunocompetence. Interrelationships between malnutrition and the immune system have been well documented in recent years.

Malnutrition appears to have a particularly important influence on cell mediated immunity; however, phagocytosis, the complement system, and the humoral response in general are all affected by the nutritional status of the host. These, acting in concert, determine the magnitude of response to infectious disease. This is now coming under intense scrutiny as a potential for enhancing or decreasing the susceptibility of mammalian systems to environmental carcinogens. Suggestions for this have come from observations about the increased susceptibility to cancer of immunosuppressed individuals.

The interactions of individual nutrients, and the response to infectious disease have been covered in great detail in a series of reviews.[248,249] In the interest of space the interactions of nutrition in immunologic function will be restricted to a very short discussion with appropriate references included to allow the reader to gain entry into the literature.

The complexities and subtle balances that exist within the immune system are only now becoming evident in a general way, and, in particular, how may these properties relate to cancer.[260-266] Many new leads have emerged in recent years, pointing the way for studies of nutrients at the molecular level and allowing a better understanding about how these influence the immune system. The knowledge about lymphocyte biology is growing rapidly; and such information is being applied in general to the field of immunological regulation of carcinogenesis. In comparison to the brief survival time of most other circulating white blood cells lymphocytes are extremely long lived. Associated with this long life is a thin outer layer of carbohydrate, referred to as a glycocalyx, which is a highly dynamic structure. The turnover of this complex layer of carbohydrate takes place from 12 to 24 times per hour requiring that surface components lost by shedding must be resynthesized on a continuing basis.

The B-lymphocyte class of white cells have about 10^5 molecules of immunoglobulin on their surface membranes which exhibit limited antigenic specificity similar to the immunoglobulin secreted by the cell after it has been stimulated. These proteins are also lost by shedding or by enzymatic degradation but can be restored completely in a matter of a few hours. Thus the lymphocyte is a highly active metabolic unit and must therefore be regarded as a consumer of the many individual nutrients required for orderly and optimum function.

It is the influence of nutrition on the T and B cell populations and subsets of these populations that may link nutrition to some observed effects on either the inhibition or enhancement of carcinogenesis. While many advances have been made in recent years regarding the potential relation of the immune function and cancer, much remains to be done. It should be pointed out here however that the interactions of nutrition, the immune system and cancer are exceedingly important to an overall understanding of the mechanisms of cancer induction. If there is indeed a surveillance by the immune system over transformed cells and their progression to malignancy, nutrition is undoubtedly involved in very important ways. As noted above those cells responsible for presumed surveillance over foreign cells and other types of foreign materials entering the body require a continuing supply of nutrients in order to carry out their physiological functions. It is in this area of research that there is great promise for progress in the prevention and treatment of cancer. Some of the considerations of perceived importance of the immune system to the occurrence of cancer are covered briefly in the following pages. The discussion is arranged into sections to include: (1) components of the immune system and their perceived role in immunocompetence; (2) immunologic aspects of primary liver cancer, an organ site studied in much detail in recent years; and (3) lipotropes as nutrient examples and their relation to immunocompetence and the induction of experimental cancer.

A. Components of the Immune System

As noted above the activity of the various components of the immune system and how they interact may be determined in part by nutrients in the diet. This segment will consider the subject in a general way and point out the various components of the immune system and their relation to the development or the rejection of tumors; it must be realized however that this area is far from well understood. The components of significance include the cellular and humoral systems involved in specific and nonspecific immune responses. These are served by the T-lymphocytes derived from the thymus. There are several subsets of cells in this category each of which is recognizable by specific antigens on the cell surface. These include cytotoxic T-cells, suppressor T cells, and helper T cells. An additional subpopulation is that derived from the bone marrow, referred to as B-lymphocytes; these make the specific antibodies providing for humoral immunity. An additional class serving the immune system is that containing the non-specific reactive cells; these are the macrophages, granulocytes, natural killer cells, activated killer cells, and antibody dependent killer cells. Finally, there are the nonspecific reactive humoral agents such as complement.

Our concept of the immune system and its surveillance over the development of tumors suggests that destruction of a transformed or developing tumor cell might be the outcome to be expected. However some members of the immune system, particularly suppressor T cells and at times humoral (blocking) antibodies, actually seem to aid in the growth of tumors. These characteristics may be important in human cancer, particularly hepatocellular carcinoma, cirrhosis and immunocompetence.[266] A few comments on the better recognized aspects of the immunological regulation of carcinogenesis follows.

The cytotoxic T-lymphocytes derived from the thymus have as a major function the elimination of virus infected cells. They are also reactive against alloantigens or cells from other members of the given species. Furthermore, the function of cytotoxic T-cells can be considerably enhanced by immunization. Cytotoxic T-lymphocytes determines the immune control of tumors such as polyoma virus-induced tumors in mice, Marek's lymphoma disease in chickens, and, probably tumors of the lymphatic system in man induced by Epstein-Barr virus, but other immune mechanisms may participate in the latter. It is established further that humans with T-cell deficiencies are more susceptible to lymphoid tumors associated with oncogenic effects of the Epstein-Barr virus.[265,266]

Suppressor T-cell lymphocytes have as their main function the down regulation of immune responses. This cell type may be related to the growth of tumors or at least associated with weak or ineffective immune responses.[265] An example where this may work in experimental systems is that of mice exposed repeatedly to ultraviolet (UV) light which induces fibro-sarcomas and epithelial tumors. Two influences appear to be operational in this case; one is direct because of the oncogenic effect of UV light and the other is indirect because mice exposed to UV light develop suppressor T-cells in their lymphoid organs. Suppressor T lymphocytes appear in UV-irradiated mice prior to the appearance of primary skin tumors and are demonstrable by an ability to prevent the immunological rejection of transplanted, highly antigenic UV-induced tumors.

T-helper lymphocytes (helper inducer lymphocytes) appear to participate in virtually all specific immune responses. This class of T-cells can also be regarded as effectors because of their capacity for the release of lymphokines. They are thus mediators of delayed type hypersensitivity responses. The rejection of skin and organ allografts in mice appears to be dependent more on one type of helper cell than on another,[267] but details regarding this aspect of response are not yet available. It would appear however that responses by T-cells to autochtonous tumors may be strongly dependent on the T-helper lymphocytes but direct evidence is yet to be provided.

B-cells and humoral antibodies may be polyfunctional in relation to the regulation of carcinogenesis. There may be a potential for either retarding or for facilitating the growth of tumors. Antibodies to antigens on the surface of tumor cells may promote complement dependent lysis of a cell but may be ineffective as shown by the response to antigens on cells of T-cell leukemia of mice. It has been pointed out that the appearance of T-helper antigens signals a qualitative change in gene expression associated with neoplasia; further-more, high titers of antibodies to T-helper antigens have been raised in some strains of mice. It has been pointed out also that antibody can actually inhibit effector responses to tumors (blocking antibody) and may act by coating antigenic sites on tumor cells or by forming immune complexes with tumor antigens. Thus, high levels of immune complexes in some forms of cancer may in fact signal an adverse prognosis.

During the past few years a great deal of attention has been focused on the so-called natural killer (NK) cells. This concept was developed to explain the ability of lymphocytes from normal individuals to kill cells of tumor cell lines spontaneously, without any prior activation. Human natural killer cells (NK) resemble to some extent T-cells in that they possess the sheep erythrocyte receptor but morphologically they are distinct and have the appearance of "large granular" cells. Because the NK cells can kill tumor cells in vitro, without any prior stimulation, they have gained a great deal of credibility as cells of special interest for immune surveillance. It is not known however whether or not such a function actually exists in in vivo systems as a line of defense.

Activated lymphocyte killer cells (ALK) are a subset of a cell type recently identified as developing during the culture of lymphocytes under the stimulatory influence of T-cell growth factor (TCGF). In humans these cells have certain of the markers of T-cells yet have the morphological appearance and cytochemical marker of NK cells. Functionally, ALK cells are of interest because they have a broad and potent cytotoxic capacity for fresh and for cultured tumor targets. The origin of these cells is at present unknown.

For many years the macrophage has been a strong candidate for immunologic defense against cancer. When innocula of tumor cells are injected in the rats, together with an adjuvant (BCG) known to augment the activity of macrophages the growth of tumor is suppressed. A single intraperitoneal injection of silica however, only 3 days before challenge with a tumor eliminated this suppressive effect. In addition there is usually a high content of macrophages in tumors in rats and mice; in some cases they may constitute more than half of the total population of the tumor. The capacity of macrophages derived from rodent tumors often fail to inhibit specifically the growth of the tumor cells in vitro. It still remains a possibility that macrophage infiltration implies that antitumor effects of this class of cells may be conditional on other influences.

The concept of immunosurveillance was proposed by Thomas and Burnett many years ago. It postulated a continuing cytotoxic control over cells the surface of which might signify premalignancy because of certain changes on the surfaces. Somewhat later however the major responsibility for surveillance was passed largely on to macrophages and, then, more recently to NK cells. It must be pointed out that evidence for surveillance has always been circumstantial. It is more likely that if surveillance does exist, and there is reason to believe that it does, it probably results from most if not all of the effector mechanisms acting in concert.

Some promising leads are now emerging relative to the possibility of amplifying immunity to cancer. If this hopeful suggestion proves to be a realistic goal it should work to the advantage of the host, perhaps by specific immunization combined with adjuvants, through mechanisms effected by expanding a population of T-cells which are immune to the tumor, or in some way reducing suppressor cell populations.

B. Immunological Aspects of Primary Liver Cell Carcinoma

Perhaps one of the best examples supporting the concept of immunosurveillance of cancer is that of hepatocellular carcinoma where hepatitis B virus (HBV), combined with perhaps other environmental exposures, such as aflatoxins, are clearly pointing the way toward development of methods for intervention in cancer. Primary hepatocellular carcinoma develops in patients with chronic inflammatory liver disease; in most cases this has progressed to cirrhosis of the macronodular type. Patients who have hepatitis B virus exposure, or alcohol excess, or hemachromatosis, are the most likely subpopulations at risk for hepatocellular carcinoma. This tumor is rare in patients with autoimmune chronic active liver disease or in those with primary biliary cirrhosis.

The etiological agents themselves are suspected of being directly carcinogenic in the high risk disease groups. HBV-DNA is found in an integrated form in the majority of HBs antigen positive tumors.[266] The association of excessive alcohol ingestion with carcinoma of the head and neck region has led to suggestions that agents such as alcohol or impurities in it, and tobacco are in fact direct carcinogens.

Diseases associated with a high risk of primary hepatocellular carcinoma most commonly affect males,[266] and usually have reached the stage of macronodular cirrhosis by the time the tumor is detected. In addition many primary hepatocellular carcinomas secrete alphafetoprotein which we now know has immunomodulatory activity. The role of this protein in the entire process must be considered. A few comments on how these various factors may relate to the immunocompetence of host and the immunology of primary hepatocellular carcinoma which may in fact reflect similar findings in other types of tumors, will be reviewed briefly. Humoral responses are considered first.

Antibodies can be demonstrated in the serum of animals bearing a wide variety of experimentally induced tumors. The antibodies are useful in serological characterization and in isolation of tumor associated antigens but the presence of the humoral response is not necessarily correlated with increased resistance to the tumor by the host. Some of the ways in which tumor specific antibodies could theoretically mediate antitumor activity are (1) complement mediated lysis; (2) opsinization and phagocytosis; (3) loss of cell adhesion; all of these could contribute to antitumor activity.

Cell mediated responses may involve several different cell types. These can include: (1)

direct lysis by T-lymphocytes; (2) antibody dependent cell-mediated cytotoxicity; (3) killing mediated by activated macrophages, the latter being one of the major effects of cell mediated immunologic responses. Cytotoxic activity of macrophages is remarkably enhanced after the donor is infected with intracellular microorganisms. The efficiency of the in vitro cytolysis of tumor target cells by macrophages is also increased by the presence of activated lymphocytes or lymphokines which activate macrophages; (4) lysis by natural killer (NK) cells. NK cells are different from immune cytotoxic T lymphocytes; they lack surface immunoglobulins and C3 receptors but appear to possess other types of receptors. They also have extremely low levels of T-cell antigen and do not exhibit phagocytic activity. NK cells do infiltrate tumors but their role in vivo has not been determined.

The effectiveness of the immune response against tumors is limited in a number of ways. (1) The tumor may be located in an area of the body where effector cells cannot reach it; (2) the tumor cells may loose antigenicity or may exhibit a change in antigen markers; (3) there may be an enhancement of blocking factors; (4) the tumor mass may overwhelm the immune capacity; (5) suppressor T-lymphocytes may be enhanced; (6) the tumor may mediate suppression by synthesizing various materials such as prostaglandins and alphafetoprotein; these in turn affect the immune response.

There are a number of immune defects in patients with chronic liver disease.[264-266] For example, hyperglobulinemia and depressed cell mediated immunity are common to most forms of chronic liver disease and probably secondary to liver damage. The anergic state seems to be related to increased activity of prostaglandin secreting monocytes which have suppressor functions.

It has been shown that many of the host defense systems are impaired when patients with chronic liver disease develop cirrhosis, effects insufficient in themselves to result in a significant rate of tumor development. However, when there are increased levels of carcinogenic stimuli such as environmental toxins, viruses, etc. the tendency is for a more severe immunodeficiency to result. This then makes malignant transformation and evasion of immune surveillance more likely. The profound influence of nutritional status on immunocompetence makes this area of nutrition and risk for cancer an attractive one for further study. A brief description of one aspect is presented here.

IV. LIPOTROPES, IMMUNITY, AND CARCINOGENESIS

An interesting correlation between animal and human hepatocellular cancer exists in an experimental animal tumor model where cirrhosis is induced by lipotrope deficiency.[20,21] This in turn depresses cell mediated immunity and other functions of the immune system and both conditions appear to act in concert to increase risk of susceptibility to cancer induction.[267] These studies have demonstrated that lipotrope deficiency leads to impaired immunocompetence the severity of which depends on the age of the animal and the adequacy of other nutrients in the diet. In addition other factors appear to be involved which contribute in one way or another to the development of cancer in the lipotrope deficient animal. For example, a number of in vitro assays have demonstrated that rats fed the lipotrope deficient diet had a reduced potential for hepatic metabolism of most carcinogens.[268] If most chemical carcinogens require metabolic activation this situation should reduce susceptibility to carcinogens. Instead, susceptibility is increased by lipotrope (methyl group) deficiency. A factor which may bear on this seeming paradox is the fact that liver S-adenyslmethionine (SAM) is decreased in lipotrope deficient rats.[269-271] On the other hand, GSH content of the liver does not appear to be appreciably affected; this latter observation requires confirmation. Some recent (unpublished) data from our laboratories have shown that SAM is decreased in lipotrope deficient rat liver; in addition, GSH is also diminished, in contrast to the findings of Poirier. Furthermore, there is a marked decrease in the methylation of cytosine, an essential link in the cell proliferation process required for hyperplasia. While this appears to be opposed to what one might expect, depressed 5-methylcytosine may result in defects in the genetic material which leads to neoplasia. It is likely through this process that the lymphatic system is unable to respond in a normal fashion to stimuli and that deranged genetic material leads to increased susceptibility to neoplasia in other organs and tissues.

SAM may influence carcinogenesis by serving as a trap for electrophilic carcinogens or by playing some role in alkylation. It is reduced by feeding the carcinogen or by the administration of certain tumor promoters.[270,271] A reduction in hepatic SAM by carcinogens clearly points to a direct interaction occurring in one carbon metabolism with depletion of SAM. This may be directly related to deranged genetic material through decreased or aberrant methylation and thus, greater susceptibility to carcinogenesis.

Since the lipotropes are directly involved in cell proliferation and tumors require cell proliferation for growth, they are sensitive to alterations in methyl group metabolism. This is the basis for antimetabolic chemotherapy of tumors. Rapid cell turnover creates a need for methyl groups over and above a normal requirement, a condition characteristic of neoplastic diseases. For rapid cell proliferation such as that required by the immune system lipotropes are important in the cell proliferation activities. It is likely through this manner that immunocompetence is depressed in lipotrope deficiency which in turn bears upon the greater susceptibility to carcinogenesis.[20] Emerging data in this field, where lipotrope deficiency alone is associated with liver tumor initiation make this an exciting area for research into nutrient-cancer interaction.

V. OTHER FACTORS INVOLVED IN RISK FOR INCREASED CANCER SUSCEPTIBILITY

A considerable effort has been expended in human and animal research in recent years relative to the carcinogenicity of drugs and other chemicals used for therapeutic measures for many diseases including cancer. Evidence that human exposure to a chemical substance may be potentially carcinogenic is provided by epidemiologic studies, long-term bioassays in animals, and a combination of short-term in vitro screening methods for detecting neoplastic transformation.[272] Table 13 lists a number of drugs and chemicals that are known to be carcinogenic in man. Despite this the risk benefit evaluation of a drug must take into account the likely outcome of the disease if it is not treated and whether or not safer alternative therapies are available.

There are more than 20,000 generically different drug products available in the United States today and those that are established as human oncogens include the cytotoxic alkylating agents, immunosuppressive agents, estrogens, oral contraceptives, androgenic anabolic steroids, phenacetin-containing analgesics, diethylstilbestrol (DES) and phenytoin. The latter two are the only known examples of human transplacental carcinogens.

Nitrate consumption has been associated with increased risk for stomach cancer.[273] There is also some concern regarding drugs that can be nitrosated.[193,197] Many of the most effective drugs are tertiary amines such as aminopyrene, chlorpromazine, methadone, and oxytetracycline all of which can be relatively easily nitrosated to form nitrosamines. Table 14 lists data regarding nitrosation and blocking of the reaction by ascorbate. Nitrosamines generally are potent toxins many of which are carcinogenic.[274] As noted earlier this process can be inhibited by antioxidants including ascorbic acid, vitamin E or glutathione, perhaps our most effective naturally occurring system for defense against cancer-inducing agents.

Additional factors which must be mentioned in passing, relative to cancer causation, include the genetics of the host, exposure to the heavy metals arsenic, cadmium and lead,[22] food additives; pesticides; polycyclic aromatic hydrocarbons; polychlorinated biphenyls; polybrominated biphenyls, mycotoxins, pyrrolizidines, plant constituent metabolites and mutagens in foods.

Table 13
CARCINOGENIC DRUGS AND CHEMICALS

Agent	Species	Neoplasm
Alkylating	Mouse	Lymphosarcoma
Antineoplastic	Rat	Mammary Tumors
		Lymphomas
	Rabbit	Bladder
	Human	Bladder
Antimetabolites	Mouse	Lymphoma
	Rat	None
	Rabbit	None
	Human	Lymphoreticular
Anticonvulsant	Mouse	Lymphoma
	Rat	Leukemia
	Human	Lymphoma, Neuroblastoma
Hormones	Mouse	Vagina, Uterus
	Women	Vagina, Cervix
Estrogen/Progestin	Rat	Hepatocellular
		Carcinoma
	Women	Hepatocellular
		Adenoma/Carcinoma

Abridged from Schottenfeld, D., *Cancer J. Clin.*, 32, 258, 1982.

Table 14
ASCORBATE AND NITROSATION OF AMINES

Amine mM	Nitrite mM	Ascorbate mM	PH	Yield (%) Plus Ascorbate	Minus Ascorbate	Blocking (%)
				Oxytetracycline		
21.6	315	475	2.8	0	44	100
21.6	315	475	3.8	0.04	63	99
				Morpholine		
25	50	100	2.0	0.48	20	98
25	50	100	3.0	0	65	100
25	50	100	4.0	0	34	100
				Piperazine		
10	10	20	1.0	0.6	35	98
10	10	20	2.0	1.2	44	98
10	10	20	3.0	0.3	56	99
10	10	20	4.0	0.2	54	99

Note: Abridged from Mirvish, S. S., Wallcave, L., Eagen, M., and Shubik, P., *Science*, 177, 65, 1972.

VI. SUMMARY

The foregoing review of current risk factors, primarily concerned with nutritional status seems to clearly point to the fact that nutrients may be one of our major hopes for prevention of cancer. The observations that migrant populations exhibit changes in incidence and frequency of certain types of cancer point to external causes for cancer with diet high on the list of suspects. As pointed out in this chapter, the foods we eat are probably the most important risk factors that human populations encounter. While this may be discouraging on the one hand, on the other it is also encouraging because of all lifestyle factors associated with cancer, diet is one over which we exercise control. Epidemiologic observations, supported by controlled animal model research on carcinogenesis strongly suggest that through dietary control, some types of cancer are preventable.

REFERENCES

1. Surgeon General Report, Ch.14, Constituents of tobacco smoke, *Natl. Cancer Inst.*, 1979.
2. **Higginson, J. and Muir, C. S.,** Environmental carcinogens: misconceptions and limitations to cancer control, *J. Natl. Cancer Inst.*, 63, 1291, 1979.
3. **Doll, R. and Peto, R.,** The causes of cancer: quantitative estimates of avoidable risks of cancer in the United States today, *J. Natl. Cancer Inst.*, 66, 1191, 1981.
4. **Wynder, E. L. and Shigematsu, T.,** Environmental factors of cancer of the colon and rectum, *Cancer*, 20, 1520, 1967.
5. **Haenszel, W. M. and Kurihari, M.,** Studies of Japanese migrants. I. Mortality from cancer and other diseases among Japanese of the United States, *J. Natl. Cancer Inst.*, 40, 43, 1968.
6. **Haenszel, N., Berg, J. W., Segi, M., et al.,** Large bowel cancer in Hawaiian Japanese, *J. Natl. Cancer Inst.*, 51, 1765, 1973.
7. **Ackerman, L. V.,** Some thoughts of food and cancer, *Nutr. Today*, 2, Jan.-Feb., 1972.
8. **Phillips, R. L.,** Role of lifestyle in dietary habits in risk of cancer among Seventh-Day Adventists, *Cancer Res.*, 35, 3513, 1975.
9. **Enstrom, J. E.,** Health and dietary practices in cancer mortality among California Mormons, in *Cancer Incidence in Defined Populations*, Cairns, J., Lyon, J. L., and Skolnick, M., Eds., Cold Spring Harbor Laboratory, New York, 1980, 69.
10. **Phillips, R. L., Kuzma, J. W., Bisson, W. L., and Lotz, T.,** Influence of selection versus lifestyle on risk of fatal cancer and cardiovascular disease among Seventh-Day Adventists, *Am. J. Epidemiol.*, 112, 296, 1980.
11. **Lynch, H. T., Guirgis, H. A., Lynch, P. M., and Lynch, J. F.,** The role of genetics and host factors in cancer susceptibility and cancer resistance, *Cancer Detection Prev.*, 1, 175, 1976.
12. **Young, V. R. and Richardson, D. P.,** Nutrients, vitamin and minerals in cancer prevention, facts and fallacies, *Cancer*, 43, 2125, 1979.
13. **Hoover, R. and Fraumeni, J. F., Jr.,** Risk of cancer in renal transplant recipients, *Lancet*, 2, 55, 1973.
14. **Harris, C. C.,** Immunosuppressive anticancer drugs in man: their oncogenic potential, *Radiology*, 114, 163, 1975.
15. **Seiber, S. M. and Adamson, R. H.,** The clastogenic, mutagenic, teratogenic, and carcinogenic effects of various antineoplastic agents, in *Pharmacological Basis of Cancer Chemotherapy*, Williams and Wilkins, Baltimore, 1974, 401.
16. **Strong, L. C., Herson, J., Osborne, B. M., et al.,** Risk of radiation related subsequent malignant tumors in survivors of Ewings sarcoma, *J. Natl.Cancer Inst.*, 62, 1402, 1979.
17. **Shafritz, D. A.,** Hepatitis B virus DNA molecules in the liver of HBsAg carriers: mechanistic considerations in the pathogenesis of hepatocellular carcinoma, *Hepatology*, 2, 35, 1982.
18. **Brechot, C., Poureel, C., Habchouel, D. A., et al.,** State of hepatitis B virus DNA in liver diseases, *Hepatology*, 2, 27, 1982.
19. **MacSween, R. N. M.,** Alcohol and cancer, *Br. Med. Bull.*, 38, 31, 1982.
20. **Rogers, A. E. and Newberne, P. M.,** Lipotrope deficiency in experimental carcinogenesis, *Nutr. Cancer*, 2, 104, 1980.

21. **Newberne, P. M., Harrington, B. H., and Wogan, G. N.,** Effects of cirrhosis and other liver insults on induction of liver tumors by aflatoxin in rats, *Lab. Invest.,* 15, 962, 1966.
22. *Diet, Nutrition and Cancer,* National Academy of Sciences, Washington, D.C., 1982.
23. **Berg, J. W.,** Can nutrition explain the pattern of international epidemiology of hormone-dependent cancer? *Cancer Res.,* 35, 3345, 1975.
24. **Hill, M., MacLennan, R., and Newcombe, K.,** Letter to the editor: diet and large-bowel cancer in three socioeconomic groups in Hong Kong, *Lancet,* 1, 436, 1979.
25. **Gregor, O., Toman, R., and Prusova, F.,** Gastrointestinal cancer and nutrition, *Gut,* 10, 1031, 1969.
26. **Armstrong, B. and Doll, R.,** Environmental factors and cancer incidence and mortality in different countries, with special reference to dietary practices, *Intl. J. Cancer,* 15, 617, 1975.
27. **Gaskill, S. P., McGuire, W. L., Osborne, C. K., and Stern, M. P.,** Breast cancer mortality and diet in the United States, *Cancer Res.,* 39, 3628, 1979.
28. **Miller, A. B., Kelly, A., Choi, N. W., Matthews, V., Morgan, R. W., Munan, L., Burch, J. D., Feather, J., Howe, G. R., and Jain, M.,** A study of diet and breast cancer, *Am. J. Epidemiol.,* 107, 499, 1978.
29. **Jain, M., Cook, G. M., Davis, F. G., Grace, M. G., Howe, G. R., and Miller, A. B.,** A case-control study of diet and colorectal cancer, *Intl. J. Cancer,* 26, 757, 1980.
30. **deWaard, F. and Baanders-van Halewijn, E. A.,** A prospective study in general practice on breast-cancer risk in post-menopausal women, *Intl. J. Cancer,* 14, 153, 1974.
31. **Mirra, A. P., Cole, P., and MacMahon, B.,** Breast cancer in an area of high parity: Sao Paulo, Brazil, *Cancer Res.,* 31, 77, 1971.
32. **Lin, T. M., Chen, K. P., and MacMahon, B.,** Epidemiologic characteristics of cancer of the breast in Taiwan, *Cancer,* 27, 1497, 1971.
33. **MacMahon, B.,** Formal discussion of breast cancer incidence and nutritional status with particular reference to body weight and height, *Cancer Res.,* 35, 3357, 1975.
34. **deWaard, F., Cornelis, J. R., Aoki, K., and Yoshida, M.,** Breast cancer incidence according to weight and heights in two cities of the Netherlands and in Aichi prefecture, Japan, *Cancer,* 40, 1269, 1977.
35. **Lew, E. A. and Garfinkel, L.,** Variations in mortality by weight among 750,000 men and women, *J. Chronic Dis.,* 32, 563, 1979.
36. **Tannenbaum, A.,** The genesis and growth of tumors. II. Effects of caloric restriction per se, *Cancer Res.,* 2, 460, 1942a.
37. **Tannenbaum, A.,** The genesis and growth of tumors. III. Effects of a high-fat diet, *Cancer Res.,* 2, 468, 1942b.
38. **Tannenbaum, A.,** The dependence of the genesis of induced skin tumors on the caloric intake during different stages of carcinogenesis, *Cancer Res.,* 4, 673, 1944.
39. **Tannenbaum, A.,** The dependence of tumor formation on the degree of caloric restriction, *Cancer Res.,* 5, 609, 1945a.
40. **Tannenbaum, A.,** The dependence of tumor formation on the composition of the calorie-restricted diet as well as on the degree of restriction, *Cancer Res.,* 5, 616, 1945b.
41. **Lavik, P. S. and Baumann, C. A.,** Further studies on the tumor promoting action of fat, *Cancer Res.,* 3, 749, 1943.
42. **Kolonel, L. N., Hankin, J. H., Lee, J., Chu, S. Y., Nomura, A. M. Y., and Hinds, M. W.,** Nutrient intakes in relation to cancer incidence in Hawaii, *Br. J. Cancer,* 44, 332, 1981.
43. **Knox, E. G.,** Foods and diseases, *Br. J. Prev. Soc. Med.,* 31, 71, 1977.
44. **Hems, G.,** The contributions of diet and childbearing to breast-cancer rates, *Br. J. Cancer,* 37, 974, 1978.
45. **Hems, G.,** Associations between breast-cancer mortality rates, child-bearing and diet in the United Kingdom, *Br. J. Cancer,* 41, 429, 1980.
46. **Gray, G. E., Pike, M. C., and Henderson, B. E.,** Breast-cancer incidence and mortality rates in different countries in relation to known risk factors and dietary practices, *Br. J. Cancer,* 39, 1, 1979.
47. **Lubin J. H., Blot, W. J., and Burns, P. E.,** Breast cancer following high dietary fat and protein consumption, *Am. J. Epidemiol.,* 114, 422, 1981.
48. **Bingham, S., Williams, D. R. R., Cole, T. J., and James, W. P. T.,** Dietary fibre and regional large-bowel cancer mortality in Britain, *Br. J. Cancer,* 40, 456, 1979.
49. **Berg, J. W. and Howell, M. A.,** The geographic pathology of bowel cancer, *Cancer,* 34, 805, 1974.
50. **Howell, M. A.,** Diet as an etiological factor in the development of cancers of the colon and rectum, *J. Chronic Dis.,* 28, 67, 1975.
51. **Haenszel, W., Berg, J. W., Segi, M., Kurihara, M., and Locke, F. B.,** Large-bowel cancer in Hawaiian Japanese, *J. Natl. Cancer Inst.,* 51, 1765, 1973.
52. **Haenszel, W., Locke, F. B., and Segi, M.,** A case-control study of large bowel cancer in Japan, *J. Natl. Cancer Inst.,* 64, 17, 1980.

53. **Bjelke, E.,** Dietary factors and the epidemiology of cancer of the stomach and large bowel, *Aktuel Ernaehrungsmed. Klin. Prax. Suppl.,* 2, 10, 1978.
54. **Graham, S., Dayal, H., Swanson, M., Mittelman, A., and Wilkinson, G.,** Diet in the epidemiology of cancer of the colon and rectum, *J. Natl. Cancer Inst.,* 61, 709, 1978.
55. **Hirayama, T.,** A large-scale cohort study on the relationship between diet and selected cancers of the digestive organs, in *Gastrointestinal Cancer, Endogenous Factors; Banbury Report 7,* Bruce, W. R., Correa, P., Lipkin, M., Tannenbaum, S. R., and Wilkins, T. D., Eds., Cold Spring Harbor Laboratory, New York, 1981, 409.
56. **Lea, A. J.,** Neoplasms and environmental factors, *Ann. R. Coll. Surg. Eng.,* 41, 432, 1967.
57. **Ishii, K., Nakamura, K., Ozaki, H., Yamada, N., and Takeuchi, T.,** (In Japanese) Epidemiological problems of pancreas cancer, *Jpn. J. Clin. Med.,* 26, 1839, 1968.
58. **Hirayama, T.,** Changing patterns of cancer in Japan with special reference to the decrease in stomach cancer mortality, in *Origins of Human Cancer, Book A: Incidence of Cancer in Humans,* Hiatt, H. H., Watson, J. D., and Winsten, J. A., Eds., Cold Spring Harbor Laboratory, New York, 1977, 55.
59. **Voegtlin, C. and Maver, M. E.,** Lysine and malignant growth. II. The effect on malignant growth of a gliadin diet, *Publ. Health Rep.,* 51, 1436, 1936a.
60. **Voegtlin, C. and Thompson, J. W.,** Lysine and malignant growth. I. The amino acid lysine as a factor controlling the growth rate of a typical neoplasm, *Publ. Health Rep.,* 51, 1429, 1936b.
61. **Engel, R. W. and Copeland, D. H.,** Protective action of stock diets against the cancer-inducing action of 2-acetylaminofluorene in rats, *Cancer Res.,* 12, 211, 1952.
62. **Gilbert, C., Gillman, J., Loustalot, P., and Lutz, W.,** The modifying influence of diet and the physical environment on spontaneous tumour frequency in rats, *Br. J. Cancer,* 12,565, 1958.
63. **Ross, M. H. and Bras, G.,** Tumor incidence patterns and nutrition in the rat, *J. Nutr.,* 87, 245, 1965.
64. **Harris, P. N.,** Production of tumors in rats by 2-aminofluorene and 2-acetylaminofluorene: failure of liver extract and of dietary protein level to influence liver tumor production, *Cancer Res.,* 7, 88, 1947.
65. **Tannenbaum, A. and Silverstone, H.,** The genesis and growth of tumors. IV. Effects of varying the proportion of protein (casein) in the diet, *Cancer Res.,* 9, 162, 1949.
66. **Carroll, K. K.,** Experimental evidence of dietary factors and hormone-dependent cancers, *Cancer Res.,* 35, 3374, 1975.
67. **Larsen, C. D. and Heston, W. E.,** Effects of cystine and calorie restriction on the incidence of spontaneous pulmonary tumors in strain A mice, *J. Natl. Cancer Inst.,* 6(1), 31, 1945.
68. **Ross, M. H. and Bras, G.,** Influence of protein under- and overnutrition on spontaneous tumor prevalence in the rat, *J. Nutr.,* 103, 944, 1973.
69. **Madhavan, T. V. and Gopalan, C.,** The effect of dietary protein on carcinogenesis of aflatoxin, *Arch. Pathol.,* 85, 133, 1968.
70. **Wells, P., Alftergood, L., and Alfin-Slater, R. B.,** Effect of varying levels of dietary protein on tumor development and lipid metabolism in rats exposed to aflatoxin, *J. Am. Oil Chem. Soc.,* 53, 559, 1976.
71. **Temcharoen, P., Anukarahanonta, T., and Bhamarapravati, N.,** Influence of dietary protein and vitamin B_{12} on the toxicity and carcinogenicity of aflatoxins in rat liver, *Cancer Res.,* 38, 2185, 1978.
72. **Engel, R. W. and Copeland, D. H.,** The influence of dietary casein level on tumor induction with 2-acetylaminofluorene, *Cancer Res.,* 12, 905, 1952b.
73. **Morris, H. P., Westfall, B. B., Dubnik, C. S., and Baumann, C. A.,** Some observations on carcinogenicity, distribution and metabolism of N-acetyl-2-aminofluorene in the rat, *Cancer Res.,* 8, 390, 1948.
74. **Walters, M. A. and Roe, F. J. C.,** The effect of dietary casein on the induction of lung tumours by the injection of 9,10-dimethyl-1,2-benzanthracene (DMBA) into newborn mice, *Br. J. Cancer,* 18, 312, 1964.
75. **Silverstone, H.,** The levels of carcinogenic azo dyes in the livers of rats fed various diets containing p-dimethylamino-azobenzene: relationship to the formation of hepatomas, *Cancer Res.,* 8, 301, 1948.
76. **Clinton, S. K., Truex, C. G., and Visek, W. J.,** Dietary protein, aryl hydrocarbon hydroxylase and chemical carcinogenesis in rats, *J. Nutr.,* 109, 55, 1979.
77. **Topping, D. C. and Visek, W. J.,** Nitrogen intake and tumorigenesis in rats injected with 1,2-dimethyl-hydrazine, *J. Nutr.,* 106, 1583, 1976.
78. **Mgbodile, M. U. K. and Campbell, T. C.,** Effect of protein deprivation of male weanling rats on the kinetics of hepatic microsomal enzyme activity, *J. Nutr.,* 102, 53, 1972.
79. **Campbell, T. C.,** Influence of nutrition on metabolism of carcinogens, *Adv. Nutr. Res.,* 2, 29, 1979.
80. **Preston, R. S., Hayes, J. R., and Campbell, T. C.,** The effect of protein deficiency on the *in vivo* binding of aflatoxin B_1 to rat liver macromolecules, *Life Sci.,* 19, 1191, 1976.
81. **Tsuda, H., Lee, G., and Farber, E.,** Induction of resistant hepatocytes as a new principle for a possible short-term *in vivo* test for carcinogens, *Cancer Res.,* 40, 1157, 1980.
82. **Appleton, B. S. and Campbell, T. C.,** Effects of dietary protein level and phenobarbital (PB) on aflatoxin (AFB₁)-induced hepatic -glutamyl transpeptidase (GGT) in the rat, *Fed. Proc. Fed. Am. Soc. Exp. Biol.,* 40, 842, 1981.

83. **Haley, H. B. and Williamson, M. B.,** Growth of tumors in experimental wounds, *Proc. Am. Assoc. Cancer Res.,* 3, 116, 1960.
84. **Babson, A. L.,** Some host-tumor relationships with respect to nitrogen, *Cancer Res.,* 14, 89, 1954.
85. **Devik, F., Elson, L. A., Koller, P. C., and Lamerton, L. F.,** Influence of diet on Walker rat carcinoma 256, and its response to X-radiation — cytological and histological investigations, *Br. J. Cancer,* 4, 298, 1950.
86. **White, F. R. and Belkin, M.,** Source of tumor proteins. I. Effect of low-nitrogen diet on the establishment and growth of a transplanted tumor, *J. Natl. Cancer Inst.,* 5, 261, 1945.
87. **White, F. R.,** The relationship between underfeeding and tumor formation, transplantation, and growth in rats and mice, *Cancer Res.,* 21, 281, 1961.
88. **Jose, D. G. and Good, R. A.,** Quantitative effects of nutritional essential amino acid deficiency upon immune responses to tumors in mice, *J. Exp. Med.,* 137, 1, 1973.
89. **Hems, G. And Stuart, A.,** Breast cancer rates in populations of single women, *Br. J. Cancer,* 31, 118, 1975.
90. **Drasar, B. and Irving, D.,** Environmental factors and cancers of the colon and breast, *Br. J. Cancer,* 27, 167, 1973.
91. **Modan, B., Lubin, F., Barrell, V., Greenberg, R. A., Modan, M., and Graham, S.,** The role of starches in the etiology of gastric cancer, *Cancer,* 34, 2087, 1974.
92. **deJong, U. W., Breslow, N., Hong, J. G. E., Sridharan, M., and Shanmugaratnam, K.,** Aetiological factors in oesophageal cancer in Singapore Chinese, *Int. J. Cancer,* 13, 291, 1974.
93. **Roe, F. J. C., Levy, L. S., and Carter, R. L.,** Feeding studies on sodium cyclamate, saccharin and sucrose for carcinogenic and tumour-promoting activity, *Food Cosmet. Toxicol.,* 8,135, 1970.
94. **Friedman, L., Richardson, H. L., Richardson, M. E., Lethco, E. J., Wallace, W. C., and Sauro, F. M.,** Toxic response of rats to cyclamates in chow and semisynthetic diets, *J. Natl. Cancer Inst.,* 49, 751, 1972.
95. **Hunter, B., Graham, C., Heywood, R., Prentice, D. E., Roe, F. J. C., and Noakes, D. N.,** Tumorigenicity and carcinogencity study with xylitol in long-term dietary administration, in *Xylitol,* Vol.20-23, F. Hoffman La Roche Company, Ltd., Basel, Switzerland, 1978a.
96. **Gershoff, S. N. and McGandy, R. B.,** The effects of vitamin A-deficient diets containing lactose in producing bladder calculi and tumors in rats, *Am. J. Clin. Nutr.,* 34, 483, 1981.
97. National Academy of Sciences, Recommended Dietary Allowances, 9th Ed. Committee on Dietary Allowances, Food and Nutrition Board, National Academy of Sciences, Washington, D.C., 1980, 187.
98. **Burkitt, D. P. and Trowell, H. C.,** Refined carbohydrate foods and disease, in *Some Implications of Dietary Fibre,* Academic Press, New York, 1975.
99. **Malhotra, S. L.,** Dietary factors in a study of cancer colon from cancer registry, with special reference to the role of saliva, milk and fermented milk products and vegetable fibre, *Med. Hypotheses,* 3, 122, 1977.
100. **MacLennan, R., Jensen, O. M., Mosbech, J., and Vuori, H.,** Diet, transit time, stool weight, and colon cancer in two Scandinavian populations, *Am. J. Clin. Nutri.,* 31, S239, 1978.
101. **Liu, K., Stamler, J., Moss, D., Garside, D., Persky, V., and Soltero, I.,** Dietary cholesterol, fat, and fibre, and colon-cancer mortality, *Lancet,* 2, 782, 1979.
102. **Lyon, J. L. and Sorenson, A. W.,** Colon cancer in a low-risk population, *Am. J. Clin, Nutr.,* 31, S227, 1978.
103. **Modan, B., Barell, V., Lubin, F., Modan, M., Greenberg, R. A., and Graham, S.,** Low-fiber intake as an etiologic factor in cancer of the colon, *J. Natl. Cancer Inst.,* 55, 15, 1975.
104. **Bjelke, E.,** Dietary factors and the epidemiology of cancer of the stomach and large bowel, *Aktuel. Ernaehrungsmed. Klin. Prax. Suppl.,* 2, 10, 1978.
105. **Dales, L. G., Friedman, G. D., Ury, H. K., Grossman, S., and Williams, S. R.,** A case-control study of relationships of diet and other traits to colorectal cancer in American blacks, *Am. J. Epidemiol.,* 109, 132, 1978.
106. **Martinez, I., Torres, R., Frias, Z., Colon, J. R., and Fernandez, M.,** Factors associated with adenocarcinomas of the large bowel in Puerto Rico, in *Advances in Medical Oncology, Research and Education, Vol. III: Epidemiology,* Birch, J. M., Ed., Pergamon Press, Oxford, 1979, 45.
107. **Glober, G. A., Klein, K. L., Moore, J. O., and Abba, B. C.,** Bowel transit-times in two populations experiencing similar colon-cancer risks, *Lancet,* 2, 80, 1974.
108. **Glober, G. A., Nomura, A., Kamiyama, S., Shimada, A., and Abba, B. C.,** Bowel transit-time and stool weight in populations with different colon-cancer risks, *Lancet,* 2, 110, 1977.
109. **Barbolt, T. A. and Abraham, R.,** The effect of bran on dimethylhydrazine-induced colon carcinogenesis in the rat, *Proc. Soc. Exp. Biol. Med.,* 157, 656, 1978.
110. **Chen, W. F., Patchefsky, A. S., and Goldsmith, H. S.,** Colonic protection from dimethylhydrazine by a high fiber diet, *Surg. Gynecol. Obstet.,* 147, 503, 1978.

111. **Wilson, R. B., Hutcheson, D. P., and Wideman, L.,** Dimethylhydrazine-induced colon tumors in rats fed diets containing beef fat or corn oil, with and without wheat bran, *Am. J. Clin. Nutr.,* 30, 176, 1977.

112. **Freeman, H. J., Spiller, G. A., and Kim, Y.S.,** A double-blind study on the effect of purified cellulose dietary fiber on 1,2-dimethylhydrazine-induced rat colonic neoplasia, *Cancer Res.,* 38, 2912, 1978.

113. **Freeman, H. J., Spiller, G. A., and Kim, Y. S.,** A double-blind study on the effects of differing purified cellulose and pectin fiber diets on 1,2-dimethylhydrazine-induced rat colonic neoplasia, *Cancer Res.,* 40, 2661, 1980.

114. **Watanabe, K., Reddy, B. S., Weisburger, J. H., and Kritchevsky, D.,** Effect of dietary alfalfa, pectin, and wheat bran on azoxymethane- or methylnitrosourea-induced colon carcinogenesis in F344 rats, *J. Natl. Cancer Inst.,* 63, 141, 1979.

115. **Watanabe, K., Reddy, B. S., Wong, C. Q., and Weisburger, J. H.,** Effect of dietary undergraded carrageenin on colon carcinogenesis in F344 rats treated with azoxymethane or methylnitrosourea, *Cancer Res.,* 38, 4427, 1978.

116. **Fleiszer, D. M., Murrary, D., Richards, G. K., and Brown, R. A.,** Effects of diet on chemically induced bowel cancer, *Can. J. Surg.,* 23, 67, 1980.

117. **Cassidy, M. M., Lightfoot, F. G., Grau, L. E., Story, J. A., Kritchevsky, D., and Vahouny, G. V.,** Effect of chronic intake of dietary fibers on the ultrastructural topography of rat jejunum and colon: a scanning electron microscopy study, *Am. J. Clin. Nutr.,* 34, 218, 1981.

118. **Lubin, J. H., Burns, P. E., Blot, W. J., Ziegler, R. G., Lees, A. W., and Fraumeni, J. F., Jr.,** Dietary factors and breast cancer risk, *Int. J. Cancer,* 28, 685, 1981.

119. **Miller, A. B., Kelly, A., Choi, N. W., Matthews, V., Morgan, R. W., Munan, L., Burch, J. D., Feather, J., Howe, G. R., and Jain, M.,** A study of diet and breast cancer, *Am. J. Epidemiol.,* 107, 499, 1978.

120. **Nomura, A., Henderson, B. E., and Lee, J.,** Breast cancer and diet among the Japanese in Hawaii, *Am. J. Clin. Nutr.,* 31, 2020, 1978.

121. **Schuman, L. M., Mandell, J. S., Radke, A., Seal, U., and Halberg, F.,** Some selected features of the epidemiology of prostatic cancer: Minneapolis-St. Paul, Minnesota case-control study, 1976-1979, in *Trends in Cancer Incidence: Causes and Practical Implications,* Magnus, K., Ed., Hemisphere Publishing Corp., Washington, D.C., 1982, 345.

122. **Higginson, J.,** Etiological factors in gastrointestinal cancer in man, *J. Natl. Cancer Inst.,* 37, 527, 1966.

123. **Graham, S., Schotz, W., and Martino, P.,** Alimentary factors in the epidemiology of gastric cancer, *Cancer,* 30, 927, 1972.

124. **Enig, M. G., Munn, R. J., and Keeney, M.,** Response to letters, *Fed. Proc. Fed. Am. Soc. Exp. Biol.,* 38, 2437, 1979.

125. **Enstrom, J. E.,** Colorectal cancer and consumption of beef and of fat, *Br. J. Cancer,* 32, 432, 1975.

126. **Bull, A. W., Soullier, B., Wilson, P. S., Hayden, M. T., and Nigro, N. D.,** Promotion of AOM-induced intestinal cancer by high fat diet in rats, *Cancer Res.,* 39, 4956, 1979.

127. **King, J. T., Casas, C. B., and Visscher, M. B.,** The influence of estrogen on cancer incidence and adrenal changes in ovariectomized mice on calorie restriction, *Cancer Res.,* 9, 436, 1949.

128. **White, F. R.,** The relationship between underfeeding and tumor formation, transplantation, and growth in rats and mice, *Cancer Res.,* 21, 281, 1961.

129. **Carroll, K. K. and Khor, H. T.,** Effects of level and type of dietary fat on incidence of mammary tumors induced in female Sprague-Dawley rats by 7,12-dimethylbenz(a)anthracene, *Lipids,* 6, 415, 1971.

130. **Carroll, K. K.,** Lipids and carcinogenesis, *J. Environ. Pathol. Toxicol.,* 3(4), 253, 1980.

131. **Wetsel, W., Rogers, A. E., and Newberne, P. M.,** Dietary fat and DMBA mammary carcinogenesis in rats, *Cancer Detect. Prev.,* 4, 535, 1981.

132. **Hopkins, G. J. and West, C. E.,** Possible roles of dietary fats in carcinogenesis, *Life Sci.,* 19, 1103, 1976.

133. **Chan, P. C., Head, J. R., Cohen, L. A., and Wynder, E. L.,** Influence of dietary fat on mammary tumors, *J. Natl. Cancer Inst.,* 59, 1279, 1977.

134. **McCay, P. B., King, M., Rikans, L. E., and Pitha, J.,** Interactions between dietary factors, mammary cancer and liver cancer, *J. Environ. Path. Toxicol.,* 3, 451, 1980.

135. **Hopkins, G. J. and West, C. E.,** Effect of dietary polyunsaturated fat on the growth of a transplantable adenocarcinoma in C3HAᵛfB mice, *J. Natl. Cancer Inst.,* 58, 753, 1977.

136. **Kidwell, W. R., Monaco, M. E., Wicha, M. S., and Smith, G. S.,** Unsaturated fatty acid requirements for growth and survival of a rat mammary tumor cell line, *Cancer Res.,* 38, 4091, 1978.

137. **Hillyard, L. A. and Abraham, S.,** Effect of dietary polyunsaturated fatty acids on growth of mammary adenocarcinomas in mice and rats, *Cancer Res.,* 39, 4430, 1979.

138. **Corwin, L. M., Varshovsky-Rose, F., and Broitman, S. A.,** Effect of dietary fats on tumorigenicity of two sarcoma cell lines, *Cancer Res.,* 39, 4350, 1979.

139. **Rogers, A. E. and Wetsel, W. C.,** Mammary carcinogenesis in rats fed different amounts and types of fats, *Cancer Res.,* 41, 3755, 1979.

140. **Newberne, P. M. and Zeiger, E.,** Nutrition, carcinogenesis and mutagenesis, in *Advances in Modern Toxicology, Vol. 5,* Flamm, W. G. and Mehlman, M. A., Eds., John Wiley and Sons, New York, 1978.

141. **Newberne, P. M., Weigert, J., and Kula, N.,** Effects of dietary fat on hepatic mixed-function oxidases and hepatocellular carcinoma induced by aflatoxin B₁ in rats, *Cancer Res.,* 39, 3986, 1979.

142. **Longnecker, D. S., Roebuck, D. B., Yager, J. D., Dobbins, W. O., Lilja, H. S., and Siegmund, B.,** Pancreatic carcinoma in azaserine-treated rats, *Cancer,* 47, 1562, 1981.

143. **Roebuck, B. D., Yager, J. D., Jr., and Longnecker, D. S.,** Dietary modulation of azaserine-induced pancreatic carcinogenesis in the rat, *Cancer Res.,* 41, 888, 1981.

144. **Nigro, N. D., Singh, D. V., Campbell, R. L., and Pak, M. S.,** Effect of dietary beef fat on intestinal tumor formation by azoxymethane in rats, *J. Natl. Cancer Inst.,* 54, 439, 1975.

145. **Reddy, B. S., Watanabe, K., and Weisburger, J. H.,** Effect of high-fat diet on colon carcinogenesis in F344 rats treated with 1,2-dimethylhydrazine, methylazoxymethanol acetate, or methylnitrosourea, *Cancer Res.,* 37, 4156, 1977.

146. **Reddy, B. S., Weisburger, J. H., and Wynder, E. L.,** Effects of dietary fat level and dimethylhydrazine on fecal acid and neutral sterol excretion and colon carcinogenesis in rats, *J. Natl. Cancer Inst.,* 52, 507, 1974.

147. **Broitman, S. A.,** Cholesterol excretion and colon cancer, *Cancer Res.,* 41, 3738, 1981.

148. **Bansal, B. R., Rhoads, J. E., Jr., and Bansal, S. E.,** Effects of diet on colon carcinogenesis and the immune system in rats treated with 1,2-dimethylhydrazine, *Cancer Res.,* 38, 3293, 1978.

149. **Reddy, B. S., Cohen, L. A., McCoy, G. D., Hill, P., Weisburger, J. H., and Wynder, E. L.,** Nutrition and its relationship to cancer, *Adv. Cancer Res.,* 41, 888, 1981.

150. **Nauss, K. M., Locniskar, M., and Newberne, P. M.,** Alterations in quality and quantity of dietary fat on 1,2-dimethylhydrazine-induced colon tumorigenesis in rats, *Cancer Res.,* 1983.

151. **Pearce, M. L. and Dayton, S.,** Incidence of cancer in men on a diet high in polyunsaturated fat, *Lancet,* 1, 464, 1971.

152. **Miettinen, M., Turpeinen, O., Karvonen, M. J., Elosuo, R., and Paanilainen, E.,** Effect of cholesterol-lowering diet on mortality from coronary heart causes and other causes: a twelve-year clinical trial in men and women, *Lancet,* 2, 835, 1972.

153. **Ederer, F., Leren, P., Turpeinen, O., and Frantz, I. D., Jr.,** Cancer among men on cholesterol-lowering diets: experience from five clinical trials, *Lancet,* 2, 203, 1971.

154. **Rose, G. and Shipley, M. J.,** Plasma lipids and mortality: a source of error, *Lancet,* 1, 523, 1980.

155. **Bjelke, E.,** Letter to the editor: colon cancer and blood-cholesterol, *Lancet,* 1, 1116, 1974.

156. **Rose, G., Blackburn, H., Keys, A., Taylor, H. L., Kannel, W. B., Paul, O., Reid, D. D., and Stamler, J.,** Colon cancer and blood-cholesterol, *Lancet,* 1, 181, 1974.

157. **Williams, R. R., Sorlie, P. D., Feinleib, M., McNamara, P. M., Kannel, W. B., and Dawber, T. R.,** Cancer incidence by levels of cholesterol, *J. Am. Med. Assoc.,* 245, 247, 1981.

158. **Kark, J. D., Smith, A. H., and Hames, C. G.,** The relationship of serum cholesterol to the incidence of cancer in Evans County, Georgia, *J. Chronic Dis.,* 33, 311, 1980.

159. **Peterson, B., Trell, E., and Sternby, N. H.,** Low cholesterol level as risk factor for noncoronary death in middle-aged men, *J. Am. Med. Assoc.,* 245, 2056, 1981.

160. **Dyer, A. R., Stamler, J., Paul, O., Shekelle, R. B., Schoenberger, J. A., Berkson, D. M., Lepper, M., Collette, P., Shekelle, S., and Lindberg, H. A.,** Serum cholesterol and risk of death from cancer and other causes in three Chicago epidemiological studies, *J. Chronic Dis.,* 34, 249, 1981.

161. **Moskovitz, M., White, C., Barnett, R. N., Stevens, S., Russell, E., Vargo, D., and Floch, M. H.,** Diet, fecal bile acids, and neutral sterols in carcinoma of the colon, *Dig. Dis. Sci.,* 24, 746, 1979.

162. **Mower, H. F., Ray, R. M., Shoff, R., Stemmermann, G. N., Nomura, A., Glober, G. A., Kamiyama, S., Shimada, A., and Yamakawa, H.,** Fecal bile acids in two Japanese populations with different colon cancer risks, *Cancer Res.,* 39, 328, 1979.

163. **Miller, S. R., Tartter, P., Papatistas, A., Slater, G., and Aufses, Jr.,** Serum cholesterol and human colon cancer, *J. Natl. Cancer Inst.,* 67, 297, 1981.

164. **Bjelke, E.,** Dietary vitamin A and human lung cancer, *Intl. J. Cancer,* 15, 561, 1975.

165. **Gregor, A., Lee, P. N,., Roe, F. J. C., Wilson, M. J., and Melton, A.,** Comparison of dietary histories in lung cancer cases and controls with special reference to vitamin A, *Nutr. Cancer,* 2, 93, 1980.

166. **MacLennan, R., DaCosta, J., Day, N. E., Law, C. H., Ng, Y. K., and Shanmugaratnam, K.,** Risk factors for lung cancer in Singapore Chinese, a population with high female incidence rates, *Intl. J. Cancer,* 20, 854, 1977.

167. **Smith, P. G. and Jick, H.,** Cancers among users of preparations containing vitamin A, *Cancer,* 42, 808, 1978.

168. **Mettlin, C., Graham, S., and Swanson, M.,** Vitamin A and lung cancer, *J. Natl. Cancer Inst.,* 62, 1435, 1979.

169. **Shekelle, R. B., Liu, S., Raynor, W. J., Jr., Lepper, M., Maliza, C., and Rossof, A. H.,** Dietary vitamin A and risk of cancer in the Western Electric study, *Lancet,* 2, 1185, 1981.

170. **Graham, S., Mettlin, C., Marshall, J., Priore, R., Rzepka, T., and Shedd, D.,** Dietary factors in the epidemiology of cancer of the larynx, *Am. J. Epidemiol.,* 113, 675, 1981.

171. **Mettlin, C., Graham, S., Priore, R., Marshall, J., and Swanson, M.,** Diet and cancer of the esophagus, *Nutr. Cancer,* 2, 143, 1981.

172. **Cook-Mozaffari, P. J., Azordegan, F., Day, N. E., Ressicand, A., Sabai, C., and Aramesh, B.,** Oesophageal cancer studies in the Caspian littoral of Iran: results of a case-control study, *Br. J. Cancer,* 39, 293, 1979.

173. **Sokula, A.,** Letter to the editor: vitamin A and lung cancer, *Br. Med. J.,* 2, 298, 1976.

174. **Cambien, F., Ducimetiere, P., and Richard, J.,** Total serum cholesterol and cancer mortality in a middle-aged population, *Am. J. Epidemiol.,* 112, 388, 1980.

175. **Wald, N., Idle, M., Boreham, J., and Bailey, A.,** Low serum vitamin A and subsequent risk of cancer — preliminary results of a prospective study, *Lancet,* 2, 813, 1980.

176. **Crocker, T. T. and Sanders, L. L.,** Influence of vitamin A and 3,7-dimethyl-2, 6-octadienal (citral) on the effect of benzo(a)pyrene on hamster trachea in organ culture, *Cancer Res.,* 30, 1312, 1970.

177. **Nettesheim, P. and Williams, M. L.,** The influence of vitamin A on the susceptibility of the rat lung to 3-methylcholanthrene, *Intl. J. Cancer,* 17, 351, 1976.

178. **Smith, D. M., Rogers, A. E., Herndon, B. J., and Newberne, P. M.,** Vitamin A (retinyl acetate) and benzo(a)pyrene-induced respiratory tract carcinogenesis in hamsters fed a commercial diet, *Cancer Res.,* 35, 11, 1975.

179. **Saffiotti, U., Montesano, R., Sellakumar, A. R., and Borg, S. A.,** Experimental cancer of the lung: inhibition by vitamin A of the induction of tracheobronchial squamous metaplasia and squamous cell tumors, *Cancer,* 20, 857, 1967.

180. **Welsch, C. W., Goodrich-Smith, M., Brown, C. K., and Crowe, N.,** Enhancement of retinyl acetate of hormone-induced mammary tumorigenesis in female GR/A mice, *J. Natl. Cancer Inst.,* 67, 935, 1981.

181. **Rogers, A. E., Herndon, B. J., and Newberne, P. M.,** Induction by dimethylhydrazine of intestinal carcinoma in normal rats and rats fed high or low levels of vitamin A, *Cancer Res.,* 33, 1003, 1973.

182. **Sporn, M. B., Dunlop, N. M., Newton, D. L., and Smith, J. M.,** Prevention of chemical carcinogenesis by vitamin A and its synthetic analogs (retinoids), *Fed. Proc. Fed. Am. Soc. Exp. Biol.,* 35, 1332, 1976.

183. **Sporn, M. B. and Newton, D. L.,** Chemoprevention of cancer with retinoids, *Fed. Proc. Fed. Am. Soc. Exp. Biol.,* 38, 2528, 1979.

184. **Newberne, P. M. and McConnell, R. G.,** Nutrient deficiencies in cancer causation, *J. Environ. Pathol. Toxicol.,* 3, 323, 1980.

185. **Peto, R., Doll, R., Buckley, J. D., and Sporn, M. B.,** Can dietary beta-carotene materially reduce human cancer rates? *Nature (London),* 290, 201, 1981.

186. **Mathews-Roth, M. M., Pathak, M. A., Fitzpatrick, T. B., Harber, L. H., and Kass, E. H.,** Beta carotene therapy for erythropoietic protoporphyria and photosensitivity diseases, *Arch. Dermatol.,* 113, 1229, 1977.

187. **Shamberger, R. J.,** Inhibitory effect of vitamin A on carcinogenesis, *J. Natl. Cancer Inst.,* 47, 667, 1971.

188. **Meinsma, L.,** (In Dutch: English summary) Nutrition and cancer, *Voeding,* 25, 357, 1964.

189. **Cook-Mozaffari, P.,** The epidemiology of cancer of the esophagus, *Nutr. Cancer,* 1, 51, 1979.

190. **Hormozdiari, N., Day, N. E., Aramesh, B., and Mahboudi, F.,** Dietary factors and esophageal cancer in the Caspian littoral of Iran, *Cancer Res.,* 35, 3493, 1975.

191. **Wassertheil-Smoller, S., Romney, S. L., Wylie-Rosett, J., Slagle, S., Miller, G., Lucido, D., Duttagupta, C., and Palan, P. R.,** Dietary vitamin C and uterine cervical dysplasia, *Am J. Epidemiol.,* 114, 714, 1981.

192. **Mirvish, S. S.,** Inhibition of the formation of carcinogenic N-nitroso compounds by ascorbic acid and other compounds, in Cancer: Achievements, Challenges, and Prospects for the 1980s, Vol. I, Burchenal, J. H. and Oettgen, H. F., Eds., Grune and Stratton, New York, 1981, 557.

193. **Mirvish, S. S., Wallcave, L., Eagen, M., and Shubik, P.,** Ascorbate-nitrite reaction: possible means of blocking the formation of carcinogenic N-nitroso compounds, *Science,* 177, 65, 1972.

194. **Mirvish, S. S., Cardesa, A., Wallcave, L., and Shubik, P.,** Induction of mouse lung adenomas by amines or ureas plus nitrite and by N-nitroso compounds: effect of ascorbate, gallic acid, thiocyanate, and caffeine, *J. Natl. Cancer Inst.,* 55, 633, 1975.

195. **Benedict, W. F., Wheatley, W. L., and Jones, P. A.,** Inhibition of chemically induced morphological transformation and reversion of the transformed phenotype by ascorbic acid in C3H/10T-1/2 cells, *Cancer Res.,* 40, 2796, 1980.

196. **Park, C. H., Amare, M., Savin, M. A., and Hoogstraten, B.,** Growth supression of human leukemic cells *in vitro* by L-ascorbic acid, *Cancer Res.,* 40, 1062, 1980.

197. **Mergens, W. J., Vane, F. M., Tannenbaum, S. R., Green, L., and Skipper, P. L.,** *In vitro* nitrosation of methapyrilene, *J. Pharm. Sci.,* 68, 827, 1979.

198. **Wattenberg, L. W.,** Inhibition of carcinogenic and toxic effects of polycyclic hydrocarbons by phenolic antioxidants and ethoxyquin, *J. Natl. Cancer Inst.,* 48, 1425, 1972.

199. **Harman, D.,** DMBA-induced cancer: inhibitory effect of vitamin E, *Clin. Res.,* 17, 125, 1969.

200. **Lee, C. and Chen, C.,** Enhancement of mammary tumorigenesis in rats by vitamin E deficiency, *Proc. Am. Assoc. Cancer Res.,* 20, 1321, 1979.

201. **Shamberger, R. J.,** Relationship of selenium to cancer. I. Inhibitory effect of selenium on carcinogenesis, *J. Natl. Cancer Inst.,* 44, 931, 1970.

202. **Haber, S. L. and Wissler, R. W.,** Effect of vitamin E on carcinogenicity of methylcholanthrene, *PSEBM,* 111, 774, 1962.

203. **Boutwell, R. K., Brush, M. K., and Rusch, H. P.,** The influence of vitamins of the B complex on the induction of epithelial tumors in mice, *Cancer Res.,* 9, 747, 1949.

204. **Kensler, C. J., Suguira, N., Young, N. F., Halter, C. R., and Rhoads, C. P.,** Partial protection of rats by riboflavin with casein, *Science,* 93, 308, 1941.

205. **Miller, J. A. and Miller, E. C.,** The carcinogenic aminoazo dyes, *Adv. Cancer Res.,* 1, 339, 1953.

206. **Rogers, A. E. and Newberne, P. M.,** Lipotrope deficiency in experimental carcinogenesis, *Nutr. Cancer,* 2, 104, 1980.

207. **Thanassi, J. W., Nutter, L. M., Meisler, N. T., Commers, P., and Chiu, J.-F.,** Vitamin B_6 metabolism in Morris hepatomas, *J. Biol. Chem.,* 256, 3370, 1981.

208. **Tryfiates, G. P.,** Effects of pyridoxine on serum protein expression in hepatoma-bearing rats, *J. Natl. Cancer Inst.,* 66, 339, 1981.

209. **Bell, E. D., Tong, D., Mainwaring, W. I. P., Hayward, J. L., and Bulbrook, R. D.,** Tryptophan metabolism and recurrence rates after mastectomy in patients with breast cancer, *Clin. Chim. Acta.,* 42, 445, 1972.

210. **Furst, A.,** Problems in metal carcinogenesis, in *Trace Metals in Health and Disease,* Kharasch, N., Ed., Raven Press, New York, 1979, 83.

211. **Sunderman, F. W., Jr.,** Metal carcinogenesis, in *Toxicology of Trace Elements,* Goyer, R. A. and Mehlman, M. A., Eds., Hemisphere Publishing Corporation, Washington, D.C., 1977, 257.

212. **Schroeder, H. A. and Mitchener, M.,** Scandium, chromium (VI), gallium, yttrium, rhodium, palladium, indium in mice: effects on growth and life span, *J. Nutr.,* 101, 1431, 1971a.

213. **Schroeder, H. A. and Mitchener, M.,** Selenium and tellurium in rats: effects on growth, survival and tumors, *J. Nutr.,* 101, 1531, 1971b.

214. **Schroeder, H. A. and Mitchener, M.,** Selenium and tellurium in mice: effects on growth, survival and tumors, *Arch. Environ. Health,* 24, 66, 1972.

215. **Schroeder, H. A., Balassa, J. J., and Vinton, W. H., Jr.,** Chromium, lead, cadmium, nickle and titanium in mice: effect on mortality, tumors and tissue levels, *J. Nutr.,* 83, 239, 1964.

216. **Schroeder, H. A., Balassa, J. J., and Vinton, W. H., Jr.,** Chromium, cadmium and lead in rats: effects on lifespan, tumors and tissue levels, *J. Nutr.,* 86, 51, 1965.

217. **Schroeder, H. A., Mitchener, M., Balassa, J. J., Kanisawa, M., and Nason, A. P.,** Zirconium, niobium, antimony and fluorine in mice: effect on growth, survival and tissue levels, *J. Nutr.,* 95, 95, 1968.

218. **Schroeder, H. A., Mitchener, M., and Nason, A. P.,** Zirconium, niobium, antimony, vanadium and lead in rats: life term studies, *J. Nutr.,* 100, 59, 1970.

219. Selenium in Nutrition Rep. Subcommittee on Selenium, Committee on Animal Nutrition, National Academy of Sciences, Washington, D.C., 1971.

220. **Shamberger, R. J. and Frost, D. V.,** Letter to the editor: possible protective effect of selenium against human cancer, *Cancer Med. Assoc. J.,* 100, 682, 1969.

221. **Shamberger, R. J. and Willis, C. E.,** Selenium distribution and human cancer mortality, *CRC Crit. Rev. Clin. Lab. Sci.,* 2, 211, 1971.

222. **Shamberger, R. J., Tytko, S. A., and Willis, C. E.,** Antioxidants and cancer. VI. Selenium and age-adjusted human cancer mortality, *Arch. Environ. Health,* 31, 231, 1976.

223. **Shamberger, R. J., Rukovena, E., Longfield, A. K., Tytko, S. A., Deodhar, S., and Willis, C. E.,** Antoixidants and cancer. I. Selenium in the blood of normals and cancer patients, *J. Natl. Cancer Inst.,* 50, 863, 1973.

224. **Schrauzer, G. N.,** Selenium and cancer: a review, *Bioinorg. Chem.,* 5, 275, 1976.

225. **Schrauzer, G. N., White, D. A., and Schneider, C. J.,** Cancer mortality correlation studies. III. Statistical associations with dietary selenium intakes, *Bioinorg. Chem.,* 7, 23, 1977a.

226. **Schrauzer, G. N., White, D. A., and Schneider, C. J.,** Cancer mortality correlation studies. IV. Associations with dietary intakes and blood levels of certain trace elements, notably Se-antagonists, *Bioinorg. Chem.,* 7, 35, 1977b.

227. **Nelson, A. A., Fitzhugh, O. G., and Calvery, H. O.,** Liver tumors following cirrhosis caused by selenium in rats, *Cancer Res.,* 3, 230, 1943.

228. **Volgarev, M. N. and Tscherkes, L. A.,** Further studies in tissue changes associated with sodium selenate, in *Symp. Selenium in Biomedicine. 1st Int. Symp.,* AVI Publishing Co., Westport, Conn., 1967, 179.

229. **Schrauzer, G. N., White, D. A., and Schneider, C. J.,** Selenium and cancer: effects of selenium and of the diet on the genesis of spontaneous mammary tumors in virgin inbred female C_3H/St mice, *Bioinorg. Chem.,* 8, 387, 1978.

230. **Harr, J. R., Exon, J. H., Whanger, P. D., and Weswig, P. H.** Effect of dietary selenium on N-2-fluorenyl-acetamide (FAA)-induced cancer in vitamin E supplemented, selenium depleted rats, *Clin. Toxicol.,* 5, 187, 1972.

231. **Ip, C. and Sinha, D. K.,** Enhancement of mammary tumorigenesis by dietary selenium deficiency in rats with a high polyunsaturated fat intake, *Cancer Res.,* 41, 31, 1981.

232. **Clayton, C. C. and Baumann, C. A.,** Diet and azo dye tumors: effect of diet during a period when the dye is not fed, *Cancer Res.,* 9, 575, 1949.

233. **Shamberger, R. J.,** Relation of selenium to cancer. I. Inhibitory effect of selenium on carcinogenesis, *J. Natl. Cancer Inst.,* 44, 931, 1970.

234. **Griffin, A. C.,** Role of selenium in the chemoprevention of cancer, *Adv. Cancer Res.,* 29, 419, 1979.

235. **Medina, D. and Shepherd, F.,** Selenium-mediated inhibition of mouse mammary tumorigenesis, *Cancer Lett.,* 8, 241, 1980.

236. **Thompson, H. J. and Becci, P. J.,** Selenium inhibition of N-methyl-N-nitrosourea-induced mammary carcinogenesis in the rat, *J. Natl. Cancer Inst.,* 65, 1299, 1980.

237. **Greeder, G. A. and Milner, J. A.,** Factors influencing the inhibitory effect of selenium on mice inoculated with Ehrlich ascites tumor cells, *Science,* 209, 825, 1980.

238. **Newberne, P. M. and Conner, M.,** Acute studies with selenium and aflatoxin B_1 liver injury, Trace Sub. Conf. VIII, University of Missouri, 323, 1974.

239. **Newberne, P. M. and McConnell, R. G.,** Dietary nutrients and contaminants in laboratory animal experimentation, Proc. 4th Int. Symp. Lab Animals, Boston, Mass., October 28-31, *J. Environ. Pathol. Toxicol.,* 4, 105, 1980.

240. **Newberne, P. M.,** Influence of selenium on response to toxins, *The Toxicol.,* 3, 140, 1983.

241. **Suphakarn, V. S., Newberne, P. M., and Goldman, M.,** The influence of vitamin A and aflatoxin (AFB_1) on enzymes of liver and colon mucosa, and of binding of AFB_1 to DNA of colon mucosa, *Nutr. and Cancer,* 5, in press, 1983.

242. **Benson, A. M., Batzinger, R. P., Os, S. L., et al,** Elevation of hepatic glutahione S-transferase activities, *Cancer Res.,* 38, 4486, 1978.

243. **Sparnins, V. L., Chuan, J., and Wattenberg, L. W.,** Enhancement of glutathione S-transferase activity of the esophagus by phenols, and benzyl isothiolyanate, *Cancer Res.,* 42, 205, 1982.

244. **VanRensburg, S. J.,** Epidemiologic and dietary evidence for a specific nutritional predisposition to esophageal cancer, *J. Natl. Cancer Inst.,* 67, 243, 1981.

245. **Gyorkey, F., Min, K. W., Huff, J. A., and Gyorkey, F.,** Zinc and magnesium in human prostate gland: normal, hyperplastic and neoplastic, *Cancer Res.,* 27, 1348, 1967.

246. **Prasad, A. S.,** *Trace Elements and Iron in Human Metabolism,* Plenum Publishing Corp., New York, 1978.

247. **Golden, M. H. N., Golden, B. E., Harland, P. S. E., and Jackson, A. A.,** Zinc and immunocompetence in protein-energy malnutrition, *Lancet,* 1, 1226, 1978.

248. **Gross, R. L. and Newberne, P. M.,** Role of nutrition in immunologic function, *Physiol. Rev.,* 60, 188, 1980.

249. **Beisel, W. R.,** Single nutrients and immunity, *Am. J. Clin. Nutr.,* 35, 417, 1982.

250. **Strain, W. H., Mansour, E. G., Flynn, A., Pories, W. J., Tomaro, A. J., and Hill, O. A., Jr.,** Letter to the editor: plasma-zinc concentration in patients with bronchogeni cancer, *Lancet,* 1, 1021, 1972.

251. **Davies, I. J., Musa, M., and Dormandy, T. L.,** Measurements of plasma zinc. II. In malignant disease, *J. Clin. Pathol.,* 21, 363, 1968.

252. **Lin, H. J., Chan, W. C., Fong, Y. Y., and Newberne, P. M.,** Zinc levels in serum, hair and tumors from patients with esophageal cancer, *Nutr. Rep. Intl.,* 15, 635, 1977.

253. **Petering, H. G., Buskirk, H. H., and Crim, J. A.,** The effect of dietary mineral supplements of the rat on the antitumor activity of 3-ethoxy-2-oxobutyraldehyde bis (thiosemicarbazone), *Cancer Res.,* 27, 1115, 1967.

254. **DeWys, W. and Pories, W.,** Inhibition of a spectrum of animal tumors by dietary zinc deficiency, *J. Natl. Cancer Inst.,* 48, 375, 1972.

255. **Fenton, M. R., Burke, J. P., Tursi, F. D., and Arena, F. P.,** Effect of a zinc-deficient diet on the growth of an IgM-secreting plasmacytoma (TEPC-183), *J. Natl. Cancer Inst.,* 65, 1271, 1980.

256. **Fong, L. Y. Y., Sivak, A., and Newberne, P. M.,** Zinc deficiency and methylbenzylnitrosamine-induced esophageal cancer in rats, *J. Natl. Cancer Inst.,* 61, 145, 1978.

257. **Gabrial, G., Schrager, T., and Newberne, P. M.,** Zinc deficiency, alcohol and a retinoid: association with esophageal cancer in rats, *J. Natl. Cancer Inst.,* 68, 785, 1982.
258. **Poswillo, D. E. and Cohen, B.,** Inhibition of carcinogenesis by dietary zinc, *Nature (London),* 231, 447, 1971.
259. **Duncan, J. R. and Dreosti, I. E.,** Zinc intake, neoplastic DNA synthesis, and chemical carcinogenesis in rats and mice, *J. Natl. Cancer Inst.,* 55, 195, 1975.
260. **James, S. P., Hoffnagle, J. H., Jones, E. A., and Stroker, W.,** Augmentation of natural killer cell function in vitro by interferon, *Hepatology,* 2, 690, 1982.
261. **Murgita, R. A., Goidl, E. A., Kontianinen, S., and Wiggell, H.,** Alphafetoprotein induces suppressor T cells in vitro, *Nature (London),* 267, 257, 1977.
262. **Edginton, T. S. and Curtiss, L. K.,** Plasma lipoproteins with bioregulatory properties including the capacity to regulate lymphocyte function and the immune response, *Cancer Res.,* 41, 3786, 1981.
263. **Harmony, J. A. K. and Hui, D. Y.,** Inhibition of membrane-bound LDL of the primary inductive events of mitogen stimulated lymphocyte activation, *Cancer Res.,* 41, 3799, 1981.
264. **Mackay, I. R.,** Immunological regulations of carcinogenesis, U.I.C.C. Workshop on Hepatic Carcinoma, International Union Against Cancer, Geneva, November 22-26, 1982.
265. **Okuda, K., Nakashima, T., Sakamoto, K., et al.,** Hepatocellular carcinoma arising in non-cirrhotic and highly cirrhotic livers, *Cancer,* 49, 450, 1982.
266. **Thomas, H. C.,** Immunological aspects of chronic liver disease: relationship to development of primary liver cell carcinoma, U.I.C.C. Workshop on Hepatic Carcinoma, International Union Against Cancer, Geneva, November 22-26, 1982.
267. **Newberne, P. M., Roger, A. E., and Nauss, K. M.,** Choline methionine and related factors in oncogenesis, in *Nutritional Factors in the Induction and Maintenance of Malignancy,* Butterworth, C. E. and Hutchinson, M., Eds., Academic Press, New York, in press, 1983.
268. **Suit, J. L., Rogers, A. E., and Jettin, N. F.,** Effects of diet on conversion of aflatoxin B_1 to bacterial mutagens by rats in vivo and by rat hepatic microsomes in vitro, *Mutation Res.,* 46, 312, 1977.
269. **Mikol, Y. B. and Poirier, L. A.,** An inverse correlation between hepatic ornithine decarboxylase and *S*-adenosylmethionine in rats, *Cancer Lett.,* 13, 196, 1981.
270. **Poirier, L. A., Grantham, P. H., and Rogers, A. E.,** The effects of a marginally lipotrope diet on the hepatic levels of *S*-adenosylmethionine and on urinary metabolites of 2-AAF in rats, *Cancer Res.,* 37, 744, 1977.
271. **Shivapurkar, N. and Poirier, L. A.,** Decreased levels of *S*-adenosylmethionine in the livers of rats fed phenobarbital or DDT, *Carcinogenesis,* 3, 589, 1982.
272. **Schottenfeld, D.,** Cancer risks of medical treatment, *Ca — A Cancer J. Clin.,* 32, 258, 1982.
273. **Armijo, R., Gonzalez, A., Orellana, M., Coulson, A. H., Sayre, J. W., and Detels, R.,** Epidemiology of gastric cancer in Chile. II. Nitrate exposures and stomach cancer frequency, *Int. J. Epidemiol.,* 10, 57, 1981.
274. **Magee, P. N., Montesano, R., and Preussmann, R.,** N-nitroso compounds and related carcinogens, in *Chemical Carcinogens,* Searle, C. E., Ed., Monograph 173, American Chemical Society, Washington, D.C., 1976, 491.

Chapter 4

THE SIGNIFICANCE OF BENEFITS IN REGULATORY DECISION-MAKING

William J. Darby

TABLE OF CONTENTS

I. INTRODUCTION

Responsible regulatory decision-making gives consideration both to risks and benefits. Risk has been described as "the probability of the occurrence of some unfavorable event".[1] Benefit, similarly, may be regarded as advantage, good, or profit. Webster's New International Dictionary defines benefit as "whatever promotes welfare; advantage; profit." Interestingly, it quotes Burke: "Men have no right to what is not for their benefit." The connotation of profit in such context is that of accession of good; valuable result; useful consequences; avail; gain, and not in the narrow pecuniary sense.

In decision-making both benefit and risk must be considered in relation to whom is exposed to the risk and who is the anticipated beneficiary-individuals? Categories or groups of persons? The general population? A percentage of the general population? If so, how large a percentage? Such concepts are easy to understand but difficult to delimit, and virtually impossible to quantitate in predicting human experience.

Regulatory agencies have long sought to identify and protect their constituency against readily recognizable risks, initially of acute ill effects or debased products and, during the last quarter of a century or so, there has been an unprecedented effort to identify and develop predictive methods for assessing subtle risks to vanishingly minimal levels. The cost in resources, human talent, and money have continuously accelerated to a cumulatively astronomical figure. Any attempt to place a monetary figure on the total costs would perforce have to include the nebulous estimate of national losses resulting from diversion of resources into this effort, resources that might have been far more beneficially deployed elsewhere. Recognizable progress has been made in risk asessment, particularly in the identification of sources of uncertainty in interpreting experimental findings in terms of human significance and in recognizing that toxicologic studies must be individually designed to take account of the differences in metabolism, diet, and eating habits of differing species, including man.

II. BENEFITS

Identification and assessment of benefits have received much less attention than have questions relating to risk. Despite 6 years of study and deliberation, the Food Safety Council, in its final report,[2] does not examine in depth the nature of benefits. It addresses instead

almost exclusively the question of risk assessment. Concerning cost-benefit analysis, the report states

. . . Cost-benefit analysis compares the costs of an action with its benefits. It uses identical units for comparison, often dollars.

When applying cost-benefit analysis to food safety, a substance's costs and benefits are compared for various alternative levels of use. For food safety, the costs are a substance's risks and the benefits can be lower prices or more food. With cost-benefit analysis, the numerical risks that are generated by quantitative risk assessment must be converted to dollars for each level of use being considered.

. . . By converting costs and benefits to dollars, cost-benefit analysis, if adhered to rigidly, provides a decision. However, the analysis first must convert the numerical risk estimates into dollars.

The problem with cost-benefit analysis can be stated in a simple question: How can one convert risks to human health into dollars and cents? How much is a certain level of risk (e.g., one in a million risk of bladder cancer) worth?

One suggested approach for handling this problem is to ask people what they are willing to pay to reduce their risk for a particular toxic effect or condition. Few people, however, consider risk when they purchase products. Even fewer people know what the probabilities associated with the very low risks typically experienced in foods might mean. Thus, the economic valuing of risk poses difficulties in reaching decisions on whether to restrict or prohibit specific substances on the grounds of their safety.

It is this conversion of risk estimates that provides the greatest problem when applying cost-benefit analysis to food safety. Human health or even life just cannot be meaningfully translated directly into economic costs. For this reason, the Council rejects the use of cost-benefit analysis.

This report continues by discussing risk-benefit analysis:

. . . risk-benefit analysis, attempts to identify and estimate all the risks and benefits of a food substance. Like cost-benefit, risk-benefit analysis compares the estimated differences of the risks and benefits (marginal risks and benefits) for each level of use. Unlike cost-benefit analysis, however, the risks and benefits are expressed in their commonly used units. The risk data are not converted into dollars, thus avoiding the major problems with cost-benefit analysis. The data can either be qualitative or quantitative, but quantitative data more clearly show the differences in risks and benefits for various use levels.

Risk benefit analysis seeks to organize information on a food substance so that more informed judgments can be made. These judgments will determine whether a food substance should be permitted and, if so, at what level of use. Risk-benefit analysis does not provide definitive answers, but may be considered a valuable aid in the making of decisions.

. . . While ideally risk-benefit analysis requires extimating and weighing all known risks and benefits, a practical application limits the nature and type of risks and benefits to be analyzed. Without limitations, a seemingly endless array of primary, secondary, and tertiary effects, particularly benefits, could be used. Such an open-ended analysis could be excessively time consuming.

. . . On the other hand, such limitations could lead analysts to underestimate a substance's potential benefits, and thus possibly exclude some potentially useful substances from the market. Further, such exclusion may remove from consideration benefits that may later be perceived by society to be important.

Decisions to limit risk-benefit analysis can be made in two stages. The first relates to the nature of risks and benefits. Should only risks to consumers be analyzed or should other risks, such as those to the environment or to processing and farm workers, be considered as well? With regard to benefits, should only those to consumers be included or should the benefits to food producers and/or processor be analyzed too? At present, several regulatory agencies have different criteria for dealing with environmental, occupational, and food risks. Because this report focuses on food safety, the Council recommends that only risks and benefits to the ultimate consumer be included in any risk-benefit analysis that leads to the establishment of a substance's acceptable or safe level of use in food.

The second decision requires listing specific consumer risks and benefits to be included in a risk-benefit analysis. Any risk to consumers should be included. Realistically, the identified risks of a substance for which quantitative risk assessment is applied will have predominant weight in the analysis. However, other identified risks should be considered in the analysis as well, even if the risk data are in a qualitative form.

Risk-benefit analysis can include several different health and economic benefits to the consumers. Health benefits are either the avoidance or reduction of risk that a food poses with the substance and/or the addition or retention of important nutrients in foods. Economic benefits include lower consumer costs, increased availability of food, convenience in storage and home preparation, and sensual characteristics such as color, flavor, and texture. Specific benefits to be weighed require social and political judgments...

Because risk-benefit analysis avoids the valuation-of-risk problem posed by cost-benefit analysis and because a decision-assisting approach is needed when quantitative risk assessment is applied, the Food Safety Council recommends that some form of risk-benefit analysis be employed when quantitative risk assessment is used. However, the Council recommends that the type of consumer benefits included in the analysis be limited.

There is wide agreement that benefits should be weighed against risks in the decision-making process. The evident problem as is clearly indicated by the Food Safety Council's report is that no units of benefit can be applied across the board. The process, therefore, remains a judgemental one at the purely subjective level, unless one is dealing with definable monetary values or units of production of essentials that can be weighted against quantitative needs.

Yudkin[3] has emphasized the importance of distinguishing between wants and needs. This is especially important in balancing health benefits and risks. There is a biological requisite or need for definable quantities of nutrients, food, and water. Until the intake of these essentials meets the biological needs, the organism's wants coincide with its need. Once the requirements are satisfied, however, the organism may, because it derives pleasure of some sort, or satisfaction from increased intake, want additional food, beverage, or nutrients. In such situation the human organism usually, mistakenly, describes his wants as "I need so and so." Needs are physiological, wants may be physiologically determined but also may be behavioral expressions of psychological origin. Nevertheless, satisfying the human wants that are in excess of human need, constitute benefits unless demonstrably harmful. The pleasure of satisfaction is interpreted by the individual as a benefit. Likewise increasing the convenience, decreasing the waste, enhancing the appeal, or increasing the pleasure derived from food through the use of flavors, colors, and other esthetic devices constitute allowable benefits unless harmful to health, economically excessively costly, or compromising of availability of food or of significant nutrient intake. The Social and Economic Committee of the Food Safety Council in its report[4] summarized such considerations as follows:

Regarding benefits, these fall into four categories, namely, health, supply, appeal, and convenience benefits. While each of these benefits is different in kind and largely incommensurable, they can be weighed against risk individually or cumulatively.

The Committee urges, whenever possible, the balancing of health risks with health benefits. But, if after offsetting health benefits against health risks there is still a net health risk, there may exist other benefits that offset this residual risk.

What is at stake is an overall balancing of values. The goal is to arrive at the maximum net social gain. That is, the reason for considering the social and economic consequences of a food safety decision is to assure that as a result of any food safety decision society comes out ahead, that the benefits offset the risks. The Committee recognizes that the estimation of both risks and benefits is difficult and controversial but we are convinced that the uncertainties surrounding the measuring of benefits are no greater than the uncertainties involved in the scientific techniques used to determine risk.

While some minimum benefit must be shown as a precondition for admitting a substance into the food supply even where there is no measurable risk, this condition can be met by showing that a substance is expected to accomplish its intended functionality which would be some level of one or more of the following as predetermined by the Standards Committee:

- Enhance nutritive value
- Avoid toxin formation or other disease incidence
- Reduce cost
- Increase availability
- Increase convenience
- Enhance appeal

The Report of the Social and Economic Committee of the Food Safety Council described four categories of benefits: (1) reduction in health risks or the generation of other health benefits such as through contributing to the general nutritional value of food — collectively "health benefits"; (2) a reduced cost or increased supply of food — supply benefit"; (3) enhanced appeal; and (4) increased convenience. Adding to or maintaining in the food supply any substance that accomplishes any of those effects produces benefits. These benefits may or may not offset the risks involved in having that substance in the food stream. This report, however, subsequently confuses terminology by discussing "negative benefits" which, in fact, are but risks and should be so termed.

It is reassuring that responsible deliberative bodies repeatedly have evolved similar principles as guidelines and have clearly set these forth. In 1957 The first report of the joint FAO/WHO Expert Committee on Food Additives[5] was published. It read in part:

The increase in the number of chemicals used or proposed for use in or on foods has imposed upon public health authorities and other governmental agencies the responsibility for deciding whether or not such substances should be employed. The socioeconomic position of a country is an important factor in arriving at such decisions. Additives can contribute greatly to the preservation of food; for example, they can help prevent wastage of seasonal surpluses. In economically underdeveloped countries, lack of modern storage facilities and the inadequacy of transport and communications may increase the necessity of using certain food additives for purposes of food preservation. Again, in tropical regions, where high temperature or humidity favor microbial attack and increase the rate of development of oxidative rancidity, a wider use of antimicrobial agents and antioxidants may be justified than in more temperate climates. In these regions possible risks associated with the increased use of food additives must be weighed against the benefits in the form of preventing wastage and making more food available in areas in which it is needed. In these circumstances, however, food additives should be used to supplement the effectiveness of traditional methods of food preservation rather than to replace these methods.

In countries which are technically and economically highly developed, the availability of adequate facilities for refrigerated transport and storage reduces, even if it does not eliminate, the need for antimicrobial agents. In these countries, however, there is an increasing demand for more attractive foods, for uniformity of quality and for a wide choice at all seasons of the year. Moreover, large quantities of many of the foods consumed have to be transported from distant producing areas, a fact which may create special transport and storage problems. For such purposes the variety of useful food additives is great and their employment promotes the better utilization of the available foods.

The extent to which food additives are likely to be needed and their nature will therefore vary considerably from region to region and even from country to country. In decisions concerning the use of an additive, attention should be given to its technological usefulness, protection of the consumer against deception, the use of inferior techniques in processing and, above all, to the evidence bearing on the safety for use of the substance.

The summary of this report spoke to benefits as follows:

The increase in the number of chemicals used or proposed for use in or on foods has imposed upon public health authorities and other governmental agencies the responsibility for deciding whether or not such substances should be employed. Food additives have a legitimate use in the food processing and distribution systems of both technologically advanced and of less well developed countries, in that they promote the better utilization of available foods.

The use of food additives to the advantage of the consumer may be technologically justified when it serves the following purposes: the maintenance of the nutritional quality of a food; the enhancement of keeping quality or stability with resulting reduction in food wastage; making food attractive to the consumer in a manner which does not lead to deception; providing essential aids in food processing.

The use of food additives is not in the best interest of the consumer in the following situations and should not be permitted: to disguise the use of faulty processing and handling techniques; to deceive the consumer; when the result is a substantial reduction of the nutritive value of the food; when the desired effect can be obtained by good manufacturing practices which are economically feasible.

Safety for use of an additive is an all-important consideration. While it is impossible to establish absolute proof of the nontoxicity of a specified use of an additive for all human beings under all conditions, critically designed animal tests of the physiological, pharmacological and biochemical behavior of a proposed additive can provide a reasonable basis for evaluating the safety of use of a food additive at a specified level of intake. Any decision to use an intentional additive must be based on the considered judgment of properly qualified scientists that the intake of the additive will be substantially below any level which could be harmful to consumers. Permitted additives should be subjected to continuing observation for possible deleterious effects under changing conditions of use. They should be reappraised whenever indicated by advance in knowledge. Special recognition in such reappraisals should be given to improvements in toxicological methodology.

Other factors must be taken into account in food additives control. When a new food additive is proposed for use, clear evidence must be available to show that benefits to the consumer will ensue. In classes of foods which constitute a considerable proportion of the diet the use of intentional additives should, in principle, be limited. The presence of harmful impurities in food additives can be excluded most effectively by the establishment of specifications of purity. Food additives should be identifiable in chemical and physical terms.

The 1973 statement of the Committee on Food Protection[6] sets forth:

Several circumstances justify the use of food additives to the advantage of the consumer. Every chemical used in food processing should serve one or more of these purposes:

- Improve or maintain nutritional value
- Enhance quality
- Reduce wastage
- Enhance consumer acceptability
- Improve keeping quality
- Make the food more readily available
- Facilitate preparation of the food.

It then cautions that

Apart from the question of safety the use of food additives in some situations is not in the best interests of the consumer and should not be employed when:

- Used to disguise faulty or inferior processes
- Used to conceal damage, spoilage, or other inferiority
- Used to deceive the consumer
- Otherwise desirable results entail substantial reduction in important nutrients
- The desired effects can be obtained by economical, good manufacturing practices
- Used in amounts greater than the minimum necessary to achieve the desired effects.

Numerous other responsible groups have arrived at similar guiding principles and agree that balance in judgments between benefits and risk must recognize the sociocultural characteristics of the system of production, distribution and consumption.

III. BALANCING BENEFITS VS. RISKS

"In evaluating benefits, health benefits carry more weight then benefits of supply and these in turn are more important than benefits of appeal or convenience."[4]

Health benefits, which are weighted most heavily in assessing risk-benefit, may be either direct or indirect. Direct health benefits are those, for example, that improve the nutrient intake of the consumer, such as enrichment where needed (with B vitamins, vitamin A and calcium, fluoride, iodine, and so on), where needed or judged protective; those which modify the composition of foods through processing in such manner as improves the health of a significant number of consumers; or those that result in removal or reduction of content of potentially toxic material inherent within general purpose foods or foods for specific groups of the population.

Indirect health benefits accrue from developments in preservation, storage, production, etc., through making particular seasonal foods available throughout the year and thereby increasing the varieties of foods available to the consumer, and changes in technology that result in increased consumption of nutritious foodstuffs not widely represented in the food supply.

IV. DECISION-MAKING

It is helpful in decision-making to view the needs and setting in relation to Maslow's theory of needs. This theory holds that man evolves through the 5 levels of survival, security, belongingness, esteem, and self-actualization.

Food benefits at the survival level are the basic essentials to avoid starvation. At the security level benefits of food translate into avoidance of obvious food-related ill health, for example prevention of disease directly or immediately recognizable as due to the lack of or poor quality of food. Examples are scurvy, pellagra, or other deficiency diseases. At subsequent levels the benefits of food are particularly subject to societal influences. Additional benefits are not biological requisites, but rather the fulfillment of wants or nonsurvival requisites. Such benefits differ with the perceived role of food in fulfilling expectations generated by the individual's life style or in overcoming a situation that is inimical to the individual. Clearly, judgment as to whether a particular measure may be beneficial and the weight to be given to the resulting "benefit", if any, varies with the level of the society or the perceived needs (wants) of individual members of the society.

Similarly the acceptable degree of risk varies with the position attained in this scale of needs between survival and self-actualization. Obviously decisions reached concerning regulatory balance cannot be based on precisely measured quantitation of benefit, particularly as one rises above the level of security where the perceived benefits are subject to varying societal influences and are not determined by the biological purpose of survival. The consumer may, however, display a quantitation of his perceived value of a "benefit" by the price he is willing to pay. When the food available is sufficient to meet nutrient needs we then perceive benefits of food as contributing to the fulfillment of our nonsurvival requisites. These requisites differ with the perception of the role of food in fulfilling expectations in keeping with individual lifestyles or overcoming situations that individually are regarded as inimical.

Benefit, as well as risk, may be perceived as voluntary or involuntary. Voluntary benefits are those in which the individual makes the conscious decision to participate; involuntary benefits are those in which participation is not determined by the individual but is imposed by the society in which the individual lives, i.e., imposed by law, regulation, social custom, religion, etc. Regulatory imposition of a benefit demands more certain knowledge of existing need and judged safety than does a voluntarily perceived benefit. This latter may carry with it a greater potential risk but does impose on the regulatory decision-maker a responsibility for assuring the presence of understandable consumer information.

Regulatory agencies have a responsibility for giving leadership to informing the public and for imposing involuntary benefits generally where there is abundant evidence of health benefits and minimal risk; for example, fluoridation of water at optimally beneficial levels. A similar responsibility pertains where there is overwhelming evidence of need, and hence benefit to the vast majority of the population even with potential risk to an exceedingly small percentage of the population. Such a situation exists in regions of prevalent severe endemic goiter as there is initiated an effective program of iodinization. In the latter type of situations, regulatory agencies have a further responsibility to assure availability of an alternate supply of the food without ingredients that may be undesirable to a minimal group of consumers; for example, noniodized salt. While such judgments can readily be made within more affluent, industrialized areas, there are situations where these alternatives are not feasible and cultural heritage must then dominate the decision as to imposition of involuntary benefit.

V. RESPONSIBILITY

The responsibility for policy decisions and enforcement resides within the regulatory agency. A corollary is that the regulatory agency must have highly competent, knowledgeable, responsible, scientific staff, adequate in judgment to make informed decisions, independent of political or commercial pressure and with sensitivity to its public's perception of needs. Regulatory decisions should be the continuing responsibility of the agency which has been empowered with the responsibility to alter existing regulations as evidences of need, effectiveness, and safety, may change. The agency should not be subjected to legislatively imposed specific scientific constraints. Such constraints minimize benefits and even increase risks despite sound scientific knowledge and evidence, proper consideration of which is precluded by preemptory legislative dictum.

The objective of science to provide man with the necessities, facilities, and pleasures that contribute to the highest quality of life which can be attained only with proper weighting of benefits in the benefit-risk decision-making process. This occurs only through mature judgment that responsibly examines the scientific evidence against societal criteria of acceptable safety and desirable benefits, in keeping with moral principles of problems related to national, regional, and world food supply resources and standards.[7]

REFERENCES

1. **Wodica,** *Nutri. Rev.,* 38, 45, 1980.
2. The Food Safety Council, A Proposed Food Safety Evaluation Process, Final Report of the Board of Trustees, June, 1982.
3. **Yudkin,** *Diet of Man: Needs and Wants,* Applied Science Publishers, Ltd. London, 1978, 358.
4. The Food Safety Council, Principals and Processes for Making Safety Divisions, Social and Economic Committee, Washington, D.C., 15, 1979.
5. First Representative Joint Fad/Who Expert Committee on Food Additives, Tech. Ser. No. 129, Geneva, 1957.
6. The National Academy of Sciences, The Use of Chemicals in Food Production, Processing, Storage, and Distribution. Committee on Food Protection, Washington, D.C., 7, 1973.
7. **Stumpf, S. E.,** The moral dimension of the world's food supply, *Annu. Rev. Nutri.,* 1, 1, 1981.

Chapter 5

MEASURING HEALTH BENEFITS

Mark S. Thompson

TABLE OF CONTENTS

I. INTRODUCTION

General types of health benefits include economic, noneconomic, and externality benefits. Specific components are many and may be grouped into the categories of life extensions, enhanced life quality, genetic benefits, gains from the better understanding of the health care system, lower costs, and environmental enhancement. Current means of measuring benefits are economic quantification, shadow pricing, willingness to pay, decision analysis, and health status indices. These measurements are systematically combined in benefit-cost, cost-effectiveness, cost-utility, and risk-benefit analysis and in multi-attribute utility theory (MAUT).

II. TYPES OF BENEFITS

A. Economic

Economic benefits are effects that have clear positive impact on the economic product of a nation. Distinction frequently is made between direct and indirect components. Direct benefits represent the reduction of expenditures for goods and services — as for instance the avoidance of a visit to a physician. Indirect benefits are additions to the economic product — such as the value of work performed by a cured individual. In recent years, some analysts have gone beyond the standard measurement of economic benefits to quantify such subtle effects as saved expenditures on transportation and the additional economic product arising from freeing up friends and relatives from care of the sick to increase their own earnings. This distinction between direct and indirect effects is specifically that of health economics. In an alternative special usage, indirect expenses are taken to apply to overhead as, for instance, the percentage allocations for office space, administrative support, and the like in connection with research contracts. In common language terms, direct and indirect refer to the immediacy of impact. Terminological confusion may arise as in the case of toxic chemical control which, as an initial effect, reduces ambient levels of the toxins and, as a consequent effect, reduces physician expenses for treatment of toxin-induced ill health. The reduction in physician expenses would be termed direct in health economics, indirect in common usage. In this chapter, the usage of health economics will generally be followed.

B. Noneconomic

Noneconomic benefits are gains sensed by individuals that are not related in a clear and immediate way to economic product. Examples include longer life among persons no longer working, less pain, better ability to function in the home in nonessential duties, more enjoyment in leisure activities, freedom from depression, and improved physical appearance. A noneconomic social benefit is the freeing up of the time of friends and relatives who can therefore devote more time to pursuits other than earning. Noneconomic effects are sometimes termed intangible or unquantifiable.

C. Externality

Externality benefits arise when actions with immediate impact on but a small number of persons (perhaps patients and their families) positively affect a broader community. Examples occur in the treatment of alcoholics which may reduce rates of fires, crime, and motor vehicle accidents in the community.[1] Another is lowered contagion hazard: successful treatment of some patients keeps them from spreading communicable diseases to others: inoculation programs benefit even the uninoculated members of a society. Externality benefits are frequently environmental: the immediate impacts of many environmental programs are costs of compliance to point polluters: benefits are sensed by a broader population — in such forms as reduced risks to plants and animals and enhanced esthetic value. Externality benefits may be genetic: for instance from prenatal screening or genetic counseling programs which not only eliminate specific cases of chromosomal or single-gene disease from the next generation but also improve the gene pool for the indefinite future. Controversy has arisen over externality effects on animals and humans in the case of policies toward antibiotics in animal feed: they are known to yield cost-effective increases in meat production but may additionally foster the emergence of antibiotic-resistant strains.[2] Externality effects have also been known as side effects, spillovers, or neighborhood effects.

III. COMPONENTS OF BENEFIT

A. Overview

Across these different types of benefit are many specific benefit components. One scheme for listing health program benefit components is presented in Table 1. Organization of the table within categories is generally from the more restricted impacts on individuals to impacts on broader populations and, in cases, on future generations. Analysts have pointed out that entirely different benefit-cost analyses may be performed, depending on whose decision-making perspective the analyst adopts. A study carried out from the point of view of the consumer would, for instance, focus on such issues as out-of-pocket charges, time lost in seeking care, inconvenience, and personal health status. From the viewpoint of the provider, professional time, staffing requirements, overhead costs, and the congeniality of the clinical encounter would be more important concerns; third-party payors would concentrate on impacts on reimbursements; a broad societal analysis would encompass all of the items of interest to the separate groups, as well as political, symbolic, and ethical concerns. A sense of how the individual health benefit components are regarded is indicated by the letters in parentheses following each component, suggesting who the primary beneficiaries are.

A listing of this type can never be definitive. There will always be different ways of dividing up the sum total of health program benefits and of ordering the pieces. Overlaps are inevitable as, for instance, among the catagories of disease, physical disability, and pain. Virtually any benefit, considered finely enough, has impacts on all of the groups indicated in parentheses.

B. Longer Life

The most evident and dramatic effects of health programs are life savings — to the extent that the major killers, cancer and coronary heart disease, it is often alleged, receive more than their fair share of attention and funding in comparison with such crippling but usually nonfatal chronic diseases as rheumatoid arthritis.[3,4] Life savings themselves were often undervalued in the past as their effects on gross economic product. Schelling[5] pointed out that this approach — which still is predominant in empirical studies and in court proceedings — overlooks the noneconomic dimensions of benefit, both to the persons with longer lives and to their acquaintances. A valuation methodology appropriate for income-producing livestock is hardly suitable for the family collie. Moreover, Schelling argues, much of the harm done by the existence of premature deaths arises for people who do not die prematurely but worry that they will. Lowering the rates of premature deaths reduces the anxiety costs imposed by the risks. Similar arguments would apply to risks of disease, disability, and pain but for brevity are not shown in Table 1. Slovic et al.[6] discuss popular attitudes toward risk and perceptions of it.

C. Higher Life Quality

Specifying the components of life quality has frustrated many quantitative analysts of the past ten years. Primary obstacles have been the number of possible aspects of life quality and the differences in attitude across people toward these aspects. To an avid tennis player,

Table 1

COMPONENTS OF HEALTH PROGRAMS BENEFITS

I.   Longer Life
 A.   Greater personal enjoyment of life (P)
 B.   Reduced risks of disease (P)
 C.   Nonmonetary benefits to acquaintances (F, H)
 D.   Increased earnings (Ec, H, P)

II.   Higher Life Quality
 A.   Less disease
  • Greater personal enjoyment of life (P)
  • More positive interactions with friends and family (F, H)
  • Fewer demands for unpaid assistance (F, H)
  • Less money spent on health care (Ec, Em, H, P)
  • Less absenteeism (Ec, Em)
  • Reduced contagion hazard (Ec, En, F, H)
 B.   Less physical disability
  • Greater personal enjoyment of life (P)
  • Greater range of possible leisure pursuits and greater gratification from them (F, H, P)
  • Fewer demands for unpaid assistance (F, H)
  • Less money spent for therapy, prosthetics, counseling and other components of care (Ec, Em, F, P)
  • Longer working life, less absenteeism, and better vocational function (Ec, Em, H, P)
 C.   Better mental health
  • Greater personal enjoyment of life (P)
  • More rewarding and meaningful interpersonal encounters (Em, F, H, P)
  • Reduced need for emotional support from friends and family (F, H)
  • Less money spent for care (Ec, Em, H, P)
  • Longer working life, less absenteeism, and better vocational function (Ec, Em, H, P)
  • A social environment of higher quality and pleasantness and fewer risks and concerns (Ec, En)
 D.   Less pain, malaise, or discomfort
  • Greater personal enjoyment of life (P)
  • Reduced need for emotional support from friends and family (F, H)
  • Longer working life, less absenteeism, and better vocational function (Ec, Em, H, P)
  • Less money spent for analgesics, therapy, and other components of care (Ec, Em, H, P)
 E.   Improved physical appearance
  • Greater personal enjoyment of life (P)
  • Better interpersonal relationships (F, H, P)

III.   Genetic Counseling, Diagnosis, and Remedial Actions
 A.   Reduced emotional burden on acquaintances of the severely handicapped (F, H)
 B.   Less need for financial support of persons with mental and physical handicaps (Ec, H)
 C.   Greater economic productivity of persons who would not otherwise have been born (Ec)
 D.   An improved gene pool (F. G.)

IV.   Better Understanding of the Health Care System
 A.   Better health care and greater satisfaction with it (Ec, Em, F, H, P)
 B.   Health care of lower cost (Ec, Em, H, P)

V.   Lower Costs of Health Care (Ec, Em, H, P)

VI.   Environmental Enhancement
 A.   Less disease with environmental etiology (Ec, En)
 B.   Reduced harm to plant and animal life (Ec, En)
 C.   More esthetically pleasing environment (En)
 D.   Prevention of harm to ecosystems that is long-lasting or irreversible (Ec, F.G.)

Note: The letters in parentheses indicate primary effects on: Ec = the economy of a region; Em = employers of persons directly affected; En = the physical and social environment of a region; F = friends of persons directly affected; F.G. = future generations; H = households of persons directly affected; and P = persons directly affected.

elbow problems are a catastrophe while, to a skier, they are a minor inconvenience. To the nonathletically inclined, many physical limitations are rated as relatively unimportant. Problems of specification and measurement are the more challenging in the area of mental health benefits which may be taken to include less neurosis and psychosis, reduced depression and anxiety, improved self-esteem and self-confidence, greater personal independence, less self-injurious behavior, less substance abuse, improved socialization skills, less antisocial and careless behavior. The enormous difficulties in classifying and calibrating these subjective dimensions of disease have hampered development of the International Classification of Disease codes and operationally hinder attempts to control costs through classification into diagnostically related disease groups. Pain, pruritus, and related ailments have defied attempts to develop interpersonally comparable scales. Improved physical appearance is a controversial health benefit, included in Table 1 because of the billions of dollars spent within health systems to achieve it.

D. Genetic Counseling, Diagnosis, and Remedial Actions

A special area of health interventions is that of genetics — where decisions can have the effects of (1) reducing the birth rates of some types of persons, (2) increasing the birth rates of others, and (3) changing the human gene pool. The current polarization over the issue of abortion in many nations illustrates the difficulties in reaching an objective analytic consensus. Analysts have shrunk from such daunting tasks as comparing the values achieved by aborting defective or unwanted fetuses and by replacing defective fetuses with healthy ones. Gains or losses to the individuals themselves are accordingly not shown in Table 1.

E. Better Understanding of the Health Care Systems

Programs of education and counseling can help patients to guide themselves toward medical care that is most appropriate for their ailments and for their personal systems of values.[7-9] In addition, patients would learn to avoid unjustified apprehensions and would better understand their own health states, their possibilities, and their limitations. One would hope that such programs would have the effect of lowering the costs of medical care as there would be less inappropriate use of emergency rooms and specialists, fewer medical visits for nonthreatening and minimally remediable complaints, and less need for placebo prescriptions. It is possible that patient education would at times increase the costs of health care as people learn that some apparently minor symptoms do merit medical attention. Providers of medical care may also benefit from a higher level of congeniality in the patient care encounter as general understanding of medicine and health is raised.

F. Lower Costs of Health Care

Several programs of recent years have had as an explicit major goal the lowering of costs of medical care with minimal compromise of patient benefit. Such programs include education, deinstitutionalization of various categories of individuals, educational programs directed toward persuading physicians to order fewer unnecessary tests, and promotion of second opinions before surgery.

G. Environmental Enhancement

Environmental programs, unlike many health programs, typically do not provide benefits to isolated individuals within populations. Their impacts are rarely as narrow even as on a town with contaminated drinking water. In this broad sense, environmental programs both extend lives and enhance life quality, thereby reaping in a more general way benefits shown in these other sections of Table 1.

IV. COSTS AND BENEFITS OF TOXIC CHEMICALS

A similar disaggregation may be made of the effects of toxic chemicals and of their control. These effects are displayed in Figure 1, where the major distinction is between the

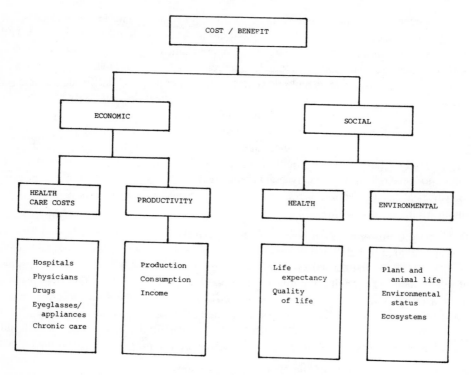

FIGURE 1.   Costs and benefits of toxic management.

economic and social dimensions. Economic effects include impacts on spending within the health care delivery system (direct costs, in health economics) and consequences for overall economic productivity (indirect costs, in health economics).

Social effects in Figure 1 encompass the noneconomic and externality dimensions of impact. For analyses of toxic chemical impact, the leading noneconomic effect will be on health: all changes in life expectancy and quality that are not translated into economic terms. The externality benefits here are overwhelmingly environmental, with impacts on plant and animal life, general environmental status, and specific ecosystems.

V. METHODS OF MEASUREMENT

A. Economic Quantification

Complementary economic methodologies for measuring the burdens imposed by various illnesses and the gains realized by combatting them have evolved. Estimation of the direct cost of illness (or conversely the direct gains realized in reducing illness) have been relatively easy. Money spent for the time of physicians, for hospitalizations, for drugs and appliances, and for other expenses of care and cure are undeniable costs of illness; their reductions achieved through health programs represent clear benefits. Good first approximations of the value of these expenditures are their current dollar costs — even though slightly improved numbers would be obtained by using the theoretically superior concept of opportunity costs.

More difficult to quantify are the indirect economic effects: effects on length of life and health status. Valuation of such effects was initiated by Sir William Petty in the late seventeenth century.[10] Petty's objective in one study was to compare the benefits and costs of better general medicine. As a strategy for valuing the lives that would thereby be saved, he took the current minimum price for slaves on the London slave market as the lower bound to the average value of a human life. While there is a modern tendency to ridicule this approach, a closer look at the leading life valuation methods of 1985 reveals that they follow the main lines of slave pricing: people are valued as one would value productive machines, as the present value of future expected output. By far the most common means for valuing the extension of a life or the enhancing of health status is by focusing on the addition to economic product thereby obtained. This method is the human capital approach to health benefit measurement.

The most extensive applications of the human capital approach have been in a series of studies quantifying the costs of illness.[11-15] These studies have yielded refined estimates of future productivity: disaggregating by age and sex, adjusting for inflation, using sensitivity analyses of the discount rates, incorporating alternative methods for capturing the worth of work performed in the home, supposing reasonable rates of productivity growth. Yet, for all these methodological refinements, the cost-of-illness techniques still follow the basic idea of Sir William Petty: valuing lives solely as the present value of their future economic product. This approach would not be pernicious if decision makers heeded the caution given by Petty and echoed by modern analysts: that such calculations are only lower bounds to life values, not the analysts' best estimates of those values. Decision makers all too frequently have concluded that health programs were justifiable if and only if benefits as calculated by the human capital approach exceeded costs.

B. Shadow Pricing

Benefits and costs are easily quantified if they pertain to market-valued goods and services. The value of an air pollution control program insofar as it increases crop yields is simply the market value of the additional crops. But what is the value to people of not having to breathe polluted air? Since this benefit is not traded in the marketplace, other methods must be applied. One approach is that of shadow pricing:[16,17] of valuing benefits and costs by reference to their near equivalents having market prices. Shadow pricers might thus value better air as the amount of money that would otherwise be spent on air conditioners, the value of reduced crime from a program of drug control as the reduction in expenditures on home security devices. But shadow pricing frequently breaks down because the near equivalents are not near enough. The values people associate with clean outdoor air are not captured by the amounts of money they might spend on air conditioners. The human capital approach to the costs of illness could be seen as a shadow-pricing methodology.

C. Willingness to Pay

Analysts recognize that the effects of health and other social programs rarely have market values and are not well quantified by shadow pricing. The approach adopted by many benefit-cost analysts[18-20] has been to monetarize such effects using willingness to pay. In this method, the value to me of better air is the amount of money I would be willing to pay for it. The value of lowering air quality would be the amount of money I would have to be paid to be reconciled to this cost to me. The technical term for either of these amounts of money is compensating variation.

While the theoretical superiority of willingness-to-pay methods has long been recognized, the practical difficulties in asking people their maximum willingness to pay in ways that elicit meaningful responses and a few disappointing experiences have prevented widespread applications. The need for an approach like that of willingness to pay is especially evident in chronic debilitating diseases such as arthritis. There, a recent study[21] has shown that many patients are reluctant to respond, but that, among responders, stated willingness to pay for arthritis cure tended to be high for patients with more concern for their health, with great ongoing medical expenses, and with more severe disease. Average willingness to pay for patients stratified over the four levels of arthritis defined by the Health Interview Survey was 17 percent of total family income, or $3900 per year. Noteworthy applications of the willingness-to-pay methodology have also been achieved in such areas as transportation[22] and environmental management.[23]

D. Decision-Analytic Quantifications

An alternative approach to valuing social program effects that are hard to quantify is that of decision analysis.[24-26] Valuation of health benefits using decision analysis has been carried out in the cases of amniocentesis[7] and electronic fetal monitoring.[27] In each of these cases, valuation had to be made of marginal changes in small risks of deaths, of genetic diseases, and of perinatally caused mental retardation — phenomena for which economic valuations are not highly reliable. This can be seen in the case of maternal death consequent to cesarean section. Economic valuation shows that the future discounted earnings of a mother are close to twice the economic value of changing a baby with severe mental retardation to perfect health. However, the median maximum risk of maternal risk acceptable to most parents and doctors in order to change neonatal status from severe mental retardation to perfect health is one in 300 — a vastly different number from the 0.50 risk that would seem acceptable on the basis of economic valuation.[27]

E. Health Status Indices

Related to the decision analytic methodology is that of health status indices.[28-30] The idea here is to quantify health states as proportions of perfect health. With such measures, to raise a person's health status from 0.75 to 0.85 of perfect health for one year would achieve 0.10 quality-adjusted life years (QALY's), as would extending the life of a perfectly healthy person for 0.10 year.[31-33] Health status indices have been used for the quantification of benefits in a broad variety of health programs including control of hypertension,[34] vaccinations against pneumococcal pneumonia,[35] and estrogen use in postmenopausal women.[36]

VI. CONCEPTS VITAL IN HEALTH BENEFIT MEASUREMENT

Concepts important in quantifying the benefits of health programs include:

(1) Consumption — the amounts of money spent for personal reasons, including expenditures on food, shelter, transportation, health care services, and recreation, among others. Arguments have been made that human life extensions are more appropriately valued as the present worth of anticipated consumption than as the present worth of future production.[37] A situation to which such arguments may apply is a vacation. Under the human capital approach, the value of vacationers' time is zero since no economic product is created by them. Others would hold however that time spent on vacation is even more valuable to the individual than time spent working, as is indicated by comparing levels of expenditures. The difference between the consumption and the production measures of life value is also graphic in the case of retired persons;

(2) Direct cost of illness — expenditures incurred for the care of patients;

(3) Health status index — a scale usually set between 0 as the value of a unconsciousness, death, or health so bad that it is considered equivalent to death and 1.0 as the value of perfect health. Some health status indices are comprehensive and have methodologies for quantifying all health states; others focus only on health states associated with particular diseases;

(4) Human capital value — calculation of human worth as the present value of all future contributions by a person to the total economic product of a society;

(5) Income — money periodically received as earnings, as return on investments, or as social support payments. Income is typically more than production by the amounts of investment return and social support and differs from consumption by the amounts of saving or dissaving and of tax;

(6) Indirect cost of illness — economic product not produced due to illness;

(7) Opportunity cost — the benefits resources would achieve in alternative employments;

(8) Production — economically valued goods and services. Earnings are usually taken as the best measure of personal production, although wage scales correlate poorly with marginal individual contributions to overall economic product;

(9) Utility — the general term used by economists and decision analysts to describe what people seek to achieve in their decisions. Near synonyms are good, happiness, satisfaction, or well being. Health analysts often use the term in a more narrow sense, as the proportion of perfect health, as measured by a health status index; and

(10) Willingness to pay — valuation of a benefit as the maximal amount the beneficiary is willing and able to pay for it.

VII. FORMS OF NORMATIVE ANALYSIS

A. Benefit-Cost Analysis

Synonymous with cost-benefit analysis, benefit-cost analysis is the systematic weighing of the advantages and disadvantages of an action or program, expressed in monetary terms. Figures summarizing benefit-cost evaluation of programs are net present value, the difference between present-valued benefits and present-valued costs, and the benefit-cost ratio, their quotient. The primary shortcoming of the benefit-cost analysis relative to other forms of normative analysis in the area of health is the extreme difficulty in quantifying monetarily health effects.

B. Cost-Effectiveness Analysis

As a means of circumventing the difficulties encountered by benefit-cost analysis in dealing with effects that are hard to value monetarily, cost-effectiveness analysis was developed. The basic insight underlying cost-effectiveness analysis is that even without knowing the value of a benefit unit - perhaps a life saved, or a case of pneumonia prevented, or a one percent decrease in levels of atmospheric ozone — decision makers know that programs achieving set benefit levels at lower cost than others are better and that, for a set budget, one should attempt to obtain as many benefit units as possible. The cost-effectiveness of programs is summarized in the cost-effectiveness ratio, the ratio of dollars spent per benefit unit achieved. Several authors[31,38,39] have argued that the most appropriate way to spend limited budgetary amounts is by funding programs in the priority order of their cost-effectiveness ratios.

C. Cost-Utility Analysis

A special form of cost-effectiveness analysis is cost-utility analysis — where utility is taken in the narrow sense of proportion of full health and the analysis focuses on comparing net dollars spent per net quality-adjusted life year gained. This seems to be the most broadly applicable of all possible cost-effectiveness ratios in the health domain.

D. Risk-Benefit Analysis

An offshoot of both benefit-cost and cost-effectiveness analysis is risk-benefit analysis. The term has been used to cover a variety of types of analysis, and generally refers to any normative analysis of situations involving mortal risk.[40]

E. Multi-Attribute Utility Theory

An extension of cost-effectiveness analysis is multi-attribute utility theory (MAUT). The cost-effectiveness approach, as generally applied, is limited to dealing with one hard-to-quantify effect. This is acceptable in analyzing programs with single, main dimensions of benefit — such as goals of many disease control programs to reduce the number of cases. The approach falters, however, when a program has many dimensions of effect — such as an environmental program that seeks at once to reduce mortal risks, to reduce morbid risks, to reduce risks to plant and animal life, to promote esthetic qualities, and to preserve the viability of ecological systems for future generations, among other goals. To reduce all these program goals to a single dimension is an impossible task. Multi-attribute utility theory is based on comparing the decision tradeoffs policy setters would make in weighing gains and losses across the different dimensions.[25,41]

VIII. CONCLUSIONS

A. Multiple Aspects and Perspectives

Application of normative analysis in social program areas has become vastly more complicated and sensitive than 20 years ago when benefit-cost analysts mainly estimated the economic values of improved waterways and the other sibling disciplines did not exist. Decision-makers realize the need for comprehensive consideration of programs — comprehensiveness that may be facilitated by schemata such as Table 1 and Figure 1, listing the many possible dimensions of concern. It is useful too to bear in mind the distinctions among such different types of health benefits as economic and noneconomic effects, externality effects, and the degree of directness or immediacy of the impacts.

B. Economic Quantification

Recent advances in health program analysis have led to downplaying the importance of economic measures. These still, however, are useful as lower-bound indicators of value. It remains effective and persuasive to show that economic benefits for a proposed program substantially exceed costs and that noneconomic or intangible effects, which may be considered too difficult to quantify, only strengthen the case for the program.

C. Noneconomic Quantification

Health program analysis is currently undergoing a revolutionary development of methods for measuring noneconomic effects and for expressing them in monetary terms. Despite the many advances of the past decade, important obstacles remain. Decision makers for a considerable time to come will have to continue relying on sensitive qualitative judgment, rather than on exact quantification of the intangibles.

D. Disciplines

The development of willingness-to-pay methods has given benefit-cost analysis an important handle on noneconomic aspects of health programs and a renewed relevance. The trend in recent years, however, has been away from the more heavily economic analysis toward alternative forms — such as cost-effectiveness, cost-utility, and risk-benefit analysis and multi-attribute utility theory — that do not require aggressive monetarization of all impacts. In this light, it seems reasonable to expect that new variants of these disciplines will continue to be formulated. Those methods gaining greatest acceptance will likely be those that combine capture of decision-maker concerns with morally palatable expressions of them.

REFERENCES

1. **Berry, R. E., Jr. and Boland, J. P.,** *The Economic Cost of Alcohol Abuse,* Free Press, New York, 1977.
2. **Marshall, E.,** Antibiotics in the barnyard, *Science,* 208, 376, 1980.
3. **Meenan, R. F., Yelin., E. H., Nevitt, M., and Epstein, W. V.,** The impact of chronic disease, *Arthritis and Rheumatism,* 24(3), 544, 1981.
4. **Policy Analysis, Inc.,** Evaluation of Cost of Illness Ascertainment Methodology, final report submitted under contract 233-79-2048 to the National Center for Health Statistics, 1981.
5. **Schelling, T. C.,** The life you save may be your own, in *Problems in Public Expenditure Analysis,* Chase, S., Ed., The Brookings Institution, Washington, D.C., 1968, 127.
6. **Slovic, P., Fischhoff, B., and Lichtenstein, S.,** Facts and fears: understanding perceived risk, in *Societal Risk Assessment: How Safe Is Safe Enough,?* Schwing, R. C. and Albers, W. A., Jr., Eds., Plenum, New York, 1980, 181.
7. **Pauker, S. P. and Pauker, S. G.,** The amniocentesis decision: an explicit guide for parents, in National Foundation — March of Dimes, *Birth Defects: Original Article Series,* Vol. XV, No. 5C, Alan R. Liss, New York, 1979, 289.
8. **McNeil, B. J., Weichselbaum, R., and Pauker, S. G.,** Fallacy of the five-year survival in lung cancer, *N. Eng. J. of Med.,* 299,1397,1978.
9. **McNeil, B. J., Weichselbaum, R., and Pauker, S. G.,** Speech and survival: tradeoffs between quality and quantity of life in laryngeal cancer, *N. Eng. J. of Med.,* 305, 982, 1981.
10. **Fein, R.,** Measuring economic benefits of health programs, in *Medical History and Medical Care,* McLachlan, G. and McKeown, T., Eds., Oxford University Press, London, 1971.
11. **Mushkin, S. J. and Collings, F. d'A.,** Economic costs of Disease and injury, *Publ. Health Rep.,* 74(9), 795, 1959.
12. **Weisbrod, B. A.,** *Economics of Public Health,* Universtiy of Pennsylvania Press, Philadelphia, 1961.
13. **Rice, D. P. and Cooper, B. S.,** The economic value of human life, *Am. J. Pub. Health,* 57, 1954, 1967.
14. **Cooper, B. S. and Rice, D. P.,** The economic cost of illness revisited, *Social Security Bull.,* 39(2), 21, 1976.
15. **Hartunian, N. S., Smart, C. N., and Thompson, M. S.,** *The Incidence and Economic Cost of Major Health Impairments,* D.C. Health, Lexington, Mass., 1981.
16. **McKean, R. N.,** The use of shadow prices, in *Problems in Public Expenditure Analysis,* Chase, S. B., Ed., The Brookings Institution, Washington, D.C., 1968.
17. **Abt, C. C.,** The issue of social costs in cost-benefit analysis of surgery, in *Costs, Risks and Benefits of Surgery,* Bunker, J. P., Barnes, B. A., and Mosteller, F., Eds., Oxford University Press, New York, 1977, 40.
18. **Mishan, E. J.,** *Cost-Benefit Analysis,* Praeger, New York, 1971.
19. **Sugden, R. and Williams, A.,** *The Principles of Practical Benefit-Cost Analysis,* Oxford University Press, London, 1978.
20. **Thompson, M. S.,** *Decision Analysis for Program Evaluation, Sage,* Beverly Hills, 1980.
21. **Thompson, M. S., Read, L. R., and Liang, M.,** Willingness-to-pay concepts for societal decisions in health, in *Values and Long-Term Care,* Kane, R. and Kane, R., Eds., D.C. Health, Lexington, Mass., 1982, 103.
22. **Mishan, E. J.,** What is wrong with Roskill?, *Journal of Transport Economics and Policy,* 4(3), 221, 1970.
23. National Academy of Sciences and National Academy of Engineering, The costs and benefits of automobile emission control, in *Air Quality and Automobile Emission Control,* prepared for Committee on Public Works, U. S. Senate, 4, Serial 93-24, Washington, D.C., 1974.
24. **Raiffa, H.,** *Decision Analysis: introductory lectures on choices under uncertainty,* Addison-Wesley, Reading, Mass., 1968.
25. **Edwards, W., Guttentag, M., and Snapper, K.,** A decision theoretic approach to evaluation research, in *Handbook of Evaluation Research,* Vol. 1, Struening, E. L. and Guttentag, M., Eds., Sage, Beverly Hills, 1975.
26. **Thompson, M. S.,** *Decision Analysis for Program Evaluation,* Sage, Beverly Hills, 1982.
27. **Thompson, M. S. and Cohen, A. B.,** Decision analysis: electronic fetal monitoring, in *Methods of Evaluating Health Services,* Wortman, P., Ed., Sage, Beverly Hills, 1981, 109.
28. **Kaplan, R. M., Bush, J. W., and Berry, C. C.,** The reliability, stability, and generalizability of a health status index, in *Proc. Am. Stat. Assoc., Soc. Stat. Sect.,* 1978, 704.
29. **Sackett, D. L. and Torrance, G. W.,** The utility of different health states as perceived by the general public, *J.Chronic Dis.,* 31, 697, 1978.
30. **Bergner, M., et al.,** The sickness impact profile: development and final revision of a health status measure, *Medical Care,* 19(8), 787, 1981.

31. **Bush, J. W., Fanshel, S., and Chen, M. M.,** Analysis of a tuberculin testing program using a health status index, *Socio-Econ. Planning Sci.,* 6, 49, 1972.

32. **Patrick, D. L., Bush, J. W., and Chen, M. M.,** Toward an operational definition of health, *J. Health and Soc. Behav.,* 14, 6, 1973.

33. **Weinstein, M. C. and Stason, W. B.,** Foundations of cost effectivness analysis for health and medical practices, *N. Engl. J. Med.,* 296,716, 1977.

34. **Weinstein, M. C. and Stason, W. B.,** *Hypertension: a policy perspective,* Harvard University Press, Cambridge, 1976.

35. **Willems, J. S., Sanders, C. R., Riddiough, M. A., and Bell, J. C.,** Cost effectiveness of vaccination aganist pneumococcal pneumonia, *N. Engl. J. Med.,* 303, 553, 1980.

36. **Weinstein, M. C.,** Estrogen use in postmenopausal women — costs, risks, and benefits, *N. Engl. J. Med.,* 303, 308, 1980.

37. **Fischer, G. W. and Vaupel, J. M.,** A Lifespan Utility Model: Assessing Preferences for Consumption and Longevity, working paper for the Center for Policy Analysis, Duke University, 1976.

38. **Torrance, G. W., Thomas, H., and Sackett, D. L.,** A utility maximization model for evaluation of health care programs, *Health Serv. Res.,* 7, 118, 1972.

39. **Neuhauser, D. and Lewicki, A. M.,** National health insurance and the sixth stool guaiac, *Policy Anal.,* 2, 175, 1976.

40. **Wilson, R. and Crouch, E.,** *Risk/Benefit Analysis,* Ballinger, Cambridge, 1982.

41. **Keeney, R. L. and Raiffa, H.,** *Decisions with Multiple Objectives: preferences and value tradeoffs,* Wiley, New York, 1976.

Chapter 6

FOOD SAFETY REGULATION

K. Skinner and S. A. Miller

TABLE OF CONTENTS

I. INTRODUCTION

In their considerations of risk assessments, government officials are keenly aware of the special relationship in our society between law and science. Law — through its subject matter and requirements — focuses resources on certain goals and the means through which they are to be achieved. On the other hand, science — through the knowledge it provides — affects the evolution of national goals and their incorporation into statutes. An interesting study in this relationship can be obtained by looking at the manner in which risk and its assessment have been addressed in the U.S. Federal Food, Drug, and Cosmetic Act.[1] This chapter attempts to offer such an examination for the purpose of providing a basis for understanding some of the existing activities involving food safety regulation and speculating about possible trends in this area.

II. "RISK" AND "RISK ASSESSMENT" IN THE FEDERAL FOOD, DRUG, AND COSMETIC ACT

A. Background on Statutory Provisions

The Federal Food, Drug, and Cosmetic Act (the Act) is an intriguing composite of sometimes overlapping provisions which have been added by amendments through the years. One of the most important aspects of the Act is that these provisions have established categories of food substances and ingredients with varying safety requirements. Before an inquiry into the statute's attitudes toward risk can be undertaken, some understanding of this structure is necessary. For that reason, a brief backgound on various statutory provisions will first be offered. To review all of the Act's sections would be a formidable task. This discussion will look at portions of only two provisions: Section 402 and 409. It is hoped that this limited study will provide insight not only into the statute's historical view of risk, but also its perspective regarding relative risk among food categories. For a more complete guide to the food safety provisions of the Act, the reader is referred to the thorough and lucid review by Merrill.[2]

1. Section 402

Section 402 is both significant and interesting because it contains the food safety standard incorporated into the original 1906 version of the Act. This standard declared food adulterated if it contained "any added poisonous or other added deleterious ingredient which many render such article injurious to health."[3] The "may render" standard was retained when the Act was revised in 1938 and subsequent years and is present in today's version of the law. The interpretation of this standard arises from a 1914 Supreme Court case, United States vs. Lexington Mill & Elevator Co. concerning flour treated with nitrogen peroxide gas. In that case, the government claimed that the flour was adulterated because it contained substances resulting from the chemical treatment. The Court rejected this argument on the grounds that the safety standard must be considered in its entirety. The issue was not if the flour contained a poisonous substance, but if it contained it in a quanity which may cause the flour to be harmful when eaten.
Said the Court:

 . . . The word "may" is here used in its ordinary and usual signification, there being nothing to show the intention of Congress to affix to it any other meaning. It is, says Webster, "an auxiliary verb, qualifying the meaning of another verb by expressing ability, . . . , contingency or liability, or possibility or probability." In thus describing the offense Congress doubtless took into consideration that flour may be used in many ways, in bread, cake, gravy, broth, etc. It may be consumed, when prepared as a food, by the strong and the weak, the old and the young, the well and the sick; and it is intended that if any flour, because any added poisonous or deleterious ingredient may possibly injure the health of any of these, it shall come within the ban of the statute. If it cannot by any possibility, when the facts are reasonably considered, injure the health of any consumer, such flour, though having a small addition of poisonous or deleterious ingredients, may not be condemned under the act.[4]

The "may render" standard of section 402 refers only to "added" substances. When congress revised the Act in 1938, it expanded the authority of the Act to cover naturally occurring foods and their constituents. Specifically, it added language which declared a food adulterated

" . . . If it bears or contains any poisonous or deleterious substance which may render it injurious to health; but in case the substance is not an added substance such food shall not be considered adulterated under this clause if the quantity of such substance in such food does not ordinarily render it injurious to health . . . [5]

This "ordinarily injurious" standard remains in current law. Its interpretation is derived from case law as well as legislative history. Traditionally, this standard has been construed to be similar to the "may render" standard in that quantity of substance must be considered in deciding whether it renders a food injurious. Merrill has noted that the primary difference between the "may render" and "ordinarily injurious" standards is that the latter is believed to require a showing of a greater probability of harm.[6]

2. Section 409

Section 409 is perhaps the most well known portion of the Act because it contains provisions for regulating the safe use of food additives. It also contains the controversial Delaney amendment which prohibits the addition of animal carcinogens to the food supply. Section 409 declares as additive to be unsafe unless there is a regulation prescribing the conditions under which the additive may be safely used.[7] Because the statute did not define safe other than to say it has reference to the health of man or animal, the FDA has found it necessary to turn to the legislative history for a standard as to what constitutes a safe tolerance. The House of Representatives Report on the 1958 food additives amendment stated: "Safety requires proof of a reasonable certainty that no harm will result from the proposed use of an additive. It does not — and cannot — require proof beyond any possible doubt that no harm will result under any conceivable circumstances."[8] Thus, FDA has embraced the concept of reasonable certainty of no harm as the general saftey standard of section 409. The only considerations to be made in establishing tolerances under section 409 are the safety of the additive and whether it can accomplish its intended technical effect at the contemplated level of use.

Although the Delaney amendment has been the subject of considerable interest surrounding section 409, the most outstanding aspect of section 409 as a whole is that it places the burden of proof upon the sponsor of a food additive to demonstrate the substance's safety before it can be marketed. This requirement is in striking contrast to the provisions of section 402 which required that the government must demonstrate that an added ingredient already in the food supply was unsafe. In this respect, the addition of section 409 to the Act represents a milestone in U.S. food law.

B. Use of Second-Order Probes for Risk

Obviously, sections 402 and 409 contain no reference to risk as such; nevertheless, it is possible to infer statutory attitudes toward risk by exploring these sections for second-order concepts embodied in the term risk. Clues as to what might serve as appropriate concepts may be found among the many definitions of risk. For example, risk has been called "the objective uncertainty as to the occurrence of an undesirable event."[9] In other definitions, the theme of uncertainty recurs, as well as probability and negative outcome.[10]

Looking at section 402 in terms of these three ideas — uncertainty, probability, and negative outcome — reveals two attitudes towards risk in that portion of the law. The first is in its approach towards added substances as a class. Because section 402 has placed the burden of proof on the government to demonstrate that a substance is unsafe, this provision tolerates a high degree of uncertainty about the toxicological properties of the substances as a class falling under its purview. The government does not have the resources to study every substance added to food in the marketplace. Presumably, because of the great uncertainty regarding these substances, there is potential for harm from these substances as a class. The second attitude is reflected in the provision's treatment of individual substances. Once it has been determined that a particular substance may cause harm in a food at a certain quantity, section 402 does not permit that substance to be added to the food supply at that level.

In its interpretation of this section, the Court, in the Lexington Mill case, appeared to recognize the concept of probability of harm as it sought to apply meaning to the term, "may render." It also seemed to acknowledge that the probability may vary with the type of consumer, as illustrated by its view that " . . . the strong and the weak, the old and the young, the well and the sick . . . " all should be considered in assessing whether harm will or will not occur. Section 402 allowed the public to be exposed to risk from added substances as a class because it tolerated a great deal of uncertainty about the biological properties of those substances. Its approach, however, for individual substances — once a risk was known — was conservative.

When compared to the "may render" standard, the "ordinarily injurious standard" for natural foods and their constituents offers insight regarding the statutory treatment of relative risk. As already noted, the "ordinarily injurious" standard has been interpreted as requiring a higher degree of proof or a greater probability of harm than the "may render" standard. Such a requirement may be viewed as allowing a greater degree of uncertainty — and, hence, risk — for substances in this category. In other words, early framers of the law apparently were willing to assume greater risks from traditional foods than from substances added to that food.

The addition of section 409 to the Act was a bold attempt to address the problem of uncertainty straightforwardly. This section attempted to reduce uncertainty by requiring that the safety of a substance be considered before it is marketed for a particular use. Moreover, this section also required that sponsors of a substance — not the government — provide the information needed for a safety evaluation. By reducing uncertainties, section 409 apparently was intended to reduce risks to health from food additives as a class. Section 409 also permitted tolerances to be established for individual additives that were not carcinogens. The standard which FDA has used for judging appropriate tolerances — "reasonable certainty of no harm" — acknowledges the possibility of risk, as indicated in the legislative history. Interestingly, this phrase can be made to apply both to the magnitude of risk that would be acceptable, as well as the surety with which the probability of harm is known. Like section 402, section 409 does not preclude use of quantitative risk assessments as a tool in evaluating safety. The Delaney amendment, on the other hand does. It prohibits any approval of a carcinogenic food additive and thus sets forth a zero risk standard for food additives that cause cancer in man or animals when ingested.

In summary, except for the Delaney amendment, sections 402 and 409 allow the use of quantitative risk estimations in evaluating safety. They do say, however, that the level of risk must be very low.

C. "Risk Assessment" as a Process

While consideration of second-order concepts is a useful way for deducing the law's attitude toward risk, this approach is not entirely applicable to understanding the implications of statutory structure upon the notion of risk assessment. Instead, considering risk assessment as a process is helpful. It also is in keeping with the view that it is multistep activity. The steps of this activity have been partitioned in various ways, sometimes including risk management, and sometimes not.[11-13] We shall view risk assessment as consisting of three stages: (1) hazard identification; (2) risk estimation; and (3) risk evaluation. For our purpose, we shall consider the first step as involving recognition of those substances which may cause harm as well as the types of injury they may induce; the second, as concerning attempts to ascertain the extent of harm which may result from exposure to the substance; and the third, as regarding society's reactions toward the information obtained in the first two steps, including the management of the risk.

When we look at the law within this framework, we see that section 402 is little concerned with focusing resources upon the identification and estimation of hazards. Instead, it is concerned primarily with the evaluation or management phase of risk assessment. More than anything else, it represents a societal response to a known hazard in the food supply. Of course, this response is very risk aversive, for it aims to reduce the probability of harm by protecting even the most susceptible segments of the population.

The ordinarily injurious standard also makes a strong statement about evaluation. Because it tolerates a higher risk than the may render standard, it indicates some societal judgment that foods (in their traditional sense) are more valuable than added substances. Merrill has observed that, apparently, the ordinarily injurious standard was meant to permit the FDA, or a district court, to weigh the relative dangers and importance of foods that naturally contain poisonous constituents.[14]

In contrast to section 402, section 409 opened the way for a much greater emphasis on the first two stages of risk assessment. By requiring premarket safety reviews and permitting the establishment of tolerances, it demanded that some consideration be given to the nature and extent of hazard posed by a substance at various levels of exposure. This provision is still, however, risk aversive in its approach to the evaluation phase. It does not explicitly permit consideration of a substance's benefits in determing what might be an acceptable level of exposure. Instead, it assumes that a threshold dose can be established below which most of the population will not experience any injury. Of course, the risk aversive response is carried to its extreme in the Delaney amendment where no amount of a carcinogenic additive is permitted.

III. CONSEQUENCES OF STATUTORY STRUCTURE REGARDING "RISK" AND "RISK ASSESSMENT" AND RESPONSES TO IT

The consequences resulting from the statutory structure regarding risk and risk assessment have been numerous and far-reaching. We have already alluded to one: the expression of generic risk-benefit judgments by the legislature through its different attitudes toward risk among various legal categories of substances.

Undoubtedly, however, the most important consequence of the present legal structure arises from the shift of the "burden of proof" from the government to the sponsor of a substance. In the early days of the Act, requiring the government to show that a substance was unsafe tended to focus attention on acute hazards where evidence was available and essentially indisputable. On the other hand, requiring that a sponsor demonstrate safety opened the way for a full inquiry into the spectrum of possible hazards — chronic as well as acute — posed by a substance. The unleashing of this authority at a time when tremendous advances were being made in the biological sciences and analytical chemistry led to a growing interest in hazard identification. This important consequence has, in turn, impinged strongly upon the regulatory system in various ways. For example, our universe of safety concerns has expanded with respect to both the sources of hazard and the types of hazard. Today, chemicals routinely can be identified and quantified at levels around 1 ppb. We must now ponder the effects of chronic, low-level exposures to numerous substances. In addition, our concerns have grown about environmental contaminants such as aflatoxins, and there is increasing evidence that many natural components of foods can be toxicants. Our spectrum of health interests has broadened to include phenomena such as mutagenesis, carcinogenesis, and behavioral effects. And to complicate matters more, problems exemplifying individual variability — food allergies, hyperkinesis, hypertension, and lipemia, to name but a few — have gained attention.

Ironically, as our sophistication about the types of hazard has increased, we have become enmeshed in what Rowe calls the information paradox.[15] For each quantum of uncertainty we have attempted to reduce in the identification stage of risk assessment, we have multiplied our uncertainty in the estimation phase because we often lack indicators of the true extent of possible hazard.

One type of response to this situation has been predictable. Interest and controversy have mounted over the most appropriate ways to estimate human risk. And the validity and interpretation of toxicological studies and other data for this purpose sometimes also have been questioned.[16,17] Individuals responding this way have focused less on the legal requirements once data have been interpreted, and more on the decision process. Among them, have been those who have called for mandatory scientific peer review of significant regulatory decisions. Others have suggested that an institution such as a science court might be useful.[18] None of the proposed solutions to assure adequate peer review are without difficulty. At the very least, they represent an added step to an already painfully laborious process. Although these issues are beyond the scope of this paper, they are important if a broad consensus in the scientific community is necessary for regulatory decisions.

FDA's responses to the various scientific developments have been efforts to interpret the Act in a manner which provides a regulatory framework for addressing contemporary problems that still remains within the basic spirit and intent of the statute. Paramount among its challenges has been the interpretation of "safe". As already observed, the standard of safety adopted by FDA for section 409 — reasonable certainty of no harm — requires interpretation with respect to the surety of data as well as the level of acceptable risk.

In addressing the issue of the certainty of data, FDA has differentiated between the concepts of "statistical significance" and "biological significance." In one of its decisions, the agency explained its views as follows:

A finding that a test result has "statistical significance" involves the use of statistical methodology to determine the probability that the observed difference, if any, in the incidence of treated animals compared to controls is associated with the test substance rather than a chance occurrence. A finding that a test result has "biological significance" involves consideration of certain biological factors which provide information about the proper interpretation of results. Together, these two criteria help scientists to decide what, if any, conclusions can be drawn from the results of a study.[19]

The biological factors which might be considered include the methodology of the study involved; the existence of a dose-response relationship; the rarity of tumors; and the presence of similar results in other studies. FDA has been reluctant to pin its final conclusions regarding the safety of a substance to any particular level of statistical significance.

For substances already legally on the market, the emergence of unresolved safety questions has presented a difficult regulatory situation. To resolve this problem, FDA has developed regulations which allow substances to remain on the market if the agency finds there is a resonable certainty that the substance is not harmful and that no harm to the public health will result from the continued use of the substance for a limited period of time while the question raised is being resolved by further study.[20] Such substances are called interim food additives and, traditionally, the agency has conditioned their continued use on a commitment by a sponsor to conduct the studies necessary for resolving the uncertainty.

In determining the level of risk for food additives showing a toxicological response having a threshold dose, FDA typically has relied upon the application of safety factors to the observed no-effect level for the substance. The safety factors generally have been a very rough function of the confidence the agency has in the data with respect to its ability to indicate the possible level of human toxicity, and often ranges from 100 to 1000.[21] If the no-effect level exceeds the projected exposure to the substance by the appropriate safety factor, the substance is considered to be safe.

Ascertaining the safe level of a carcinogen has been a much more difficult task for the agency. Traditionally, FDA has taken the position that safe levels of carcinogens could not be established and has taken action to prohibit exposure to them. This position has been based on uncertaintites about biological mechanisms associated with carcinogenesis. Recently, however, FDA has applied a policy which uses risk assessment as a tool for determining if carcinogenic substances not covered by the Delaney amendment can be considered safe. The agency has stated that the acceptable maximum level of exposure would be calculated based on a suitable risk standard by extrapolating from an upper confidence limit of bioassay data using procedures that would include an evaluation of pharmacodynamic data, including any human studies available, epidemiological evidence, and studies of the mechanism of action of the carcinogen.

Scientific developments have presented challenges to FDA not only in interpreting "safe" but also in sorting out the substances to which the various safety standards should apply. The discovery of low levels of carcinogenic chemicals in food additives is one example of such a problem. As previously noted, the agency historically has not allowed additives containing such chemicals to be marketed, even if the additive as a whole had not been found to cause cancer. The agency, however, has felt obliged to reconsider this approach. In the same proposal addressing risk assessment for carcinogens, FDA acknowledged that if it continued its policy of prohibiting additives containing trace amounts of carcinogenic constituents, it might be forced to deny or terminate the use of many additives. As a way out of this dilemma, the agency proposed to define an additive in such a way that the carcinogenic constituents would be interpreted as falling under the general safety provisions of section 409, rather than the Delaney clause of that section. FDA then went on to say that the Delaney clause would apply only when the additive itself has been shown to cause cancer.[22]

At the outset, we observed that laws embody society's goals and objectives for achieving those goals, but that science can influence the nature of those goals and objectives. We have seen that the introduction of section 409 into the Federal Food, Drug and Cosmetic Act encouraged greater investigations into toxicity and potential toxicants. The results of these investigations, in turn, have caused us to explore the meaning of safe, and application of safety standards to various substances.

IV. TRENDS

The developments in respone to the results of increased scientific inquiry indicated that we are entering a new phase of risk evaluation. We are in the process of reconsidering our attitudes toward risk and what our response to those attitudes should be.[23] Future activites in food regulation will represent attempts to facilitate that process of reconsideration, and, ultimately, will be based upon its outcome.

Certainly, during the next decade, we will be groping for institutional means for ascertaining and expressing our society's attitudes toward risk. It may be that we will see a greater and more explicit statement of public values through Congressional hearings and passage of new laws; however, the tendency of Congress to be general rather than specific in its laws, as well as its past reliance on administrative expertise both would appear to temper the possibility that this will be the initial approach.[24] The value judgements regarding risks most probably will involve specialized, susceptible segments of the population, such as hypertensives. Too, they may involve highly technical questions such as how to respond to the risk from a carcinogenic promoter compared to a carcinogenic initiator.

What probably will be more likely as an initial approach is that the regulatory agency will serve as the primary instrument for ferreting out opinions about risks through public comments and judical oversight of its decisions. Then, through a slow, incremental process, as public values and scientific fact become more certain, Congress may incorporate new values into statutes through amendments, just as it has done in the past.

To accomplish this scenario, however, it will be necessary to establish an environment which permits the regulatory agency flexibility not only in gathering information about risks but also in acting upon the data. In the past, the emphasis of food safety policy has been protection through prohibition or reduction of exposure to certain substances. If the agency is to be exploring attitudes about risk and its control, additional policy emphasis may be needed. Laws and regulations commanding knee-jerk responses are undesirable for several reasons. First, they focus attention and resources on those risks which may not be the most significant; second, they emphasize risk management through a sequential rather than a global approach; and third, they sometimes — as with the Delaney clause — short-circuit full scientific inquiry by dictating a regulatory response based on limited findings.

Although it will always be important for a regulatory agency to have the authority to limit risks through prohibitory means, an equally important emphasis during a transition period concerned with risk evaluation would be simply developing, gathering, and disseminating scientific information, particularly for those substances that cannot be practially controlled through the establishment of tolerances. Ideally, the regulatory agency would have the authority and resources to carry on research activities related to hazards posed by substances not having a proponent, such as environmental contaminants, and by substances which are difficult or infeasible to restrict, such as traditional foods. With such information available, the public may begin to develop a better understanding of relative risks from all of the various hazards associated with diet and articulate their judgements about those hazards more clearly. In addition, individuals may become more aware of those hazards over which they have control and begin to attempt to reduce them.

Thus, the first step in our new phase of risk evaluation may involve statutory modification to increase the flexibility of regulatory agencies so that society will have a better institutional means for ascertaining and expressing attitudes towards risks. That this is indeed a trend is evidenced by the nature of the calls for food safety reform in recent years. A close examination of the changes seriously considered reveals efforts to increase the flexibility of regulatory responses to permit them to be tailored to the findings of broader scientific inquiries.

The ultimate outcomes of the new phase of risk evaluation we are now entering are extraordinarily difficult to predict. They will depend not only on competing factors for resources, but also on the preceptions of the efficacy of government in addressing societal problems.[25] A few trends, however, already are underway and it seems that they shall continue.

The first of these involves defining a threshold of regulation: sorting out potential risks that are so trivial that resources should not be devoted to discovering more about them or controlling them. A 1979 decision by a U.S. Court of Appeals concerning the regulation of migrants from food packaging has provided impetus for this activity. This decision declared: "There is latitude in the statutory scheme [of the Federal Food, Drug and Cosmetic Act] to avoid literal application of the statutory definition of 'food additive' in those de minimis situations that, in the informed judgement of the Commissioner, clearly present no public health or safety concern."[26]

The definition of a de minimis level requires insight into the upper boundaries of carcinogenic potencies, the hazards posed by persistent, low levels of exposures, and a knowledge of relative risks from food-related hazards. Most certainly, to the maximum extent permitted by statute, regulators will rely upon the results of toxicological risk assessments as they search for ways to state and refine this definition.

A second trend concerns developing systems which delineate the amounts and types of information needed for identifying and estimating risks. Such systems do, in essence, represent value judgments since they reflect the amount of uncertainty tolerated in risk determinations as well as priorities for expending societal resources in reducing those uncertainties.

Interest in such systems is apparent from the literature.[27] The Bureau of Foods has been developing its own system based upon the concept of concern levels. In the Bureau's approach, exposure estimates and inferred toxicity based on a substance's chemical structure are used to establish an initial level of concern about the safety of a substance. If the level is low, less testing would be required than if it were higher. When it becomes available, known toxicity data may cause adjustment of the level of concern up or down at any time. Decision rules for this process are being developed and under review.[28] The Bureau hopes to use this system in several ways: (1) to define better its decision-making process; (2) to focus limited resources on those substances believed to pose the greatest potential for harm; and (3) to perform sensitivity analyses concerning the effects of various types of information on hazard estimates and regulatory responses.

Finally, a third trend relates to growing interest in nutritional toxicology. Scientific advances now are begining to reveal that nutrition can greatly influence the toxicity of ingested chemicals,[29] whether they be environmental contaminants, drugs, additives, or even constituents of natural foods. Moreover, we are becoming much more aware that diet may enhance or retard certain diseases, such as cancer and circulatory ailments. The interest in nutrition is so great that improved nutrition has been targeted as a national objective for improving health.[30] If the pursuit of this objective is diligently undertaken, it is likely that risk assessments will take on new meanings. They will become even more complex, taking into account the levels of nutrients in the diet, synergisms, specific characteristics of the consumer, dietary patterns, and a host of other factors. No longer will they solely be concerned with the effects of particular poisons. Instead, they will begin to address health risks from the absence of particular dietary factors and conditions as well.

The results of these assessments could, in turn, cause our food safety laws to be broadend in scope to specifically include nutritional concerns. The objective of our food laws focusing on health promotion through activities aimed at understanding the importance of diet could become as important as the objective of health protection through regulatory activites limiting exposures to harmful substances.

In terms of the three stage risk assessment process, we may well find that we will have traveled full circle. The new phase of risk evaluation we are now entering will bring us into yet another stage in which the law stimulates intensive scientific investigations related to identifying and estimating the influences of the total diet on health.

ACKNOWLEDGMENT

The authors would like to thank Mr. Robert Lake, Regulations Staff Coordinator in the Bureau of Foods, Food and Drug Administration, for his insights and helpful discussions during preparation of this chapter.

REFERENCES

1. Federal Food, Drug and Cosmetic Act, 21 U.S. Code (1980).
2. **Merrill, R. A.,** Regulating carcinogens in food: a legislator's guide to the Federal Food, Drug, and Cosmetic Act, *Mich. Law Rev.* 77, 171, 1978.
3. Federal Food and Drugs Act of 1906, ch. 3915, §7, 34Stat. 767 (repealed 1938.)
4. United States vs. Lexington Mill & Elevator Co., 232 U.S. 399.
5. Federal Food, Drug and Cosmetic Act of 1938, §402(a)(1) (codified at 21 U.S. Code §342(a)(1)(1980)).
6. **Merrill, R. A.,** Regulating carcinogens in food: a legislator's guide to the Federal Food, Drug, and Cosmetic Act, *Mich. Law Rev.,* 77, 193, 1978.
7. Federal Food, Drug and Cosmetic Act of 1938 codified at 21 U.S. Code §348(a)(1980).
8. House Report No. 2284, 85th Congress, 2nd session (1958).
9. **Willet, A. H.,** *The Economic Theory of Risk and Insurance,* University of Pennsylvania Press, Philadelphia, 1951, 2. Cited by Miller, E. G., and Rubin, H. W., The changing qualitative and quantitative approach to risk analysis, *Risk Management,* 26(11), 29, 1979.
10. **Miller, E. G. and Rubin, H. W.,** The changing qualitative and quantitative approach to risk analysis, *Risk Management,* 26(11), 30, 1979.
11. Committee on the Institutional Means for Assessment of Risks to Public Health (Commission on Life Sciences, National Research Council), *Risk Assessment in the Federal Government: Managing the Process,* National Academy Press, Washington, D.C.,1983, chap. 1.
12. Royal Society Study Group, *Risk Assessment,* The Royal Society, London, 1983, 22.
13. **Rowe, W. D.,** *An Anatomy of Risk,* John Wiley and Sons, New York, 1977, 25.
14. **Merrill, R. A.,** Regulating carinogens in food: a legislator's guide to the Federal Food, Drug, and Cosmetic Act, *Mich. Law Rev.,* 77, 188, 1978.
15. **Rowe, W. D.,** *An Anatomy of Risk,* John Wiley and Sons, New York, 1977, 18.
16. **Gori, G. B.,** Regulation of cancer-causing substances: Utopia or reality, *Chemical and Engineering News,* 60(36), 25, 1982.
17. **Bloom., B. R.,** News about carcinogens: what's fit to print? *Hastings Center Report,* 9(4), 5, 1979.
18. **Kantrowitz, A.,** The science court experiment: criticisms and responses, *Bullentin of Atomic Scientists,* 33(4), 43, 1977.
19. Food and Drug Administration, Cyclamate, (cyclamic acid, calcium cyclamate, and sodium cyclamate), Commissioner's decision, Federal Register, 45, 61474 1980.
20. Food and Drug Administration, Food additives permitted in food on an interim basis or in contact with food pending additional study, *21 Code of Federal Regulations,* §180.1, 320, 1981.
21. Food and Drug Administration, Safety factors to be considered, 21 Code of Federal Regulations, §170.22, 12, 1981.
22. Food and Drug Administration, Policy for regulating carcinogenic chemicals in food and color additives; advance notice of proposed rulemaking, *Federal Register,* 47, 14464, 1982.
23. **Miller, S. A. and Skinner, K.,** Science and the law: the basis for new food safety legislation; speech presented at the Food and Nutrition Committee of the American Health Foundation International Conference on Enviromental Aspects of Cancer: The Role of Macro and Micro Components of Foods, New York, April 1, 1982.
24. **Mashaw, J. L.,** Regulation, logic, and ideology, *Regulation,* 3(6), 44, 1979.
25. **Mashaw, J. L.** Regulation, logic, and ideology, *Regulation,* 3(6), 51, 1979.
26. Monsanto vs. Kennedy, 613 F. 2d 947 (D.C. Cir. 1979).
27. **Weinstein, M. C.,** Decision making for toxic substances control; cost-effective information development for the control of environmental carcinogens, *Public Policy,* 27, 333, 1979.
28. Food and Drug Administration (Bureau of Foods), *Toxicological Principles for the Safety Assessment of Direct Food Additives and Color Additives Used in Food* (the "Redbook"), Rep. No. Pb 83-170696, National Technical Information Service, Springfield, Va., 1098.
29. **Parke, D. V. and Ioannides, C.,** The role of nutrition and toxicology, in *Annual Review of Nutrition,* Vol. 1, Darby, W. J., Broquist, H. P., and Olsen, R. E., Eds., Annual Reviews, Palo Alto, Calif., 1981, 207.
30. **McGinnis, J. M.,** Targeting progress in health, *Prevention,* 97, 295, 1982.

Section B—Case Studies

Chapter 7

CASE STUDY — ASBESTOS

M. E. Meek, H. S. Shannon, and P. Toft

TABLE OF CONTENTS

I. INTRODUCTION

Asbestos is a general term for fibrous silicate minerals of the serpentine or amphibole mineral groups, which are widely distributed throughout the earth's crust. Six minerals are commonly characterized as asbestos, namely chrysotile, the only member of the serpentine group, and the amphiboles: crocidolite, amosite, tremolite, anthophyllite, and actinolite. The chemical nature and crystalline structure of asbestos impart a number of desirable characteristics such as high tensile strength, durability, flexibility, and heat and chemical resistance, which make this mineral useful for a large number of applications. More than 3000 commerical uses of asbestos have been identified[1,2] and current annual world production is about 4.5 million metric tonnes, of which greater than 95% is chrysotile.[3] The other commercially important varieties are amosite and crocidolite.

Evidence linking the development of a number of diseases to the inhalation of asbestos has accumulated over the past century; increased risks have been observed not only in occupationally exposed persons, but also in families of the workers and in the residents of neighborhoods in the vicinity of asbestos industries. There has also been concern that exposure to asbestos in the general environment may pose a risk to public health.

The numerous beneficial uses of asbestos and as yet, the unavailability of suitable substitute materials for many products dictate the need to assess the nature and extent of the hazards associated with exposure, and the possibilities for control. There is an extensive data base available for such analysis. In fact thousands of articles concerning the medical, epidemiological, and toxicological aspects of asbestos exposure have been published in the scientific literature. However, there are still gaps in our knowledge concerning dose-response relationships and mechanisms of toxicity that complicate the quantitative estimation of risk. In this chapter, the available data will be discussed in the context of the historical development of regulations to limit exposure.

II. HISTORICAL ASPECTS

A. Production and Use

The beneficial characteristics of asbestos have been recognized since early times; indeed the use of this mineral dates back thousands of years. It has been found in pottery, believed to have been produced in the Stone Age.[4] One of the earliest historical references to asbestos tells of its use in the wick of a gold lamp in a statue of the goddess Athene, made by an Athenian sculptor who lived in the fifth century B.C.[4] There have been numerous additional historical references to the use of asbestos thoughout the ages;[4,5] often, historians described the astonishment of the public concerning the seemingly magical fire-resistant properties of this mineral.

The modern asbestos production industry dates from 1870; between that time and the middle of this century, there was great interest in the exploitation of deposits and development of applications for this unique mineral.[4] Initially, asbestos was used in building materials such as sealings, packings, roofing felt, cement, and insulation. Around the turn of the century the asbestos textile and asbestos cement pipe industries were developed, and the use of asbestos as a friction material in brake linings was introduced shortly thereafter. There was a tremendous increase in the demand for and uses of asbestos during World War II, and several new applications, including the use of asbestos filters for clarification of beverages, were introduced following the war.

World production of asbestos peaked in the mid-1970s and is presently decreasing due partly to a recession in the building industry, but primarily to public concern about the potential health hazards associated with exposure to asbestos in the workplace or general environment.[6]

B. Health Effects

As early as the first century A.D., Pliny the Elder is said to have commented on the sickness of the lungs of slaves who worked with asbestos.[7] The first recognition of disease associated with the modern asbestos production industry was reported in England in 1906 when Montagu-Murray described to a Departmental Committee on Industrial Diseases, a case of pulmonary fibrosis (asbestosis) that he had observed in 1899 in an autopsied worker.[8] In the ensuing years, many additional cases of asbestos related pulmonary fibrosis were reported and by 1930, there were more than 75 cases recorded in the literature.[9] These early case reports stimulated public concern and the first epidemiological studies of asbestos workers were conducted in the U.K. in 1928 and in the U.S. in 1937.[4,9]

The first suggestion that asbestos might be a human carcinogen came in 1935 when Lynch and Smith[10] in the U.S. and Gloyne in the U.K.[11] independently observed lung cancer in three asbestos workers who had died of asbestosis. There were similar reports in the following years; however the first epidemiological studies that confirmed the association between occupational asbestos exposure and lung cancer were not conducted until the mid-1950s.[13] This association has been observed repeatedly in numerous subsequent studies.*

Pleural abnormalities such as thickening, calcification, and effusion are often associated with asbestosis and several cases of mesothelial tumors of the pleura and peritoneum were noted in the literature between 1931 and 1960.[15] For example, in 1955, Bonser et al. reported 4 primary peritoneal tumors in 72 asbestos textile workers.[171] However, there was no conclusive evidence of the relationship between occupational exposure to asbestos and mesothelioma until 1960 when Wagner et al. reported 33 pleural mesotheliomas among miners and residents of a crocidolite mining region in South Africa in a 4-year period.[16]

An increased incidence of gastrointestinal neoplasms in populations of workers occupationally exposed to asbestos was first observed by Mancuso.[17] Although the association has not been observed in all studies,[187] there have been excesses of deaths from cancers of the gastrointestinal tract in several cohort studies of populations of asbestos workers.[18] The weight of the evidence suggests a causal relationship. However, the possiblity exists that at least some of the reported gastrointestinal tumors in these studies were actually misdiagnosed peritoneal mesotheliomas.[18] The balance of epidemiological data indicates that there is also a causal association between occupational exposure to asbestos and laryngeal cancer. An excess of laryngeal cancer in asbestos workers has been observed in three out of five cohort studies[19-23] and three out of four case-control studies.[24-28] Although the excess was not observed in the best described case-control study.[28] it was noted in the inherently more sensitive cohort studies.[21-23]

Goldsmith has recently reported that based on an analysis of the results of 11 cohort studies, observed to expected ratios for cancer of several nonpulmonary sites ranged from 0.97 to 2.78 and were similar to those observed for gastrointestinal cancer.[190] An increased risk of cancer of the oral cavity and pharynx,[29] pancreas,[22] kidney,[22] and brain[172] have also been observed in a cohort of asbestos insulation workers. There was a statistically significant excess of cancers of the ovary in a high exposure group of women in a London asbestos factory,[49] and there have been isolated reports of adenocarcinoma of the rete testis in a man with asbestosis,[188] and pericardial mesotheliomas in three asbestos workers.[189] However these data have not been confirmed and such unconfirmed associations may have been due to chance or to misdiagnosis. For example, analyses using additional histological data have shown that most of the excess of pancreatic cancers observed upon the analysis of death certificate data was actually due to misdiagnosis of peritoneal mesotheliomas.[22]

There is an increased risk of developing mesothelioma among the families of asbestos

---

* In a recent review of the available epidemiological data on the subject conducted by one of the authors, there was an increase in lung cancer mortality in 18 out of 22 populations of workers examined in 37 papers.[14]

workers due presumably to dissemination of fibers in the home from contaminated work clothes. In 1965, Newhouse and Thompson first observed nine cases of mesotheliomas in household contacts of asbestos workers in a London hospital case series.[69] By 1976, a total of 37 mesotheliomas attributable to household asbestos exposure had been observed in studies from nine countries;[62] additional cases have been reported in recent years.[62-74,191] In a case-control study of 52 females with mesothelioma in New York State between 1967 and 1977,[63] the relative risk of having an asbestos worker in the household in cases compared to controls was 10. Increased prevalence of predominantly pleural X-ray abnormalities which are characteristic of asbestos exposure have also been observed in the families of asbestos workers.[62,65]

C. Regulatory Developments

1. The Occupational Environment

As information became available concerning the relationship between occupational asbestos exposure and the incidence of disease, governments imposed standards to limit the exposure of workers; the most important developments in this area are summarized in Table 1. The results of an epidemiological study conducted in the asbestos textile industry by Mereweather and Price[31] prompted the introduction of Asbestos Industry Regulations in 1931 under the U.K. Factory and Workshops Act. Although no numerical limit for occupational exposure to asbestos was specified in the regulations which came into effect in 1933, requirements for exhaust ventilation in various portions of the industry were established.[4]

A positive dose-response relationship between asbestos exposure and clinical symptoms of asbestosis was found in an epidemiological study conducted by the U.S. Public Health Service in 1938.[32] Based on the results of this study, a standard of 5 million particles per cubic foot (measured by the impingement method) was adopted by most states in the U.S., by the American Conference of Governmental Industrial Hygienists (ACGIH), and eventually by many countries throughout the world.[33]

There were no further noteworthy developments until 1968 when the ACGIH proposed a new standard of 2 million particles per cubic foot (measured by the impingement method) or 12 fibers* per milliliter.[4,33] About the same time in Britain, the Committee on Hygiene Standards of the British Occupational Hygiene Society (BOHS) concluded that "early clinical signs will be less than 1% for an accumulated exposure of 100 fiber years/mℓ."[34] This conclusion was based on a log-normal dose response relationship between prevalence of crepitations and cumulative exposure observed in a study of asbestos textile workers. On the basis of this recommendation, a standard for chrysotile asbestos of 2 f/mℓ was incorporated into the U.K. Asbestos Regulations; this limit was also adopted in several countries throughout the world. In the period since these standards were introduced, serious flaws in the design of the study which was the basis of the BOHS recommendation have been noted.[58]

The ACGIH standard of 12 f/mℓ was adopted in 1970 by the U.S. Secretary of Labor as an interim standard for all workplaces. However. based on epidemiological evidence of cases of asbestosis associated with exposure to asbestos concentrations below this level, in 1971, the U.S. Occupational Safety and Health Administration (OSHA) issued a temporary emergency standard of 5 f/mℓ. This became a permanent standard in 1972.[35] In 1976, a 2 f/mℓ standard was adopted by OSHA to "prevent asbestosis and more adequately guard against asbestos-induced neoplasms."

Recently, there have been several as yet unadopted recommendations in the U.S. to reduce the occupational exposure levels, based predominantly on the mounting evidence of the carcinogenicity of asbestos.[33] In 1975, OSHA recommended a standard of 0.5 f/mℓ based on three lines of evidence: the results of an additional study with extended follow-up which

---

* For regulatory purposes, fiber is generally defined as having an aspect ratio 3:1 and lengths > 5 μm (measured by phase contrast microscopy).

Table 1
HISTORY OF DEVELOPMENT OF STANDARDS FOR OCCUPATIONAL EXPOSURE TO ASBESTOS IN THE U.K. AND U.S.[4,33,35,36]

Year	Standard[a,c]	Basis
1933	U.K. Factory and Workshops Act—requirements for exhaust ventilation	Results of an epidemiological study of asbestosis[31]
1938	U.S. Public Health Service — 5 mppcf	Results of an epidemiological study of asbestosis[32]
1968	A.C.G.I.H. — 2 mppcf — 12 f/mℓ	
1968	BOHS — 2 f/mℓ (adopted under the U.K. Factory and Workshops Act)	Estimated that lifetime exposure to this level would result in a 1% risk of developing asbestosis
1970	ACGIH standard of 12 f/mℓ adopted by U.S. Secretary of Labor as interim standard	
1971	OSHA issues temporary emergency standard — 5 f/mℓ	
1972	OSHA makes 5 f/mℓ permanent standard	
1975[b]	OSHA recommends standard of 0.5 f/mℓ	Abnormal X-ray findings for those exposed to <2 f/mℓ; X-ray changes in family members of asbestos workers; increasing evidence of carcinogenicity
1976	OSHA adopts 2 f/mℓ standard	To "prevent asbestosis and more adequately guard against asbestos-induced neoplasms"
1979[b]	NIOSH/OSHA Committee recommends standard of 0.1 f/mℓ	Limit of detection of analytical techniques
1979	U.K.-Advisory Committee on Asbestos recommends standards of 1, 0.5, and 0.2 f/mℓ for chrysotile, amosite and crocidolite, respectively	"Reasonably practicable" levels
1980	A.C.G.I.H. — 2 f/mℓ — chrysotile; 0.5 f/mℓ — amosite; 0.2 f/mℓ — crocidolite	
1983	U.K. — recommendations of the Advisory Committee adopted by the H.S.E.	

a   for regulatory purposes, fibres have been defined as particles having an aspect ratio of 3:1 and length >5um.
b   these proposals have not been adopted
c   mppcf measured by impingement; f/mℓ by phase contrast microscopy

showed abnormal X-ray findings of the lung and pleura amoung the British textile workers of the BOHS study exposed to levels below 2 f/mℓ; the results of another study in which a high proportion of family contacts of former asbestos workers showed X-ray changes characteristic of asbestos exposure and the increasing evidence of the carcinogenicity of asbestos. In 1979, a joint National Institute of Occupational Safety and Health (NIOSH)/OSHA Committee recommended a 0.1 f/mℓ standard for all types of asbestos. This recommendation was not based on available epidemiological data but rather is the limit of detection of analytical techniques which are routinely used in the workplace.

In the U.K., standards for occupational exposure to chrysotile, amosite, and crocidolite of 1, 0.5, and 0.2 f/mℓ, respectively, were recommended by an Advisory Committee on Asbestos, convened by the Health and Safety Executive. These control limits were considered to be the minimum levels that were reasonably practicable[36] and became effective January 1, 1983.

The scenario for development of occupational exposure standards for asbestos in most industrialized countries in the world has been similar to that described for the U.K. and the U.S.; the existing regulations in most countries have been presented in Table 2. The development of standards in Canada is of particular interest due to the fact that a rather large population of approximately 33,500 persons are employed directly and indirectly in the asbestos industry in this country.[3] Canada is second only to the Soviet Union in world production of asbestos; the Canadian industry was first established in 1870 and 1.5 million tonnes are now mined principally in the province of Quebec.[3]

The use of crocidolite has been restricted to those plants which possess suitable handling facilities, and a 2 f/mℓ standard for all forms has been adopted as a guideline or regulation by all of the provinces based on recommendations made in 1976 by a Federal-Provincial Working Group on Asbestosis.[38] Some provinces have instituted more stringent values for chrysotile or other forms of asbestos.

2. The General Environment

Regulations to limit the exposure of the general public to asbestos in the environment have been introduced in several countries throughout the world; these are presented in Table 3. By far the most extensive regulations have been introduced in the U.S.; these are summarized in Table 4. Airborne emissions standards and effluent guidelines for asbestos industries were introduced in the U.S. in the early 1970s; several consumer products containing asbestos and one manufacturing process have been banned; there are record-keeping requirements for manufacturers and processors, and requirements have been specified for packaging of asbestos during transport.[35,39]

Slightly different regulations have been adopted in Canada to minimize exposure of the general public to asbestos. In 1978, an outdoor emission standard of 2 f/mℓ for asbestos mining and milling operations was promulgated under the Clean Air Act. In 1975, regulations to prohibit the use of asbestos in children's toys and modeling clays were introduced under the Hazardous Products Act. Suitable labeling is required for asbestos-containing products and there are regulations prohibiting asbestos in products such as drywall joint cements, wall patching compounds, and artificial decorative ash for fireplaces. In 1979 manufacturers voluntarily replaced the asbestos liners in hair dryers and during that year, a new standard was issued by the Canadian Standards Association which specified elimination of asbestos from all electrical hairdressing equipment.

II. RECENT DEVELOPMENTS IN RISK ASSESSMENT

A. Epidemiological Data
1. Occupational Environment

In the period since the early studies mentioned above were conducted, there have been numerous epidemiological studies of various types conducted to investigate the relationship between occupational exposure to asbestos and the incidence and prevalence of disease. These studies have further characterized the risk of development of asbestos-related disease in relation to occupation, fiber-type and other factors such as cigarette smoking.

Asbestosis generally results from exposure to high concentrations of asbestos for prolonged periods of from 7 to 20 years. The clinical, radiological, and other signs on which the diagnosis is based are nonspecific and subject to wide variations in interpretation.[187] However, the incidence of asbestosis is generally considered to be directly proportional to the cumulative dose.[42,187] All types of asbestos can cause asbestosis.[18] and the results of a recent study suggest that if exposure has been sufficiently high, the disease progresses even when the worker leaves the industry.[43] A substantial proportion (9.3%) of 86 Quebec chrysotile miners showed an increase in parenchymal opacities after withdrawal from asbestos exposure. The

Table 2
OCCUPATIONAL EXPOSURE LIMITS FOR ASBESTOS[1,30,35,121-123,41]

Country	Limit (f/mℓ)	Year	Comments[b]
Australia[a]	Chrysotile, amosite — 1 crocidolite — 0.1 (4 hours)	1982	Lowest practicable levels''; importation and mining of crocidolite and use in new work prohibited
Austria[a]	1	1980	
Belgium	Crocidolite — 0.2 other — 2 (4 hours)	1979	
Canada	2 (8 hours)	1976	Acceptable level of asbestosis; some provinces have more stringent standards for various forms; use of crocidolite restricted to special facilities
Denmark[a]	Crocidolite — 0.1; other — 1	1980	General prohibition of crocidolite use
Finland[a]	2	1977	
France[a]	2 — breathing zone; 1 — workroom air (working day)	1977	
Germany (D.R.G.)	2	1972	Prohibition of production of crocidolite-containing products
Germany[a] (F.R.G.)	1	1982	"Technical threshold value"
Ireland (Republic of)	Crocidolite — 0.2; other — 2	1972	
Israel	1	1980	
Italy	2	1979	
Japan	Crocidolite — 0.2; other — 2	1976	
Netherlands[a]	Crocidolite — 0.2; other — 2 (4 hours)	1978	General prohibition of crocidolite use
New Zealand	Crocidolite — 0.2; other — 1		
Norway	2 (15 minutes)	1977	Crocidolite use prohibited
Singapore	2	1980	
South Africa	5 (mines) 2 (manufacturing)	1973	
Sweden[a]	0.5	1978	Crocidolite use prohibited
Switzerland	2 (8 hours)	1973	
U.K.	Crocidolite — 0.2; chrysotile — 1; amosite — 0.5 (4 hours)	1983	"Reasonably practicable limits''; importation of crocidolite ceased in 1970
U.S.[a]	2 (8 hours); 10 (15 minutes)	1976	"To prevent asbestosis and more adequately guard against asbestos-induced neoplasms"
U.S.S.R.	Asbestos-containing dust — 8 mg/m^3 Amphiboles and A/C dust —5 —6 mg/m^3 Fine dust — 2 mg/m^3	1974	

[a] Spraying of asbestos for insulation banned in these countries.
[b] Several countries also specify ventilation requirements and require medical surveillance of employees.

Table 3
PRODUCT AND EMISSION[a] REGULATIONS FOR ASBESTOS[30,122,41]

Country[b]	Regulations
Canada	Emmision standard of 2 f/mℓ for asbestos mining and milling operations; prohibition of use in children's toys and modeling clays; labeling requirements for asbestos-containing products; prohibition in dry-wall joint cements,; wall patching compounds and artificial decorative ash; no asbestos-containing liners in hairdryers; Ontario guideline for asbestos in ambient air — 0.04 f/c.c.
Denmark	Ban of manufacture, import and use of asbestos or asbestos-containing materials with a few exceptions.
Finland	Labelling of asbestos waste and products.
France	Emission standards of 0.1 mg/m^3 for asbestos dust and 0.5 mg/m^3 if asbestos content <50%.
Germany (Federal Republic)	Emission standards of 20 mg/m^3 for all asbestos plants, 75 mg/m^3 if asbestos content <25%; 0.1 to 0.5 ng/m^3 proposed for all point sources; proposed guideline for asbestos in ambient air — 1 f/ℓ. Labelling of asbestos products; proof that substitutes are not available is required; ban of A/C with density <1g/c.c.
Ireland (Republic)	Mandatory labeling for shipment.
Japan	Mandatory labeling of products containing ≥5% asbestos; limitations on the use of certain products.
Netherlands	Prohibition of asbestos products used for insulation, sound proofing and decorative purposes; prohibition of household products and consumer goods in which asbestos fibre is not firmly bound.
Norway	Prohibition of use in carpet underlay and floor tiles; prohibition of use for certain insulation pruposes.
Sweden	Asbestos may only be used where no acceptable substitute exists; prohibition in flooring underlay, paint, glue, jointing compounds, and A/C products (excluding sewage and water pipes); mandatory labeling of asbestos products.
U.K.	Proposal of Advisory Committee on Asbestos to ban spraying and use of asbestos materials for new or acoustic insulation.
U.S.S.R.	Emission limits of 0.15 mg/m^3 (24 hr) and 0.5 mg/m^3 as maximum concentration in total dusts.

[a]   asbestos emissions may be indirectly reduced in other countries due to enforcement of particulate emission standards.
[b]   regulations in the U.S. are presented in Table 4.

total dust exposure of these men had been considerably greater on average (up to 498 million particles per cubic foot-year) than that for the men with no radiological changes.

For mesotheliomas of the pleura and peritoneum, a history of asbestos exposure has been ascertained for 22 to 100% of the cases in various series.[44] It is possible that the proportion associated with asbestos exposure in some series may have been underestimated due to recall bias; however, there is some evidence that mesotheliomas are associated with exposure to agents other than asbestos. In addition, a small number have occurred in children where age was less than the latency period of asbestos-induced mesothelioma.[44] The latency period between first exposure and appearance of the tumor is from 20 to 40 years, irrespective of the age at first exposure.[45] There is some evidence from studies of workers that the risk of pleural mesothelioma increases with both intensity and duration of exposure;[19,47-49] however, exposures in some cases have been as short as 6 weeks.[46] On the basis of the few available data, several authors have suggested that the dose-response relationship may be linear.[18,58] The risk of development of pleural mesothelioma varies depending upon the nature of exposure; the risks for chrysotile miners and millers are apparently considerably less than those for insulation workers, for example.[50] There are insufficient data to draw any conclusions concerning the shape of the dose-response curve for peritoneal mesotheliomas and gastrointestinal cancers; however for both, the relationship with asbestos exposure is somewhat irregular, suggesting that a second causative factor may be involved.[187]

Table 4
HISTORY OF DEVELOPMENT OF REGULATIONS TO LIMIT EXPOSURE OF THE GENERAL PUBLIC TO ASBESTOS — U.S.[35,39]

Date	Regulations	Legislation
July 1972	Asbestos-containing garments banned for general use	Food, Drug and Cosmetics Act
April 1973 (amended in 1974, 1975, 1977 and 1978 to include additional processes)	National Emissions Standard[b] "No visible emissions" from mining and processing industries No spray application of friable materials containing >1% asbestos (1978-prohibition of use of friable sprayed-on materials)	Clean Air Act
February, 1974 (amended in 1975 to include additional processes)	Effluent guidelines for existing asbestos manufacturing point sources and pre-treatment standards for new sources	Water Pollution Control Act
March, 1975	Reduction of fibre contamination of parenteral drugs due to release from asbestos filters	Food, Drug and Cosmetics Act
July, 1976	Use of electrolytic diaphragm treatment for salt disallowed due to resulting asbestos contamination of the product	Food, Drug and Cosmetics Act
January 1977	Asbestos-containing patching compounds and artificial emberizing materials banned	Consumer Product Safety Act
Early 1979	C.P.S.C. negotiated voluntary agreements with manufacturers to replace the asbestos linings in hand-held hairdryers	
August 1979	Requirements for packaging of asbestos during transport	Hazardous Materials Transportation Act
October 1979 (amended Dec. 1979)	Proposal[a] for complete or partial bans of asbestos-containing products due to emissions during commercial life cycle	Toxic Substances Control Act
April 1980	C.P.S.C.[a] grants petition to ban asbestos paper	
September 1980	Requirement to identify asbestos-containing products in schools	Toxic Substances Control Act
November 1980	Limit value of 400,000 f/ℓ for asbestos in raw water sources	Clean Water Act
December 1980	C.P.S.C. issues general order requesting information on asbestos use in certain asbestos-containing consumer products	
January 1981	Reporting and record-keeping requirements for asbestos manufacturers and processors	Toxic Substances Control Act

[a] final rules have not been issued.
[b] State of Connecticut — guideline for asbestos in ambient air — 30 ng/m³

There has been considerable debate concerning the mesothelioma risk associated with exposure to the various forms of asbestos. On the basis of the available data, it has often been concluded that crocidolite is more dangerous than chrysotile in this regard and that amosite might be intermediate between the two.[18] However, there is disagreement concerning this issue and the data cannot be considered to be conclusive largely because the evidence is somewhat indirect (often the proportion of all deaths diagnosed as mesotheliomas is determined), exposure to just one fiber type in the occupational environment is rare, and mesotheliomas are difficult to diagnose. It is also possible that the high proportion of mesotheliomas associated with crocidolite exposure may have been due to higher airborne concentrations of this reputedly "dustier" material in the occupational environment.

Bronchial carcinoma has also been associated with exposure to all three commercially produced varieties of asbestos. There seems to be no interrelationship between the occurrence of bronchial carcinoma and asbestosis.[2] Based on available evidence, it is generally assumed that the relationship between the incidence of bronchial carcinoma and cumulative exposure to asbestos at low doses is linear. Consideration of risk in terms of cumulative dose produces an estimate which ignores the lapsed period since that exposure occurred. However, there is some evidence that mortality from lung cancer increases with dose when lapsed period is held constant.[187] The risk is increased at least additively and in some cohorts, multiplicatively by cigarette smoking;[52] it has been reported that the risk of developing lung cancer for asbestos workers who are smokers is about ten times that of nonsmoking asbestos workers.[51,52] The risk also varies depending on the nature of the occupational exposure, with risks apparently increasing from mining to production to application industries.

Available epidemiological data are insufficient to permit a determination of the cancer risks associated with exposure to asbestos concentrations in the workplace at or below present occupational limits. Recently, there have been several estimates of risk for bronchial carcinoma derived through extrapolation of data from historical prospective mortality studies on populations exposed to high levels of mainly chrysotile. Examination of the available data on this subject illustrates several of the limitations of the use of the epidemiological data in quantitative assessment of risks associated with exposure to asbestos.

McDonald et al.[19] compared the mortality experience of 11,379 men born between 1891 and 1920 and employed for at least one month in the Quebec chrysotile asbestos mines and mills, with that of the Quebec population. Cumulative exposure was assessed separately for each worker and reported in million particles per cubic foot-years (mp/cf-yr), based on measurements made from 1949 onwards by an early monitoring technique called midget impingement. Estimates were made for the period before that, based on "interviews with long-term employees".

The relationship between the standardized mortality ratios (SMR)* for lung cancer and cumulative exposure by age 45 appeared to be linear. Based on the reported relationship and assuming that the dose-response relationship is indeed linear, a 1% increase in total mortality from lung cancer would be associated with exposure to an average concentration of just over 1 mp/cf for a lifetime period of 50 years. A conversion factor of 1 to 5 f/mℓ for each mp/cf was employed in order to express this risk in terms of levels of exposure measured by the currently accepted monitoring technique of phase contrast microscopy. Therefore, exposure to an average level of between 0.25 f/mℓ for a 50-year period would lead to an estimated excess mortality from lung cancer of 1%.

There were several limitations of the study. Follow-up of the oldest subjects was poor and it was assumed that those that were not traced were alive; as a result, mortality rates may have been underestimated by as much as 25%. We have attempted to counterbalance this possible underestimate by deliberately rounding off dust levels on the low side in the estimation of risk; however, the accuracy of the exposure data is also questionable and correlations between exposures measured in mp/cf and fibers per milliliter are poor.[53] The authors also note that the dose-response relationship in a subgroup of the cohort (those with 20 or more years of experience) was not linear.

In another study conducted by Peto,[54] 95% follow-up was achieved for a population of men employed for at least 10 years in a British factory following introduction of the Asbestos Regulations in 1933. Airborne concentrations of asbestos were measured between 1952 and 1960 in particles per cubic centimeter (p/cc) and from 1960 onwards, in fibers per milliliter (f/mℓ) using long running thermal precipitators and membrane filtration. Yearly mean dust levels for 16 processes were available and comparison of data for the years immediately

---

*   Ratio of the number of deaths observed to the number of deaths expected, if the study population had the same structure as the standard population. The standard population is usually the national, state, or provincial population.

before and after the change in measurement technique enabled calculation of a reasonably good conversion factor (from p/cc to f/mℓ) for each process. In this manner, cumulative exposures were estimated for each individual. Assuming a linear model for the dose-response relationship, Peto calculated that a 1% excess in total mortality due to lung cancer would be associated with lifelong exposure to 0.5 f/mℓ.

In another study conducted by Dement et al,[55] the follow-up was 97% for a population of 768 asbestos textile workers who had been employed for at least 6 months with one month between the years of 1940 and 1965. The authors reported that "only an insignificant quantity of asbestos fiber other than chrysotile was ever processed" at this facility. Before 1965, sampling was conducted by midget impinger (mp/cf); from 1965 to 1971 both midget impinger and membrane filter (f/mℓ) samples were taken concurrently; from 1971 onwards, only the membrane filter method was used. Conversion factors were calculated using data from the middle period when sampling was conducted by both methods. The authors adopted a more reproducible approach to estimate early exposures than has been used in other studies; they used a multiple regression model, taking into account changes in processes and engineering controls.

Based on the three data points that were plotted, the dose-response relationship appeared to be linear. Assuming that the linear model is correct, it can be estimated that a 1% excess in total mortality would be associated with a lifetime (50 years) exposure to an average dust level of 0.06 f/mℓ. There is some question, however, about whether the control group of U.S. white males was appropriate. The rate of lung cancer in the local population was 1.75 times the national average and if this group had been used for comparison, the risk estimated would vary by this factor, with a 1% excess in total mortality due to lung cancer being associated with a lifetime exposure to approximately 0.1 f/mℓ. However; the authors argued that the local population should not form the comparison group since the increase in local lung cancer rates was probably attributable to asbestos exposure in the plant and in shipyards in the area.

Recently, over 95% of a cohort of 241 men who were hired before 1961 and employed in a Canadian asbestos-cement factory for at least 9 years, were traced.[56] The mortality of this cohort was compared with that of two groups: (1) the Ontario male population; and (2) workers at the same company exposed to rockwool rather than to asbestos. Both chrysotile and crocidolite were used at the plant and the methods of sampling and the use of historical data to provide estimates of cumulative exposure for each employee were described in detail. The SMRs for three categories fluctuated with cumulative exposure, perhaps due to the small number of subjects included in each group. Nevertheless, the author described a risk assessment, based on the assumption that the relationship detween the dose and response is linear. A best estimate was that a 1% excess in total mortality due to lung cancer would be associated with exposure to an average concentration of 0.2 f/mℓ. The author commented that the reported dust levels were believed to be accurate to within a factor of three to five. Using the upper limit of this range, a 1% excess in lung cancer mortality would be associated with lifetime exposure to a concentration of 1 f/mℓ.

Henderson and Enterline[57] examined the mortality experience of 1075 employees of a U.S. asbestos company who retired with a pension between 1941 and 1967; however, it is not clear how many subjects were not traced. Airborne concentrations of asbestos were measured by midget impingers and cumulative exposure (mp/cf-yr) was computed for each individual on the basis of his experience in each of six categories of exposure. However no information was provided on how the estimates were made. It was simply noted that "it was not always possible to estimate precisely the dust levels for each job, particularly for past periods".

Respiratory cancer SMRs were calculated for five categories of cumulative exposure. In an early paper, the dose-response relationship was represented by a hand-drawn cumulative normal curve; this characterization was criticized[58] and a linear model was adopted. Based on this linear model, a 1% excess in total mortality due to lung cancer is associated with cumulative exposure of 15 mp/cf-yr or exposure to an average concentration of 0.3 mp/cf for a 50-year period. Since no conversion factor was included, this concentration cannot be related to a level in f/mℓ.

One potentially serious source of bias in this study is the choice of retirees as subjects. Although only one mesothelioma was reported in Henderson and Enterline's study, Borow and colleagues[59] reported 72 cases in the vicinity of the plant. A reason for part of the discrepancy is that many of the subjects with mesothelioma may have died prior to retirement or may have had too little service to receive a pension. It is possible that lung cancer results may be similarly biased. However, if the relative risk is constant for different age groups, the SMRs and hence risk assessment should not be affected.

The mortality of all men who worked for at least one month before 1970 in either of two asbestos cement plants in Louisiana was examined by Hughes and Weill.[60] One of the plants began operation in the early 1920s, the other in 1942. Person-years of exposure 20 years after initial employment were reported in relation to mp/cf-yr. No airborne fiber concentrations were described but it was reported that midget impinger sampling had been conducted between 1952 and 1969. In one factory, estimates of exposure before 1952 were made by interviewing employees of long service to compare recent dust levels with those in the past. The author discussed the asbestos fiber content of each product but provided no further details.

The observed relationship between dose and response was not linear, and therefore, it is difficult to extrapolate the data to assess the risk associated with exposure to low level concentrations. Two points are worth noting: no excess respiratory cancer was detected for those with cumulative exposure of less than 100 mp/cf-yr (equivalent to 50 years of exposure at 2 mp/cf), and although the patterns were unclear, a greater excess was found for those with mixed fiber exposure (crocidolite and chrysotile).

There are, however, several limitations of this study. The follow-up was poor; office workers who presumably had no direct exposure were included in the cohort; the midget impinger dust data were poorly described and there was no information included on airborne fiber counts. (The authors use the term fiber exposure inappropriately and also refer to mp/cf-months.)

Recently there has been emphasis on the relationship between asbestos-related disease in populations and numbers and types of asbestos fibers present in lung tissue examined by electron microscopy. However, relatively few studies have been conducted to date, and variations in sample preparation and analysis, and differences in persistence of various fiber types in lung tissue complicate interpretation of the results. In addition, fibers which induce malignant changes may be cleared from the body prior to analysis.

The chrysotile content of lungs in a group of occupationally exposed cases before the U.K. Pneumoconiosis Medical Panels was similar to that of controls;[198] however, the amphibole content was approximately 100 times greater in cases than controls. In the Panel cases, amphibole fibers were more numerous than chrysotile despite the fact that chrysotile had been used in industry in much greater quantities. Chrysotile fibers were, however, more numerous than amphibole fibers in controls. There was a clear association between the grade of asbestosis and amphibole content of the lungs, but there was no correlation with chrysotile content. Similarly in two studies of series of mesothelioma patients in the U.K.[199] and in the U.S. and Canada,[200] there were no differences in the distribution of chrysotile between cases and controls. However, amphibole fibers were present in larger quantities in cases than in controls in both series; whereas amosite was most predominant in the North American series, crocidolite was most prevalent in the U.K. series.

Size distribution of fibers in lung tissue has been determined only in a few studies. Caution must be exercised in interpretation of the data since fiber sizes may change in tissue over

Table 5
POPULATION EXPOSURE TO ASBESTOS IN AMBIENT AIR AND DRINKING WATER

Drinking Water

Location	Concentrations (f/ ℓ)[a]	Analysis[b]	Ref.
Samples of raw, treated and distributed tap water from 71 municipalities across Canada	Amosite: n.d. — 13×10^6 (7% of 336 samples had detectable levels) Chrysotile: n.d. — 1800×10^6 (5% of the population drink water with levels $> 10 \times 10^6$; 0.6% drink water with levels $> 100 \times 10^6$)	T.E.M. (E.P.A. interim procedure)	117
Compendium of surveys of 1500 samples from 406 U.S. cities	n.d. — 600×10^6 (majority of the population drink water with levels $<10^6$)	T.E.M. (various methods of sample preparation)	124

Ambient Air

Location	Concentrations (ng/m³)[c]	Analysis[b]	Ref.
Compendium of 5 surveys conducted between 1972 to 1975 of >60 samples from 14 U.S. cities and rural locations	Rural: 0.01 to 0.1 urban: 0.09 to 70 (50% of the urban population inhale $<110\mu g/yr$ and 10% inhale $>230\mu g/yr$; rural populations inhale 0.05 to 0.5 $\mu g/yr$.)	E.M.	125
20 samples from Montreal	2 (1 to 10)	E.M.	73
187 samples from 49 U.S. cities	2 to 4 (0.1 to 100)	E.M.	73
Asbestos mining communities in Quebec: 7 samples from Thetford Mines; 14 samples from Black Lake	1157 (170 to 3467) 2392 (160 to 11,000)	E.M. E.M.	73

[a] N.d. = non detectable
[b] T.E.M. = transmission electron microscopy; E.M. = electron microscopy.
[c] Concentrations in ambient air determined by electron microscopy are generally reported in terms of mass; factors vary from 100 to 10,000 f/ng for conversion to E.M. — visible fibres

time; however, there is generally a greater proportion of long fibers in lung tissue than in airborne asbestos samples.[98]

2. General Environment

Asbestos is ubiquitous in the environment due to the extensive industrial use of this mineral and to dissemination of fibers from natural sources. A summary of concentrations found in ambient air and water and estimates of population exposure to asbestos in the environment are presented in Table 5. It has been estimated that exposure of the public to asbestos in ambient air is two to three orders of magnitude smaller than that of workers in a well controlled industry.[1] However, there is concern that exposure to asbestos in ambient air or the ingestion of fibers in food and drinking water may pose a risk to the health of the general population.

Increased prevalence of pleural calcification has been observed in a Finnish population residing in the vicinity of an anthophyllite mine,[66] and similar observations have been made for populations exposed to various types of asbestos in several countries.[65,67,68] However, pleural calcification has been observed in areas where there is no identifiable asbestos exposure.[173]

There is also some evidence mainly from case series and case-control studies that the risk of mesothelioma may be increased for individuals who live near asbestos mines or factories; however, the proportion of mesothelioma patients with neighborhood exposure to asbestos varies markedly in different series. In Wagner's early review of 33 cases of mesothelioma in the Northeast Cape province of South Africa, approximately half (16) were individuals with no occupational exposure who had lived in an area of crocidolite mining.[16] Newhouse and Thompson observed 11 otherwise unexposed cases (30.6% of patients in the series) that had lived within 1/2 mile of an "asbestos factory" in London.[69] (Although the type of asbestos was not specified, it has more recently been reported that this was an amphibole plant[192]) Bohlig and Hain reviewed data on mesotheliomas observed in neighborhoods of shipyards and reported 38 cases of "non-occupational" mesothelioma occuring in a 10-year period in residents of a neighborhood near a Hamburg asbestos plant.[70] In a study conducted in Canada, however, excluding those individuals with occupational or household exposure to asbestos, only 2 out of the 254 (0.75%) cases of mesothelioma recorded in Quebec between 1960 and 1978 lived within 20 miles ($\approx$33 km) of the chrysotile mines and/or mills.[64]

There are few data available on the length of residence of the patients in the vicinity of the plants in these studies. Out of 413 notified cases of mesothelioma in the U.K. in 1966 and 1967, 11 (2.7%) individuals who were not asbestos workers who did not have household exposure had lived within 1 mile of an asbestos factory for periods of 3 to 4 years. In a review of cases of mesothelioma in 52 female residents of New York State diagnosed between 1967 and 1968, three otherwise unexposed patients (5.8%) lived within 3.6 km of asbestos factories for 18 to 27 years.[63] In most of the studies there were few data presented concerning the type of asbestos to which neighborhood residents were exposed.

Three ecological epidemiological studies have been conducted to investigate the relationship between exposure to asbestos in the environment and disease.[72,73,174,175] Based on the analysis of cancer incidence data from the Quebec Tumor registry, residents of asbestos mining communities had risks from 1.5 to 8.08 times those in rural Quebec counties for ten different cancer sites among males and for seven sites among females. The higher risks in males were attributed in part to occupational exposure. There was increased risk of cancer of the pleura in both sexes, which decreased with distance of residence from the asbestos mines. The authors emphasized the limitations of their study and recommended that information concerning other exposures and lifestyle factors be considered in more powerful case-control studies.

Although such studies have not been conducted to date, an additional ecological study has recently been completed.[72,73] Mortality between 1966 and 1977 in agglomerations (several municipalities) around the asbestos mining communities of Asbestos and Thetford Mines were compared to that of the Quebec population. There was a statistically significant excess of cancer among males in these agglomerations which was attributed to occupational exposure (a telephone survey indicated that 75% of the men in these communities had worked in the mines[73]). For women, whose exposure would have been confined to the environment or in some cases to environmental exposure and family contact, there were no statistically significant excesses of mortality due to all causes (SMR = 0.89), all cancers (SMR = 0.91), digestive cancers (SMR = 1.06), respiratory cancers (SMR = 1.07), or other respiratory diseases (SMR = 0.58). Similarly there were no significant excesses when mortality at age less than 45 was considered or when the reference population was confined to towns of similar size. Unfortunately, very few causes of mortality were examined in this study and the classes were fairly broad. The authors concluded, however, that the results were consistent with the hypothesis of no excess risk, although a small excess in SMR (30 to 40% for lung cancer) could not be ruled out. In another ecological study conducted in the U.S. in which there was some attempt to control for the urban effect, geographic gradient, and socioeconomic class, there was no correlation between general cancer mortality and the location of asbestos deposits.[174]

Ecological studies such as those described above are considered to be insensitive because of the large number of confounding variables which are difficult to eliminate. In addition, true excess cancer risk is probably underestimated in such studies due to population mobility over a latent period of several decades.[176] Case-control and cohort studies are much more powerful than ecological studies because exposure and outcome are assessed for individuals rather than populations. One relevant cohort study has been conducted. Mortality data for men who lived within 1/2 mile of an amosite factory in Paterson, N.J. in 1942 were compared to that of 5206 male residents of a similar Paterson neighborhood with no asbestos plant. All men who worked in the factory were excluded. Approximately 780 (44% of the exposed population) and 1753 (46% of the unexposed population) died during the 15-year period 1962 to 1976. With respect to total deaths, deaths from cancer (all sites combined) and lung cancer, mortality experience was slightly worse in the "unexposed" population during this period. One death from pleural mesothelioma was observed in the "exposed" population; surprisingly, it occurred in the first 5 years of the 15-year period.[74]

Although few data concerning exposure of the populations in these studies were presented, it is likely that in the past, airborne fiber levels near asbestos facilities were probably significantly higher than they are today. For example, Bohlig and Hain mention that before the second World War, there was visible snowfall-like air pollution from the Hamburg asbestos factory. It is also claimed that 20 years ago in Quebec mining communities, snowfall-like films of asbestos accumulated regularly.[70,73]

There has also been concern that the ingestion of asbestos in food and drinking water may play a role in the etiology of cancer of the gastrointestinal tract or other organs, based on speculation that the increased incidence of G.I. neoplasms and peritoneal mesotheliomas observed in occupationally exposed populations is a consequence of swallowing asbestos-contaminated sputum. The results of the available epidemiological studies on this subject are presented in Table 6.

Studies have been conducted in several areas with relatively high concentrations of asbestos in their drinking water supplies including Duluth, Connecticut, Canadian cities in areas of asbestos deposits, The San Francisco Bay area, and Seattle.[99] The Duluth[75,76] and Connecticut[81,82] studies both have the disadvantage of relatively recent onset of exposure (1955 in Duluth, mostly since 1955 in Connecticut) and actual asbestos fiber concentrations in Connecticut water supplies were very low ($< 10^6$ f/ℓ). The Canadian studies[77,78] included localities with longstanding high concentrations of asbestos in drinking water (100×10^6 f/ℓ) but there were relatively small populations at risk, and cancer incidence data were not available. The San Francisco Bay area studies[79,80] involved longstanding relatively high, but variable concentrations of asbestos in water supplies, a large population at risk, and population-based cancer incidence data. However, the San Francisco Bay area population is highly mobile, an influence which would tend to reduce the chance of observing real associations between the risk of cancer and asbestos exposure. The Seattle area study[83] is the only case-control study available and included data on individual exposures based on length of residence and water source.

Although there has been evidence of an association between ingested asbestos and gastrointestinal cancer incidence in one out of several ecological studies, the association has not been observed in the inherently more sensitive case-control study.

B. Toxicological Data

For a pollutant such as asbestos where there is a great deal of information on the human health effects associated with exposure, the results of toxicological studies are important

Table 6
EPIDEMIOLOGICAL STUDIES — ASBESTOS IN DRINKING WATER

Study Population	Study Design	Results	Ref.
Duluth: ≈ 100,000 ingesting drinking water with ≈ 50 × 10⁶ f/ℓ[206,207] from amphibole mine tailings (1955—1977)	Ecological: cancer incidence data	No convincing evidence of increased risks for G.I. cancer due to presence of fibres in drinking water	75, 76, 202
Canada: ≈ 30,000 in Asbestos and Thetford Mines ingesting drinking water with ≈ 200 × 10⁶ f/ℓ chrysotile. Longstanding contamination (≈ 80 years)	Ecological: cancer mortality data		77
≈ 25,000 in Thetford Mines and 100,000 in Sherbrooke ingesting drinking water with ≈ 100 × 10⁶ f/ℓ chrysotile. Longstanding contamination (≈ 80 years);	Ecological: cancer mortality data	No consistent, convincing evidence of increased cancer risks attributable to ingestion of drinking water contaminated by asbestos	203, 204
San Francisco: ≈ 3,000,000 ingesting drinking water containing ≈ 36 × 10⁶ f/ℓ chrysotile. Longstanding contamination (≈ 60 years)	Ecological: cancer incidence data	Evidence of positive associations and exposure response relationships between asbestos concentrations in drinking water and cancer incidence	79, 80
Connecticut: ≈ 575,000 in 165 towns in Connecticut exposed to ≈ 10⁶ f/ℓ chrysotile mainly from asbestos cement pipe	Ecological: cancer incidence data	Authors attributed largely negative results to low concentrations of asbestos fibres in drinking water	81, 82
Seattle: population of Seattle; Everett-Tacoma metropolitan area exposed to concentrations of chrysotile asbestos as high as 207 × 10⁶ f/ℓ. Longstanding contamination (≈ 60 years)	Case-control: cancer incidence (1974 to 77) and mortality (1955 to 75) data	Authors conclude that study did not provide evidence of a cancer risk due to the ingestion of asbestos in drinking water	205

not only to verify that associations observed in epidemiological studies are indeed causal, but also to elucidate mechanisms of toxicity and to define hypotheses for futher epidemiological study. Although the mechanisms by which asbestos causes fibrosis, lung cancer, and mesothelioma are not well understood, one important development relevant to the assessment of risk to man has been the observation in experimental studies, that fiber dimensions are important determinants in the pathogenesis of asbestos-related disease.

1. Cytotoxicity — In Vitro Studies

In vitro systems do not model in vivo conditions in several respects; factors which affect deposition and clearance are not taken into consideration and immunological systems and renewable cell populations are absent. However, in vitro studies of the biological effects of asbestos and other mineral fibers have been conducted for purposes of elucidating mechanisms of disease or developing screening tests for detection of pathogenic fibers. Some of the results have correlated well with in vivo observations.

The macrophage plays a fundamental role in the pathogenesis of asbestos-related disease. Short fibers deposited on the alveolar surface are phagocytosed by macrophages and removed by the mucocilary escalator. Fibers longer than 10 μm are often surrounded by groups of macrophages which may fuse to form multinucleated giant cells.[84] During the formation of giant cells, a small proportion of the fibers becomes coated with mucopolysaccharide which becomes impregnated with ferritin or hemosideran to form asbestos bodies. Some of the dust-containing macrophages become incorporated into the lung parenchyma and die, and the fibers are rephagocytosed by new populations of macrophages. The presence of these asbestos fibers in distal airways stimulates excess deposition of collagen and reticulin fibers.

There have been numerous in vitro studies of the effects of asbestos on peritoneal or alveolar macrophages of various species, generally mice, rabbits, hamsters, or guinea pigs. Effects on some nonphagocytic cell lines as well as on red blood cells have also been investigated in in vitro studies. With few exceptions, chrysotile has been more cytotoxic to macrophages and to some nonphagocytic cell lines than crocidolite or other forms of asbestos. Although hemolysis of red blood cells in vitro is not considered to be a particularly good predictive assay for in vivo pathogensis of mineral dusts,[177] chrysotile is also more cytotoxic in this system. Furthermore there appears to be a correlation between hemolytic potency, macrophage cytotoxicity, and magnesium concentration for the various fibers.[50,86] The exact mechanism of cytotoxicity is unknown but it has been suggested that it may result from damage caused by the interaction of magnesium with the plasmalemma and lysosomal membranes.[86]

Recently, there has been emphasis on the significance of fiber size in in vitro cytotoxicity. Several groups have reported that fiber size was an important factor in the cytotoxicity of various fibers in Chinese hamster lung cells, human type II alveolar cells, rat ascites tumor cells, and P388D1 cells, a mouse macrophage-like cell line.[177] The size of fibers which are active in vitro are similar to those that are most potent in inducing mesotheliomas following intrapleural administration.[177,179] Generally, it has been concluded that primary macrophages and macrophage-like cell lines in tissue culture are useful in the prediction of the potential fibrogenic effects of mineral dusts[177] and that in vitro effects in some nonphagocytic cell lines may be useful in predicting carcinogenicity.[180]

2. Inoculation and Implantation

In 1962, Wagner first reported that it is possible to produce tumors which appear to be arising from the mesothelial cells of the pleura by inoculating certain dusts into the pleural cavities of rats.[148] Since that time, numerous studies involving the injection or implantation of asbestos into the pleural or peritoneal cavities of various species have been conducted; the results of such studies are summarized in chronological order in Table 7.

Although introduction into body cavities does not simulate the route of exposure of man to fibrous dusts such as asbestos, such studies have permitted clarification of a number of questions that could not feasibly be investigated using the inhalation model since insufficient numbers of mesotheliomas occur following exposure by this route. Malignant neoplasms have resulted following intrapleural and intraperitoneal implantation of all of the commercially important varieties of asbestos and in several of the studies, crocidolite was more potent in induction of malignant neoplasms than equal doses of chrysotile.[119,169,181,157] Incidence of malignancies was increased following simultaneous administration of asbestos with radiation,[166,170] carageenan[168] or 3-methylcholanthrene.[170] There was also evidence of a dose-response relationship for malignant tumor incidence following exposure to both chrysotile and crocidolite in one of the studies;[119] however, occasional tumors were observed even at the lowest dose levels (0.5 to 1 mg.)

However, the most important contribution of the intrapleural injection studies has been to focus attention on the importance of fiber size and shape in the pathogenesis of asbestos-associated diseases. The results of several of the studies indicated that longer fibers are more fibrogenic than short fibers.[155,156] In 1972, based on their study involving intrapleural implantation of 17 fibrous materials in rats, Stanton and Wrench first hypothesized that "the simplest incriminating feature for both carcinogencity and fibrogenicity seems to be a durable fibrous shape, perhaps in a narrow range of size".[87] In later experiments, it was observed that fibers with maximum carcinogenic potency were generally longer than 8 μm and less than 1.5 μm in diameter.[119,162,163] Pott has more recently hypothesized a model in which there is a gradual transition in potency rather than a sharp demarcation between sizes.[197] These observations concerning the importance of fiber size and shape in tumor induction have given rise to speculation that mesotheliomas may be caused by physical irritation by fibers which are carried to the pleural surface by both lymphatic transport within macrophages or by direct penetration of free fibers,[40,84] and have focused a great deal of attention on this "carcinogenic subset" of fibers.

However, there are still several unanswered questions concerning the relative importance of fibers with dimensions in the critical range in mesothelioma induction.[90] The possible role of submicroscopic fibers cannot be ignored;[90] fiber number may also be important.[84,90] In addition, in one of the studies reviewed here, acid leaching of asbestos significantly altered the carcinogenic potency after intrapleural injection;[169] however, it is conceivable that this could be a function of fiber size change rather than chemical modification since the authors reported that maximally leached fibers are shorter and thicker than unmodified fibers. In several other studies described in Table 7, acid leaching[119,151] and variation in the trace metal content[181] had no effect on carcinogenic potency. In addition, there is still some controversy concerning the histological nature of malignant tumors induced by intrapleural and intraperitoneal inoculation in animals.[90]

There are disadvantages of using a nonphysiological route of administration such as intrapleural implantation to study mechanisms of asbestos-related disease in man. Specifically, neither aerodynamic factors which affect fiber deposition nor defense mechanisms which determine the retention of fibers within the lung, nor factors which determine penetration of fibers from the alveolar space to the pleura are taken into consideration in this experimental model. However the results of implantation studies can be integrated with the observations from other investigations that finer fibers are more likely to penetrate to the periphery of the lung, and that short fibers ($<5\mu$m) are more effectively cleared from the lungs by macrophages than are long fibers that cannot be phagocytosed by single cells,[90] However, the need for caution in extrapolation of the results of intrapleural injection studies to predict the potency of various fiber samples with respect to induction of mesotheliomas and other types of cancer, such as lung cancer, cannot be overemphasized. In a recent study,[144] there were significant differences in tumor incidence following intrapleural injection and inhalation of the same sample of chrysotile. The authors suggested that problems of aggregation of fibers in solution for intrapleural injection might result in different size distributions.

Although factors which affect deposition are not taken into consideration in intratracheal (i.t.) injection studies, this route simulates somewhat more closely actual exposure to airborne asbestos than does intrapleural or intraperitoneal injection. However, the greater incidence of infection following exposure by this route often complicates interpretation of the results.[182] Generally speaking, the results of studies involving i.t. administration of asbestos have been similar to those described above for investigations involving intrapleural inoculation. It was first reported that ferruginous bodies developed following exposure not only to asbestos, but also to "biologically inert" particles in a study involving i.t. administration conducted by Gross et al.[209] Simultaneous i.t. administration of asbestos with benzo(a)pyrene or nitrosodiethanolamine resulted in increased lung tumor incidence in several studies[183,193] and results of the i.t. administration studies have confirmed that longer fibers are more fibrogenic.[184,194,196]

The significantly increased incidence of both mesotheliomas and lung cancer in a recently completed study involving i.t. administration of crocidolite to *dogs* is noteworthy.[185] Nine animals received yearly instillations of 66 mg/kg body weight of crocidolite (fiber lengths 2 to 90 μm; diameters 0.2 to 2.0 μm) from 3 months of age and nine of these dogs were exposed to the smoke of nine cigarettes per day, 5 days/week for 6 years. All of the animals died 5 to 8 years after the onset of exposure. Malignant pleural and/or peritoneal mesotheliomas developed in six of the dogs and adenocarcinomas of the lung developed in four. There were no tumors in three of the six dogs exposed both to asbestos and cigarette smoke.

Table 7

INTRAPLEURAL, INTRAPERITONEAL AND SUBCUTANEOUS ADMINISTRATION STUDIES

Species	N	Protocol	Results	Ref.
Wistar rats	11 groups of 10 animals each	Intrapleural injection of 50 mg of 3 samples of crocidolite from South African mines, 3 samples from mills in the same region, 2 samples of chrysotile from mines, 1 sample of amosite, 99.9% pure silica dust or pure carbon black	At 30 mo. after exposure, pleural mesotheliomas in 2 crocidolite-treated rats, 1 chrysotile-treated rat and 1 rat receiving pure silica; authors conclude "it is possible to produce tumors which appear to be arising from the mesothelial cells of the pleura by inoculating certain dusts into the pleural cavities of rats"	148
Syrian golden hamsters	15 animals in each exposed group; 15 untreated controls	Intrapleural injection of 25 mg of soft chrysotile (av. fibre length 67 μ), harsh chrysotile (36 μ) and amosite (18 μ); also 1% soft chrysotile in diet of chrysotile-treated animals; 1% amosite in amosite-treated animals	Granulomatous inflammation and fibrosis in hamsters receiving all 3 types; 5 tumors, possibly mesotheliomas; 2 in harsh chrysotile-treated hamsters and 3 in amosite-treated hamsters	149
Male white mice	100 animals in chrysotile-exposed group; 50 in heated chrysotile-exposed group and 50 in talc-exposed "control"	Intraperitoneal injections of 0.5 mℓ of a 50% suspension of asbestos powder (≈50% Quebec chrysotile, av. fibre diameter <0.5 μ) or asbestos powder heated at 1000°C for 3 hr.	Animals sacrificed at periods up to 343 days following administration; at 48 hours after administration of heat-treated chrysotile, 30 of the 50 animals had died; remaining mice subsequently recovered; fibrous tissue reaction which was proliferative, granulomatous and invasive; no unequivocal neoplastic growth	150
Female CBA mice	Groups of 20 animals each	3 subcutaneous injections in the flank of 10 mg crocidolite, amosite, chrysotile, or solvent-extracted crocidolite and amosite	Transport of asbestos fibres to submesothelial tissues of thorax and abdomen with extensive inflammatory and proliferative changes; 10 mesotheliomas (4-chest; 4-abdomen; 2-both cavities); 6 injection site sarcomas (appeared later than mesotheliomas) All 3 types induced both types of tumors and removal of oils from amosite and crocidolite did not abolish carcinogenic potency	151

Table 7 (continued)
INTRAPLEURAL, INTRAPERITONEAL AND SUBCUTANEOUS ADMINISTRATION STUDIES

Species	N	Protocol	Results	Ref.
SPF Wistar rats and "standard" rats (presumably non-SFP Wistar rats)	96 animals in each of exposed groups; 48 males and 48 females	Intrapleural injection of 20 mg of Transvaal amosite (91% <5 μ in length) superfine grade of Canadian chrysotile (92% <5 μ), North West Cape crocidolite (70% <5 μ), extracted crocidolite (86% <5 μ), silica or saline	Appreciable proportion of animals treated with all types of asbestos developed a mesothelioma; large number of tumors found in animals receiving crocidolite (SPF: 55/94, standard: 62/91); fewer tumors in amosite-treated group (SPF: 38/96, standard: 26/84) and period between inoculation and development of mesothelioma much longer than with the other 2 types; authors note "the high incidence of these neoplasms following the inoculation of chrysotile was unexpected" (SPF: 61/96, standard: 62/90)	152
Guinea pigs rats and mice	—	Intrapleural injection of crocidolite and chrysotile — 25 mg for guinea pigs and rats, 10 mg for mice	Both dusts produced large granulomas in all species but histological patterns varied; granulomas in guinea pigs contained more giant cells, were surrounded by fibrous tissue capsules and remained cellular for 6 to 12 mo.; by 18 mo. fibrosis complete, often calcification; in mice fibrosis at 12 to 15 mo.; in rats most lesions consisted of old acellular collagen by 3 to 4 mo.	153, 154
Wistar albino rats	10 rats in each of exposed groups	Intrapleural injection of 50 mg UICC crocidolite (av. fibre length 11.45 μm), milled UICC crocidolite (3.43 μm), UICC chrysotile (16 to 18 μm) and milled UICC chrysotile (2.52 μm)	Animals sacrificed at 60 and 120 days after administration; both types induced a typical foreign body reaction and progressive fibrosis which was more prominent in long fibre groups; no malignant tumors but in all groups, a few connective tissue stroma from the pleura and pericardia penetrating the heart and lung; authors comment that if these develop into mesotheliomas, "it would appear that mean fibre length or chemical composition is not of primary importance in induction of mesothelioma"	155

Balb/C mice	10 in each exposed group	Intrapleural injection of 10 mg of short and long fibre brucite, chlorite, chromite, forsterite, magnetite, olivine, pyroxene, serpentine, talc, glass fibre, silica fibre, 3 man-man insulation fibres, chrysotile, synthetic chrysotile, fragmented chrysotile	Animals killed 2 weeks to 18 mo. after exposure; long fibre dust specimens produced widespread cellular granulomas which formed firm adhesions, gradually replaced by fibrous tissue; short fibre dust specimens produced smaller granulomas without adhesions; nonfibrous mineral rocks when finely ground also produced small non-adherent granulomas; final degree of fibrosis correlated with initial cellularity of lesions	156
Female pathogen-free Osborne-Mendel rats	Total of 1200	40 mg of 17 materials applied on a fibrous glass vehicle to the pleura including 3 types of asbestos in 7 forms, 6 types of fibrous glass, 2 types of silica, etc.	Two year observation period: amosite, chrysotile and 4 different types of crocidolite produced equally high incidences (58 to 75%) of mesotheliomas; hand milled crocidolite not exposed to extraneous oils or metallic mining yielded dose-related tumor responses comparable to those of a standard reference milled crocidolite; standard crocidolite caused fewer tumors (20 to 32%) when reduced to submicroscopic fibrils; pulverized fragments of mill and nickel metal and fibrous glass vehicle alone did not induce tumors; 2 forms of fine fibreglass milled to approach length of asbestos fibre produced moderately high incidence (12 to 18%); authors conclude "the simplest incriminating feature for both carcinogenicity and fibrogenicity seems to be a durable fibrous shape, perhaps in a narrow range of size"	87
SPF Wistar rats	12 in each of exposed groups 32 in SFA-exposed group: 48 controls and 16 in all other exposed groups 24 in UICC Canadian chrysotile-exposed	Intrapleural injections of 0.5, 1, 2, 4 or 8 mg of SFA chrysotile and crocidolite (from Northeast Cape mine) Intrapleural injection of 20 mg of Canadian chrysotile samples, SFA chrysotile or saline (control) Intrapleural injection of 20 mg of the 5 UICC	The risk of developing a mesothelioma at a given time after injection was proportional to dose for both SFA chrysotile and crocidolite. Of the UICC standard reference samples, crocidolite was the most carcinogenic and removal of oils by benzene extraction did not alter the carcinogenicity of these	119

Table 7 (continued)
INTRAPLEURAL, INTRAPERITONEAL AND SUBCUTANEOUS ADMINISTRATION STUDIES

Species	N	Protocol	Results	Ref.
	group; 32 in all other groups	samples, brucite or barium sulphate	samples; results were consistent with the hypothesis that finer fibres are more carcinogenic	
	Up to 36 in each of exposed groups	Intrapleural injections of ceramic fibre, fibreglass, glass powder, aluminum oxide and 2 samples of SFA chrysotile		
Female albino rats	2 groups of 10	Intraperitoneal injection of 50 mg of UICC crocidolite or chrysotile	Observation over the lifespan; peritoneal mesotheliomas in 90% of crocidolite-exposed group and in 30% of the chrysotile-exposed group; authors suggest crocidolite might be more carcinogenic due to greater mechanical irritation.	157
Balb/C mice	25 in all exposed groups	250 mg Cassiar chrysotile heated to 1000°C and injected into the peritoneal cavity; intrapleural injection of long and short fibre samples of normal chrysotile and chrysotile heated to 400, 600, 800 or 1000°C; intrapleural injection of 10 mg of aut. brake lining dust	Results of ref. 150 confirmed; by 48 hours after administration of 250 mg dose of chrysotile heated to 1000°C, 7 animals died; authors suggested that deaths could be due to acute silica poisoning. Chrysotile heated to 800°C and brake dust lining were more cytotoxic to macrophages than normal chrysotile. No tumors were found in the animals injected with either heated chrysotile or brake lining dust but the experiments were terminated after only 1 year; authors also note that comparatively fewer mesotheliomas are produced by intrapleural injection of asbestos in mice than rats	158
Male Wistar rats	100 in both exposed groups	Intraperitoneal injection of 50 mg of UICC crocidolite or chrysotile B	Crocidolite produced foreign body giant cell granulomas; no giant cells in chrysotile-induced granulomas; amount of fibrosis similar in both groups but atypical mesothelial proliferation more prevalent in chrysotile-exposed groups; malignant peritoneal tumors in 8 of 33 chrysotile-exposed rats and in 9 of 45 crocidolite exposed rats at 14 mo.	159

Rats	—	3 intrapleural injections of 20 mg of chrysotile from filters at 2 U.S.S.R. mines (99% fibres <5 μ in length)	Authors comment that the initiation of mesothelial proliferation leading to neoplasia is associated with a small number of asbestos fibres without significant inflammatory reaction at the site. Mesotheliomas in 46% of the exposed rats	136
Duncan Hartley guinea pigs, Wistar rats and Balb/C mice	2 groups of 50 guinea pigs 2 groups of 25 mice 2 groups of 50 mice 1 group of 25 rats and 1 group of 60 mice	Intrapleural injection of 25 mg of samples of chrysotile or crocidolite prepared at Reading University Intrapleural injection of 10 mg Reading chrysotile or crocidolite Intrapleural injection of UICC crocidolite or chrysotile Intraperitoneal injection of 25 mg in guinea pigs and 10 mg in mice of UICC crocidolite	Observation over the lifespan; only 1 pleural tumour in the 100 guinea pigs and 2 in the 150 mice administered chrysotile and crocidolite intrapleurally; all tumors occurred in crocidolite-treated animals; tumors much more frequently observed in animals given intraperitoneal injections-12/25 in rats and 20/60 in mice. Author concludes that there is evidence that asbestos-induced tumors originate from areas of chronic inflammation (contrary to ref. 159)	160
Female Wistar albino rats	10 groups of 10 rats	Intraperitoneal injections of 50 mg of either standard or acid-extracted UICC crocidolite, UICC crocidolite (50 mg) in combination with iron oxide (20 mg) or silica (20 mg), 50 mg of long or short fibre crocidolite or 50 mg of long or short fibreglass	Animals killed at periods up to 330 days after injection. All suspensions except iron oxide caused an initial foreign body reaction followed by fibroblast infiltration, fibrosis and a reactive mesothelium. Malignant mesotheliomas developed only in the crocidolite-exposed groups; the carcinogenicity of acid-extracted and standard UICC crocidolite was similar and carcinogenicity of long and short fibres similar; authors conclude "mechanical irritation does not contribute to the induction of mesotheliomas"	161
Osborne-Mendel rats	30 in each exposed group	Pleural implantation on a fibrous glass vehicle of 40 mg of 17 samples of fibrous glass of diverse type or dimensional distribution for more than 1 year	Fibres <1.5 μ in diameter and >8 μ in length yielded highest probability of pleural sarcomas	162
Osborne-Mendel rats	—	Pleural implantation on a fibrous glass vehicle of 40 mg of 37 samples which were	Percentage probability of pleural sarcomas ranged from 0 to 100%; lesions in groups	163

Table 7 (continued)
INTRAPLEURAL, INTRAPERITONEAL AND SUBCUTANEOUS ADMINISTRATION STUDIES

Species	N	Protocol	Results	Ref.
		variations of 7 fibrous materials; fibre size distributions similar to asbestos	with low probability of tumors were highly cellular and fibres were completely contained within macrophages; lesions in high tumor probability groups were relatively acellular with an abundance of collagen and free, long fibres in interstitial tissue	
Han rats	48	Intrapleural injection of 25 mg UICC crocidolite	Tumors developed in 33% of the rats between 5 and 22 mo.; histological structure variable	164
Rats	—	3 intrapleural injections of 20 mg of magnesian–arphvedsonite commercial anthophyllite or anthophyllite dust from mine filters, or synthetic hydroxamphibole	"Pretumorous lesions" and pleural tumors in all groups; 77% incidence in magnesian–arphvedsonite-exposed group, 41-commercial anthophyllite, 64.9-anthophyllite dust from mine filters and 54.5-synthetic hydroxyamphibole. Author concludes natural asbestos more carcinogenic than its synthetic analogue. Pleural schlerosis, asbestos granulomas, diffuse and focal hyperplasia of mesothelial cells in monkeys dying at periods up to 38.5 mo.	165
Barrier-protected Caesarian-derived Wistar rats	48 in each exposed group; 48 in control group	Intrapleural injection of 20 mg of SFA, UICC Canadian or Grade 7 chrysotile	Allowing for different survival times, SFA was about 2 × as carcinogenic as grade 7 which was 3 × as carcinogenic as UICC sample; results did not correlate well with results of an inhalation study with these materials	144
Rats	8 10	Inhalation of 3000 WLM radon 222 over one mo; at 71 days, intrapleural injection of 2 mg chrysotile. Whole body irradiation-230 rads for 1 day; at 125 days, intrapleural injection of 2 mg chrysotile or 1% chrysotile diet for 6 mo. at 35 days	All animals developed lung tumors including 7 mesotheliomas; authors conclude "synergistic effect obvious" Extrapulmonary tumors in irradiated controls and in rats receiving asbestos orally and by intrapleural injection; no specific localization in asbestos-exposed animals	166

Animal	Number	Treatment	Results	Ref.
4 strains of rats European hamsters	50 in each exposed group 20	Intraperitoneal injection of long fibrous glass Intraperitoneal injection of 25 mg UICC crocidolite	Wistar rats had significantly lower tumor incidence than animals of 2 other strains (SIV, Sprague-Dawley); fibrosis not a necessary prerequisite for tumor development; fewer hamsters and rabbits developed tumors; author concludes "rats more sensitive for testing carcinogenicity of fibres"	167
Rabbits	8	Intraperitoneal injection of 2 × 25 mg of UICC chrysotile A or glass fibre		
Barrier-protected Caesarian-derived Wistar rats	total of 256	Intrapleural injection of 20 mg of UICC crocidolite in combination with BCG, crystalline silica, talc, carageenan or saline (control)	No significant increase in mesothelioma rate in groups exposed to BCG crystalline silica or talc in addition to crocidolite but a 3 × increase in group receiving carageenan and crocidolite	168
SPF male Sprague-Dawley rats	16 in HCl treated chrysotile-exposed group; ≥ 32 in all other exposed groups; 32 control animals	Intrapleural injection of 20 mg untreated UICC chrysotile A or 4 samples leached to various extents (10 to 90% Mg removed) by oxalic acid or HCl; also crocidolite or glass fibre	In lifetime observation period, a total of 68 pleural mesotheliomas, 1 lung cancer and 9 peritoneal mesotheliomas in the total of 304 animals; proportion of cancer lower than expected due to early deaths from infection; carcinogenicity of chrysotile decreased as Mg content increased; carcinogenicity of crocidolite greater than that of chrysotile with 44% Mg removed; authors conclude "size is not the only factor involved in the induction of pleural cancers by mineral fibres"	169
NEDH rats	—	Intrapleural, intraperitoneal and intratracheal administration of 2 mg of UICC Canadian or Rhodesian chrysotile with or without ancillary radiation treatment (1000 rads-whole body) or injection of 1 mg 3-MC	A highly significant incidence (3.8%) of mesotheliomas in 159 rats treated with asbestos alone. This incidence increased to 11.8% in animals also receiving radiation treatment (borderline statistical significance) and 25.5% in animals also administered 3-MC (significant increase); early tissue responses were similar to asbestos reactions without specific pathological changes attributable to radiation or 3-MC	170
Female Wistar rats	3 groups of 20 animals each	Intraperitoneal injection of 50 mg milled UICC crocidolite (fibre lengths 3 to 5 μm), amorphous UICC crocidolite, or saline	Lifetime observation; 8 mesotheliomas (40%) in "amorphous crocidolite"-exposed group; 3 mesotheliomas (15%) in fibrous croci-	37

Table 7 (continued)
INTRAPLEURAL, INTRAPERITONEAL AND SUBCUTANEOUS ADMINISTRATION STUDIES

Species	N	Protocol	Results	Ref.
			dolite-exposed group and none in saline group; statistically significant difference; author questions the fibrous structure of asbestos as the predominant cause of peritoneal mesothelioma and suggests that submicroscopic particles may be important in induction of tumors	
AF/HAN Wistar rats	7 groups of 32 animals each	Intraperitoneal injection of 25 mg of 5 samples of UICC and factory chrysotile, 2 samples of UICC and factory amosite collected from airborne asbestos clouds of inhalation study	Production of mesothelial tumors in 94 to 100% of the animals in 6 groups; chrysotile more carcinogenic than amosite; heated chrysotile (850°C) least carcinogenic; some correlation of carcinogenicity with fibre length; good correlation of carcinogenicity with in vitro cytotoxicity	12

3. Inhalation

The exposure conditions in inhalation experiments do, of course, approach most closely the circumstances of human exposure to asbestos and are most relevant for the assessment of risk to man. The results of available studies are presented in chronological order in Table 8. Although fibrosis has been observed in several animal species following inhalation of different types of asbestos, a consistently increased incidence of bronchial carcinomas and pleural mesotheliomas has been observed only in the rat.[84,91] The incidence of both fibrosis[139,142] and bronchial carcinomas[93] was dose-related in several of the inhalation studies, and similar to the observation made in human populations, progression of fibrosis has been observed in animals following the cessation of inhalation exposure.[93,144]

That fiber size is important in pathogenesis was recognized as early as 1937, when Gardner first suggested a mechanical theory for asbestosis, based on his observation in inhalation experiments that longer fibers were more fibrogenic.[186] This finding has been confirmed in subsequent inhalation studies.[92,147]

The interaction between exposure to asbestos and cigarette smoking has been investigated in several of the inhalation studies. Surprisingly, the incidence of laryngeal lesions and malignant tumors was significantly lower in a group of hamsters inhaling asbestos and smoke than in a group inhaling asbestos only,[137,139] However, this was probably attributable to the shorter lifespan of the asbestos-exposed animals. Shabad et al. reported that the incidence of "precancerous lesions" was greater in groups of rats exposed in inhalation studies to chrysotile and cigarette smoke than in those exposed to chrysotile alone.[136]

The results of the inhalation studies also indicate that similar to observations made in human populations, amosite and crocidolite are more persistent in lung tissue than chrysotile. However, although fibrosis and lung cancer have been observed in animals inhaling all of the commercially important varieties of asbestos, the results of the early studies regarding the relative pathogenicity of various fiber types were confusing and contradictory[91] due predominantly to the fact that generally only the mass of the dust clouds was measured; numbers or size distribution of fibers were not considered.[50,91] Variations in methods of generation of the dust clouds also complicate interpretation of the results of the early studies. In addition, numbers of malignant bronchial carcinomas and mesotheliomas in these studies were often small.

In a recent study,[91,94-96] fiber size distribution has been taken into consideration and the results indicated that inhalation of chrysotile dust caused more lung fibrosis in rats than either crocidolite or amosite even when the fiber numbers ($>5\mu$m) in the dust clouds were similar; all malignant pulmonary neoplasms occured in chrysotile-treated animals. The authors suggested that the greater fibrogenicity and carcinogenicity of chrysotile might be related to the fact that the chrysotile clouds contained many more fibers over 20 μm long.[95] In another inhalation study conducted recently to compare the pathological effects of UICC* reference dust samples used in animal studies with those collected from a factory environment, the authors concluded that while fibrogenicity and carcinogenicity both are a function of the presence of relatively long fibers in dust clouds, different lengths are involved in each process, and tumor production requires the longest fibers.[97]

Insufficient numbers of mesotheliomas have been produced in inhalation experiments to draw conclusions concerning dose-response relationships; however it has been noted that most mesotheliomas have been found in animals that received a high total dose of asbestos.[96] In contrast to the observation in human populations that mesotheliomas are most often associated with exposure to crocidolite, small numbers of mesotheliomas have been found in animals inhaling all types of asbestos.[93] Several reasons have been suggested for this seeming discrepancy between the results of the toxicological and epidemiological studies.[84] The differences in the risk of development of mesothelioma in humans may be related to

* UICC — International Union Against Cancer.

Table 8
INHALATION STUDIES

Species	N	Protocol[a]	Results	Ref.
Mice	388 in exposed groups; 388 controls	Exposure for either 17 or 22 mo. to ≈ 500 × 10⁶ ppcf "Asbestos Floats 7–TF–2" (predominantly chrysotile, fibrous content low, length > 10 μ for 21% of the fibres)	Asbestos bodies; no fibrosis; small insignificant difference between incidence of nonmalignant lung tumors in exposed and control groups killed at 19 mo; increased incidence of multiple nonmalignant lung tumors in exposed groups; authors conclude "a possible accentuation by the inhalation of asbestos dust of an existing tendency to develop lung tumors"	126
Guinea pigs, rabbits & Vervet monkeys	16 to 24 guinea pigs, 2 to 4 rabbits and 3 to 4 Vervet monkeys in exposed groups	Exposure to ≈ 30,000 p/mℓ of either chrysotile (7 to 10% fibres > 10μ), amosite or crocidolite from South African mills for varying periods of time (e.g. up to 24 months for guinea pigs exposed to chrysotile; lifetime for rabbits and Vervet monkeys exposed to chrysotile but only 14 months for these species when exposed to amosite)	Asbestos bodies present in all 3 species exposed to all 3 types; chrysotile exposure caused fibrosis in guinea pigs and monkeys but not in rabbits; amosite caused asbestosis in all 3 species. It is difficult to draw conclusions concerning the relative pathogenicity of the various fibre types due to the varying periods of exposure and lack of characterization of fibre sizes	127
Guinea pigs, rats	9 to 28 animals in exposed groups	Series of experiments in which rats were exposed to Rhodesian chrysotile and guinea pigs were exposed to chrysotile, amosite, anthophyllite and crocidolite in a specially designed apparatus. The concentration of chrysotile was reported to be ≈5000 p/mℓ (86% fibers <5μ). The animals were exposed for periods which varied from 900 hours over a 50-day period to about 100 hours over a 5-day period	Description of the pathological changes in asbestosis; bronchiolitis as little as 3 days after the start of exposure with extension of the inflammatory reaction to the adjacent alveoli. As the length of exposure increased, changes became more generalized and by 6 to 8 weeks, widespread and progressive fibrosis often followed by complete lung consolidation in some areas; asbestos bodies observed within seven days of exposure. The authors concluded "the reactions induced by each (type of asbestos) are remarkably similar. If the fibrosis is induced chemically, it must be due to the silicate moiety of the structure"	128—131
Guinea pigs	—	Series of experiments in which animals were exposed in the apparatus of the previous studies[128,131] for periods of 24 hours to anthophyllite, crocidolite, chrysotile and amosite and for ≈800 hours (in 6 weeks) to chrysotile	Description of the formation and ultrastructure of asbestos bodies; the rate of development fastest for chrysotile, then anthophyllite, crocidolite and amosite	132—134

Male white rats	61 to 70 in exposed groups; 15 to 30 in control groups	Exposure for up to 60 weeks to 86 mg/m³ (42 to 146) ball-milled, hammer-milled Canadian chrysotile; in addition one group received 0.05 mL 5% NaOH i.t. to impair lung clearance	For those animals surviving at 16 months, 48% of the chrysotile and NaOH-exposed group, 24% of the chrysotile-exposed group and 0% of the control group had malignant lung tumors; the authors concluded that the trace metal content of the dust from hammer milling may have played a role in the aetiology of the lung tumors	135
White rats	—	Group 1 exposed to 230 mg/m³ of dust collected from the chrysotile mill in Kazakhtan, U.S.S.R. for 2 hours 5 times a week for 13.5 months Group 2 similarly exposed to chrysotile for 3 days, then received 1 mg BaP i.t. for remainder of exposure period Group 3 — from 10th day onwards until the end of exposure period, rats exposed to cigarette smoke (1.5 cigarettes/rat/mo.)	"Precancerous lesions" in 20% of rats exposed to chrysotile alone, in 57% of group exposed to chrysotile and BaP and in 38% of group exposed to chrysotile and cigarette smoke; no reference to control groups	136
SPF Wistar rats	Total of 1013 rats; group sizes of 19 to 58	Groups exposed to 9.7 to 14.7 mg/m³ of either UICC amosite, anthophyllite, crocidolite, chrysotile, (Canadian) and chrysotile (Rhodesian) for periods of 1 day, 3, 6, 12, or 24 mos.	Proportional increase of lung dust with dose and much greater retention of amphiboles; less retention of chrysotile and no clear increase with dose; less asbestosis for amosite than for the other dusts; progression of asbestosis following cessation of exposure for all dusts; higher incidence of tumors with 12 mo. exposure than with 6 mo. but little difference following 12 and 24 mo. exposure; of 20 tumors which metastasized, 16 were in chrysotile-exposed groups, 3 in crocidolite-exposed groups and 1 in anthophyllite exposed groups; of the 11 mesotheliomas, 4 occurred following exposure to crocidolite and 4 following exposure to Canadian chrysotile; 2 mesotheliomas occurred following 1 day exposures; positive association between the incidence of asbestosis and lung cancer; no association between exposure and gastrointestinal cancer incidence	93
Syrian golden hamsters	102 animals exposed; 102 controls; group sizes ≈ 50	Group 1 exposed to 23 μg/ℓ UICC Canadian chrysotile for 11 months Group 2 also exposed for 10 min. 3 times a day to cigarette smoke for duration of their lifespan	Asbestosis in all exposed animals; authors report that of 12 lung adenomas found in 510 hamsters, 10 occurred among the 102 animals of the asbestos-exposed groups, indicating an early neoplastic response; incidence of	137—139

Table 8 (continued)
INHALATION STUDIES

Species	N	Protocol[a]	Results	Ref.
			laryngeal lesions and malignant tumors significantly lower in asbestos and smoke-exposed group than in smoke-exposed control group probably due to significantly shorter lifespan in asbestos-exposed animals	
SPF rats	—	Exposed to 12.5 mg/m³ UICC Rhodesian chrysotile for 15 weeks	Grade 3 asbestosis consistently apparent at 30 and 50 weeks exposure; increase in number and size of free cells in the lung after 7 and 15 weeks; returned to normal by 30 weeks; increase in pulmonary surfactant levels — 10× after 15 weeks and 2× after 30 weeks; authors suggest that this is a primary response of the lung to prevent cytoxic effect of the dust	140
SPF Wistar rats		Same exposure as in previous experiment for periods of 3 days to 15 weeks	No significant change in total protein or fatty acid composition of pulmonary surfactant; increase in activity of PGT in lung homogenates and free cell population; authors suggest that the increase in pulmonary surfactant could be due to an increase in synthesis without corresponding alteration in degradation	141
SPF white Wistar rats of the Han strain	Group sizes of 48 animals	Experiment designed so that both mass and fiber number could be examined; 5 groups exposed to either 10 mg/m³ UICC chrysotile, crocidolite or amosite (550 f/c.c. amosite 390 f/cc. UICC chrysotile or 430 f/cc. UICC crocidolite (<5μ) for 12 mos.	Amphiboles deposited and retained in the lung to a much greater extent than chrysotile, retention reduced as density of cloud increased; chrysotile caused far more fibrosis than either amphibole even when the fibre numbers were similar; all malignant pulmonary neoplasms occurred in chrysotile-exposed animals; the authors suggested that the greater pathogenicity of chrysotile may be due to greater number of fibres >20μ in length	91, 94—96
Male Syrian golden hamsters	Total of 96 exposed and 96 control animals	2 groups exposed for 3 hours/day, 5 days/wk. to either 1 μg/ℓ (5 to 13 f/cc. >5 μ) or 10 μg/ℓ (30 to 118 f/cc. >5μ) A/C aerosol (chrysotile content 10.5%) for up to 15 mo.	"Slight pulmonary fibrosis only in the 15 mos. exposure group; higher incidence and severity with increased dose after 6 mo. exposure to 10 μg/ℓ dose; increased incidence of slight emphysema after exposure to the	139, 142

			10 μg/ℓ dose for 6 months — after 15 months, no difference between exposed and control groups; authors suggested that the minimal response may be due to changes during processing of the fibres that decrease the toxicity; however exposure probably quite low and the carcinogenicity of asbestos in hamsters has not been well characterized	
Rats	Group sizes of 48 animals	Designed to compare the pathological effects of exposure to UICC sample to those of factory samples; 4 groups exposed to either 10 mg/m³ UICC amosite, UICC chrysotile, factory amosite or factory chrysotile for 12 months; animals permitted to complete lifespan	Greater accumulation in the lungs of factory samples than UICC samples; factory chrysotile produced similar levels of lung pathology to UICC sample, with the exception that a slightly smaller number of bronchial carcinomas was produced by the factory dust; factory amosite more fibrogenic than UICC sample; little carcinogenicity of both amosite samples; based on the analysis of fibre sizes in each of the samples, authors concluded that "while fibrogenicity and carcinogenicity both depend upon the presence of relatively long fibres in dust clouds, different lengths are involved in each process and tumor production requires the largest fibres"	97
SPF Wistar rats of the AF/HAN strain	Group sizes of 48 animals	Experiment designed to compare the pathological effects of peak dosing to those of even dosing; 2 groups exposed to either 10 mg/m³ UICC amosite or 2 mg/m³ UICC chrysotile 7 hr/day, 5 days/wk for 1 year and 2 groups exposed to either 50 mg/m³ amosite or 10 mg/m³ chrysotile 1 day each wk for 1 year	Lung dust levels of both chrysotile and amosite after 12 mo. similar whether "peak" or "even" dosing; levels of peribronchial fibrosis generally lower for "peak" dosing groups than for "even" dosing groups; levels of interstitial fibrosis slightly higher following "peak" dosing; no differences in the incidence of pulmonary neoplasms between "peak" dosing groups and "even" dosing groups; the authors conclude that "no indication that short periods of high dust exposure in an asbestos factory would result in significantly greater hazard than would be indicated by the raised overall dust counts for the day in question (there were, however 2 bronchial carcinomas in the "peak" dosing amosite group and none in the "even" dosing group)	143

Table 8 (continued)
INHALATION STUDIES

Species	N	Protocol[a]	Results	Ref.
Barrier-protected Caesarian-derived Wistar rats	Group sizes of 24 (6 and 12 mo. exposure) and 48 (3 mo. exposures)	Exposure for periods of either 3, 6 or 12 mo. to 10.8 mg/m³ SFA chrysotile (430 f/mℓ >5μ), Grade 7 chrysotile (1020 f/mℓ >5μ) or UICC chrysotile (3150 f/mℓ >5μ)	Deposition and retention of dust low for all chrysotiles-poor correlation with length of exposure; progression of fibrosis after end of exposure for groups inhaling all types for 6 or 12 mo; UICC produced at least as much fibrosis as other two samples in all 6 groups; tumor yield significantly greater with UICC chrysotile than with Grade 7	144
SPF Fischer rats	Group sizes of 2; 2 control animals	Exposure to 10 mg/m³ UICC chrysotile B for 50 weeks; animals killed at 4 mo. and 50 wks.	Fibrosis more marked following inhalation of chrysotile than following inhalation of various man-made mineral fibres; authors suggest that this may be due to the fact that there are more fibres in the respirable fraction of chrysotile than in a similar quantity of mmmf	145
Fischer rats		Exposure for periods up to 12 mo. to 3.07 mg/m³ NIEHS short-range chrysotile, (192 × 10³ f/m³) or to 9.40 mg/m³ NIEHS intermediate-range chrysotile (13 × 10⁸ f/m³)	Greater lung injury involving both the alveolar epithelium and the interstitium were found in the animals exposed to intermediate-range fibres	147
Charles River, Caesarian derived Sprague Dawley rats, albino male guinea pigs and hamsters	Group sizes 34 (hamsters) 35 (guinea pigs) and 46 (rats)	Exposure to 0.3 ng/ℓ UICC amosite (3.1 × 10⁶ f/ℓ >5 μ) for up to 90 days	Body weights significantly less than control values for 15 wks. after exposure; based on consideration of mass and fibre concentration asbestos considered to be 10 × more fibrogenic than mmmf, Fybex; the fibrogenic pulmonary reaction greater in rats than in guinea pigs or hamsters; two mesotheliomas developed in hamsters following exposure to Fybex	147

[a]   Unless otherwise specified, exposures were for 6 to 8 hours per day, 5 days per week.

differences in past airborne concentrations and fiber size distributions in occupational environments (crocidolite is reputedly a more dusty material than chrysotile).[98] The greater potency of crocidolite in mesothelioma induction in humans may also be due to the observed persistence of crocidolite fibers in biological tissues; chrysotile fibers may persist in the body for a sufficiently long period to induce tumors in rats but not in man.

4. Ingestion

There is no conclusive evidence from toxicological studies conducted to date (see Table 9) that ingested asbestos is carcinogenic or cocarcinogenic in animal species.[99] The results of the early studies[100-105] must be considered to be inconclusive due to shortcomings in study design; many of the investigations were conducted for relatively short periods of time with insufficient numbers of test and control animals. In addition, the administered asbestos was often not well characterized. In the more extensive studies conducted recently,[106-110] the incidence of gastrointestinal tumors in the asbestos-fed groups has not been statistically significantly greater than that in the control groups.[99] Some increases in gastrointestinal tumor incidence were observed in some of the test groups in some of the studies; however these statistically insignificant increases were not observed consistently and were not confirmed by the results of the most extensive study conducted to date in which doses which were 2 million times the likely maximum human exposure on a body weight basis, were administered over the lifetime to large groups of animals.[109,110] There was no evidence of a dose-response relationship in any of the studies.

IV. EVALUATION

Regulations limiting the exposure of workers to asbestos are based on epidemiological data. At the present time, the standard for occupational exposure in many countries in the world is 2 f/mℓ (>5 μm), time-weighted average, and although control of exposure almost certainly lessens the risk of all asbestos-related disease, this limit was originally proposed, based on an acceptable level of risk for the development of asbestosis.

It has been difficult to derive standards for occupational exposure based on acceptable cancer risks due to the limitations of and variation in quantitative estimates of this risk derived through extrapolation of epidemiological data on high level exposures. Based on the results of some of these estimates, an outright ban of the use of all forms of asbestos may be the only acceptable course of action; on the other hand the results of other analyses suggest that presently enforced standards or slightly reduced limits may afford an acceptable degree of protection.

These disparities are due partially to problems inherent in using retrospective epidemiological studies for risk extrapolation, and partially to the complexity of asbestos itself. Often exposure data are inadequate, and measurements of airborne fiber concentrations in the occupational environment have been made only in recent years. In addition, accepted methods for sampling and analysis have changed over the years and as a result, it is difficult to compare exposure data. The potential to cause disease may also be underestimated or occasionally overestimated due to methodological limitations of mortality studies such as misdiagnosis, or inadequate comparison groups or observation periods. Also, very little is known about the differences in size distribution and properties of asbestos fibers in different industries which may be responsible for variations in risk. In addition, although there is epidemiological evidence of linearity of the dose-response relationship, the use of the linear nonthreshold model may lead to overestimation of risks at low exposure levels. Data obtained in future epidemiological studies of disease incidence in populations exposed to present day levels of well characterized asbestos in the occupational environment may impart more definitive information about quantitative risks associated with exposure.

Table 9

CARCINOGENICITY STUDIES — INGESTED ASBESTOS

Species	No. test animals	Protocol	Results	Ref.
Sprague Dawley rats	40	1500 mg/kg crocidolite in the diet, 12 weeks of age until death	No tumors or lesions in the intestinal wall	100
Rats	10	5% chrysotile in the diet for 21 months		101
	30	10 mg Rhodesian chrysotile in the diet once weekly for 6 weeks		
	30	5 mg crocidolite in the diet once weekly for 6 weeks	No tumors or lesions	
	30	10 mg Transvaal crocidolite in the diet once weekly for 6 weeks		
	30	10 mg North West Cape crocidolite in the diet once weekly for 6 weeks		
Baboons	?	"Ate food and drank water heavily contaminated with asbestos" for periods up to 5 years	No tumors	102
Wistar rats	50	25 mg/kg chrysotile extracted from a commercial filter pad fed in the diet over the lifetime	Statistically significant increase in the number of malignant tumors in tissues other than G.I. tract; incidence not specified	103
SPF Wistar rats	32	100 mg/day "super-fine" chrysotile in the diet for 6 months	1 gastric leiomyosarcoma, diagnosis uncertain; authors conclude "may be a consequence of the feeding"	105
Rats	40	1% chrysotile in the diet, 2 years	2 tumours of the G.I. tract; no tumors of the G.I. tract in control group of 40 animals; authors conclude "results inconclusive"	104
Syrian golden hamsters	60	0.5 mg/ℓ amosite in drinking water over the lifetime	No tumours	106
	60	5 mg/ℓ amosite in drinking water over the lifetime	3 malignant tumors including a peritoneal mesothelioma, 2 early squamous cell carcinomas of the forestomach	
Syrian golden hamsters	60	50 mg/ℓ amosite in drinking water over the lifetime	1 malignant tumor; authors conclude "tumors not treatment-related"	
Male Wistar rats	25	250 mg amosite per week in dietary margarine supplement for periods up to 25 months	1 malignant tumor in gastric muscle layer	107

	25	250 mg chrysotile per week in dietary margarine supplement for periods up to 25 months	1 pleural histiocytic tumor; significant increase in incidence of benign tumors in tissues other than the G.I. tract; authors conclude unlikely that these benign tumors are treatment-related due to lack of evidence of widespread penetration or dissemination of fibres	
	25	250 mg crocidolite per week in dietary margarine supplement for periods up to 25 months	No primary malignant lesions of the G.I. mucosa	
F 344 Rats	500	10% chrysotile in the diet over the lifetime	5 tumors including 1 mesothelioma; incidence not statistically significantly greater than in control group; authors conclude asbestos may play a role in carcinogenesis based on additional data concerning cAMP levels, non-neoplastic lesions and transmigration	108
Syrian Golden Hamsters	500	1% amosite in the diet fed to nursing mothers and over the lifetime of the pups	No adverse effects on body weight gain and survival; no statistically significant increase in tumor incidence	109
	500	1% short range chrysotile in the diet fed to nursing mothers and over the lifetime of the pups	Significant increase in adrenal cortical adenomas in males; not considered to be treatment-related	110
	500	1% intermediate range chrysotile in the diet fed to nursing mothers and over the lifetime of the pups	Significant increase in adrenal cortical adenomas in males and females; not considered to be treatment-related	
Sprague-Dawley Rats	22 to 30	Filtered or unfiltered Duluth drinking water or drinking water containing Lake Superior sediment, taconite plant tailings, chrysotile, amosite or diatomaceous earth over the lifetime	No significant excess in tumor incidence in experimental groups	208

Several countries have recently adopted an alternative regulatory approach; for example, in the U.K., minimum practicable levels rather than limits based on an acceptable level of cancer risk have been specified as occupational standards for the various forms of asbestos; quantitative estimates of risk were examined only to insure that the standards were not unjustifiably lax.[36] On the other hand, the U.S. has proposed the "limit of detection" of analytical techniques which are routinely used in the workplace (0.1 f/ℓ) as an occupational standard for all types of asbestos.

Several quantitative estimates of the lung cancer and mesothelioma risks associated with exposure to asbestos in consumer products,[111] drinking water,[112] and ambient air[113] have also been made, based on epidemiological data. The quantitative risks associated with exposure to fibers from sprayed friable asbestos-containing surfaces in schools have also been assessed.[114-116] Sprayed asbestos was used extensively between 1946 and 1973 on structural surfaces in public buildings to retard collapse during fire. There is some evidence that the eventual relative risk from lung cancer increases slightly as age at first exposure falls and it has been argued that the effect of childhood exposure might be greater than that for those first exposed as adults.[116]

However, quantitative estimation of the risks associated with exposure to the much lower levels of asbestos in the general environment is even more problematic than assessing risks associated with occupational exposure and as a result, there has been considerable controversy concerning the results of these analyses. For example, in October, 1979 based on the results of case-control and cohort studies of gastrointestinal cancer in occupationally exposed populations, the U.S. Environmental Protection Agency estimated that a lifetime risk of gastrointestinal cancer and peritoneal mesothelioma of 10^{-5} would result from the ingestion of drinking water containing 0.4 × 10^6 asbestos f/ℓ.[112] (Levels in Canadian drinking water supplies range up to 1800 × 10^6 f/ℓ.[117]) The validity of this estimate has been assessed to some extent using available Canadian epidemiological data. Although the differences were not tested for statistical significance, application of the EPA estimate to the mortality experience of certain Canadian localities with relatively high concentrations of asbestos in their drinking water supplies yields predicted excess numbers of gastrointestinal cancer deaths which deviate in both directions from observed values.[99]

Toxicological data have rarely been used as a basis for the development of regulations to control exposure to asbestos due to the availability of epidemiological data and to limitations of the animal bioassay data for quantitative risk estimation. Many of the animal studies have involved nonphysiological routes of administration such as intrapleural implantation, and direct extrapolation of the results of such studies to estimate risk for man is not justified.

Although exposure conditions in inhalation experiments resemble more closely the circumstances of human exposure to airborne asbestos, species differences in the tissue response and technical problems associated with generation and characterization of fibrous dusts in such studies complicate extrapolation of the data. For example, although pulmonary fibrosis has developed following prolonged inhalation of asbestos in numerous species, a consistently increased incidence of pulmonary tumors, occasional pleural mesotheliomas, and a single peritoneal mesothelioma have been observed following asbestos inhalation only in the rat.[95] Even in this species, however, the asbestos tissue response varies in one respect from that of man; asbestos bodies are commonly present in human tissue but rarely occur in the rat.[95] In addition, there are difficulties in extrapolation because some fibers persist in tissue for a greater relative proportion of the lifespan of animals than of man.

Electrostatic charges on the generated aerosols in inhalation studies also affect deposition of fibers in the lungs of the experimental animals;[118] it is unknown whether fibers in occupational settings are similarly charged. In addition, standardized UICC asbestos samples have been used extensively in inhalation studies and although their use has facilitated comparison of results, the distribution of fiber shapes and sizes is not similar to those in various occupational settings.[97]

Probably the single most important contribution of the results of the toxicological studies to our understanding of asbestos-related disease is the observation that fiber size and shape are important determinants of deposition in tissues and pathogenic potential. There is some preliminary evidence in human populations that exposure to natural fibers in the critical size range may play a role in the etiology of mesothelioma.[90,201] The present regulatory definition of fibers (length > 5 μm, aspect ratio 3:1, measured by the membrane filter method) was based on early biological evidence that fibers of length 10 to 50 μm were the most fibrogenic.[98] One author has concluded that this definition should be revised, based on the available evidence that the most hazardous fibers are those within the approximate size limits:[98]

1. length = 5 to 10 μm, < 100 μm
2. diameter = 1.5 to 2 μm
3. aspect ratio = 5:1 to 10:1

Another author has suggested that it may be appropriate to retain the 5 μm lower limit but also to specify an additional count of fibers longer than 20 μm.[95] Pott has suggested that fiber dimensions in asbestos analyses should be reported in four length categories: 2.5 to 5, 5 to 10, 10 to 20, and 20 to 100 μm and three width categories: 0.1 to 0.5, 0.5 to 1, and 1 to 3μm.[120] Although there is some disagreement concerning the precise sizes of fibers that should be measured, it is clear that in the future, the fiber definition specified in regulations must be closely examined.

Although results indicate that fiber size and shape play an important role in the pathogenesis of asbestos-related diseases, it can be concluded on the basis of available data that effects may also be a function of the relative persistence of fibers in biological tissues. Durability is probably most closely related to chemical characteristics of the fibers and it is likely therefore that physical and chemical properties are both important determinants of the potential of fibers to cause disease.

V. CONCLUSIONS

The health effects associated with exposure to asbestos have been studied extensively and based on the available toxicological and epidemiological data, it can be concluded with some certainty that there is a dose-response relationship for all of the main types of asbestos-related disease. Introduction of increasingly more stringent occupational standards and improvement of working conditions over the years should undoubtedly result in a decrease in the incidence of fibrosis and cancer resulting from exposure to asbestos in the workplace; however available epidemiological data are insufficient to quantify the magnitude of this reduction in risk or to determine the adequacy of present occupational standards. At present, although there are few data available, there is no convincing evidence of increased risks of disease in the general population associated with exposure to asbestos in ambient air or with ingestion of asbestos in food and drinking water.

Regulatory decisions concerning the use of any chemical substances for various applications in society should be based on an equitable weighting of benefits against risks. For asbestos, the issue can be simply stated: Do the levels of asbestos present in the occupational and general environments present sufficient risks to health to outweigh the valuable commercial and sometimes lifesaving properties of this material? This is a difficult question to answer even though there is great deal of scientific evidence concerning the health effects of exposure to asbestos in human populations. Quantitative risks associated with exposure cannot be determined, and as is the case with all risk benefit analyses, there are problems associated with the weighting and comparison of unlike factors such as economic gains and health risks. In addition, regulatory options are limited since suitable substitute materials are not yet available for most applications of asbestos; substitutes that have been developed are generally less effective, more expensive and have not been adequately tested in toxicological studies.[3] An additional complicating factor that affects the regulatory process is the increasing public concern about health risks associated with the production and use of asbestos. These concerns are largely based upon widely publicized, divergent, and often speculative or inaccurate statements of scientific experts, the asbestos industry and its opponents, and the press. The magnitude of the public concern over asbestos exposure has also been underscored by numerous Congressional hearings and by massive litigation.

Although there are many unanswered questions concerning quantitative risks associated with exposure to asbestos and mechanisms of disease, the tremendous effort directed towards epidemiological and toxicological study of asbestos-related health effects over the past 75 years has not been without benefit. Such study has not only provided the basis for regulations to limit the exposure of workers and the general public to asbestos, but has also alerted us to potential hazards that might be associated with exposure to other natural and man-made mineral fibers. Collaboration of scientists in programs similar to that conducted recently by the man-made mineral fiber industry under the auspices of the World Health Organization[195] may be instrumental in the prevention in other fiber production industries of the hard won lessons of the asbestos industry.

REFERENCES

1. *Public Health Risks of Exposure to Asbestos*, Report of a Working Group of Experts Prepared for the Commission of the European Communities, Zeilhuis, R. L., Ed. Pergamon Press, Oxford, 1977.
2. *IARC Monogr. Evaluation of Carcinoginic Risk of Chemicals, Vol. 14, Asbestos*, International Agency for Research on Cancer, Lyon, 1977.
3. Information provided by the Department of Energy, Mines and Resources, Ottawa, 1982.
4. **Selikoff, I. J. and Lee, D. H. K.**, *Asbestos and Disease*, Academic Press, New York, 1978.
5. **Lee, D. H. K. and Selikoff, I. J.**, Historical background to the asbestos problem, *Environ. Res.*, 18, 300, 1979.
6. **Penney, W.**, Evolution in the uses of asbestos, presented at the World Symp. Asbestos, Montreal, May 24 to 27, 1982.
7. Asbestos: an informational resource, Report Prepared under Contract by SRI International, Levine, R. J., Ed., National Cancer Institute, Bethesda, 1978.
8. Report of the Department Committee on Compensation for Industrial Diseases, Her Majesty's Stationary Office, London, 1907.
9. Teaching Module on Asbestos Related Disease, Report Prepared under Contract by American College of Radiology, National Cancer Institute and National Institute of Occupational Safety and Health, Bethesda, 1981.
10. **Lynch, K. M. and Smith, W. A.**, Pulmonary asbestosis. III. Carcinoma of lung in asbestos-silicosis, *Am. J. Cancer*, 24, 56, 1935.
11. **Gloyne, S. R.**, Two cases of squamous carcinoma of the lung occurring in asbestosis, *Tubercule*, 17, 5, 1935.
12. **Bolton, R. E., Davis, J. M. G., Donaldson, K. and Wright, A.**, Variations in the carcinogenicity of mineral fibres, in Inhaled Particles V, *Ann. Occup. Hyg.*, 26, 569, 1982.
13. **Doll, R.**, Mortality from lung cancer in asbestos workers, *Br. J. Ind. Med.*, 12, 81, 1955.
14. **Shannon, H. S.**, Characterization of Statements on Asbestos Related Diseases, unpublished data, 1982.

15. **McDonald, J. C. and McDonald, A. D.,** Epidemiology of mesothelioma from estimated incidence, *Prev. Med.,* 6, 426, 1977.
16. **Wagner, J. C., Sleggs, C. A., and Marchand, P.,** Diffuse pleural mesothelioma and asbestos exposure in the North Western Cape Province, *Br. J. Ind. Med.,* 17, 260, 1960.
17. **Mancuso, T. F. and El Attar, A. A.,** Mortality pattern in a cohort of asbestos workers — a study based on employment experience, *J. Occup. Med.,* 9, 147, 1967.
18. **Acheson, E. D. and Gardner, M. J.,** The ill effects of asbestos on health, in *Asbestos — Final Report of the Advisory Committee,* Vol. 2, Health and Safety Executive, London, 1979.
19. **McDonald, J. C., Liddell, F. D. K., Gibbs, G. W., Eyssen, G. E., and McDonald, A. D.,** Dust exposure and mortality in chrysotile mining, 1910—1975, *Br. J. Ind. Med.,* 37, 11, 1980.
20. **Newhouse, M. L., Berry, G., and Steidmore, J. W.,** A mortality study of workers manufacturing friction materials with chrysotile asbestos, in Inhaled Particles V; *Ann. Occup. Hyg.,* 26, 569, 1982.
21. **Newhouse, M. L. and Berry, G.,** Asbestos and laryngeal carcinoma, *Lancet,* ii, 615, 1973.
22. **Selikoff, I. J., Hammond, E. C., and Seidman, H.,** Mortality experience of insulation workers in the United States and Canada, 1943—1976, *Ann. N.Y. Acad. Sci.,* 330, 91, 1979.
23. **Rubino, G. F., Piolatto, G., Newhouse, M. L., Scansetti, G., Aresini, G. A., and Murray, R.,** Mortality of chrysotile asbestos workers at the Balangero Mine, Northern Italy, *Br. J. Ind. Med.,* 36, 187, 1979.
24. **Stell, P. M. and McGill, T.,** Asbestos and laryngeal carcinoma, *Lancet,* ii, 416, 1973.
25. **Shettigara, P. T. and Morgan, R. W.,** Asbestos, smoking and laryngeal carcinoma, *Arch. Environ. Health,* 30, 517, 1975.
26. **Morgan, R. W. and Shettigara, P. T.,** Occupational asbestos exposure, smoking and laryngeal carcinoma, *Ann. N.Y. Acad. Sci.,* 271, 308, 1976.
27. **Hinds, M. W., Thomas, D. B., O'Reilly, H. P.,** Asbestos, dental X-rays, tobacco and alcohol in the epidemiology of laryngeal cancer, *Cancer,* 44, 1114, 1979.
28. **Newhouse, M. L., Gregory, M. M., and Shannon, H.,** Etiology of cancer of the larynx, in *Biological Effects of Mineral Fibres,* Vol. 2, Wagner, J. C., Ed., IARC Scientific Publication 30, Lyon, 1980.
29. **Selikoff, I. J.,** Cancer risk of asbestos exposure, in *Origins of Human Cancer,* Book C, Cold Spring Harbor Laboratory, New York, 1977.
30. **Steel, J.,** Asbestos control limits, in *Asbestos-Final Report of the Advisory Committee,* Vol. 2, Health and Safety Executive, London, 1979.
31. **Mereweather, E. R. A. and Price, C. V.,** *Report on Effects of Asbestos Dust on the Lungs and Dust Suppression in the Asbestos Industry,* Her Majesty's Stationary Office, London, 1930.
32. **Dreesen, W. C., Dallavalle, J. M., Edwards, J. I., Miller, J. W., And Sayers, R. R.,** A study of asbestosis in the asbestos textile industry, *Public Health Bull.,* 241, 1938.
33. **Enterline, P. E.,** Epidemiological basis for the asbestos standard, presented at the 2nd Annu. Symp. Environmental Epidemiology, Pittsburg, April 28, 1981.
34. Hygiene standards for chrysotile asbestos dust, British Occupational Hygiene Society Committee on Hygiene Standards, *Ann. Occup. Hyg.,* 11, 47, 1968.
35. Federal Register Citations Pertaining to the Regulation of Asbestos, U.S. Environmental Protection Agency, Washington, D.C., 1979.
36. **Simpson, W. J.,** What should be the rationale for regulations on asbestos, presented at the World Symp. on Asbestos, Montreal, May 24—27, 1982.
37. **Kolev, K.,** Experimentally induced mesothelioma in white rats in response to intraperitoneal administration of amorphous crocidolite asbestos: preliminary report, *Environ. Res.,* 29, 123, 1982.
38. Report of the Asbestosis Working Group, Environmental Health Directorate, Department of National Health and Welfare, Ottawa, 1976.
39. Chronology of Major Federal Actions on Asbestos, Asbestos Information Association, Arlington, 1982.
40. **Craighead, J. E. and Mossman, B. T.,** The pathogenesis of asbestos-associated diseases, *N. Engl. J. Med.,* 306, 1446, 1982.
41. Summary of Main Features of Asbestos/Health Regulations, Asbestos Information Association, Arlington, 1982.
42. **Casey, K. R., Rom, W. N., and Montamed, F.,** Asbestos-related diseases, *Clin. Chest Med.,* 2, 179, 1981.
43. **Becklake, M. R., Liddell, F. D. K., Manfreda, J., and McDonald, J. C.,** Radiological changes after withdrawal from asbestos exposure, *Br. J. Ind. Med.,* 36, 23, 1979.
44. **Kannerstein, M. and Churg, J.,** Mesothelioma in man and experimental animals, *Environ. Health Perspect.,* 34, 31, 1980.
45. **Wagner, J. C. and Elmes, P. C.,** The mineral fibre problem, in *Recent Advances in Occupational Health,* McDonald, J. C., Ed., Churchill Livingstone, Edinburgh, 1981.
46. **Newhouse, M. L., Berry, G., Wagner, J. C., and Turok, M. E.,** A study of the mortality of female asbestos workers, *Br. J. Ind. Med.* 29, 134, 1972.

47. **Whitwell, F., Scott, J., and Grimshaw, M.,** Relationship between occupation and asbestos-fibre content of the lungs in patients with pleural mesothelioma, lung cancer and other diseases, *Clin. Chest Med.,* 2, 179, 1981.

48. **Newhouse, M. L. and Berry, G.,** Predictions of mortality from mesothelial tumours in asbestos factory workers, *Br. J. Ind. Med.,* 33, 147, 1976.

49. **Newhouse, M. L. and Berry, G.,** Patterns of mortality in asbestos workers in London, *Ann. N.Y. Acad. Sci.,* 330, 53, 1979.

50. **Gibbs, G. W., Arhirrii, M. I., and Doganoglu, Y.,** Effects of Inhaled Particles on Human Health, Report prepared for the Associate Committee on Scientific Criteria for Environmental Quality, National Research Council of Canada, Ottawa, 1982.

51. **Selikoff, I. J. and Hammond, E. C.,** Asbestos and smoking, *J. Am. Med. Assoc.,* 242, 458, 1979.

52. **McDonald, J. C.,** Asbestos-related disease: an epidemiological review, in *Biological Effects of Mineral Fibres,* Wagner, J. C., Ed., IARC Scientific, Publication 30, Lyon, 1980.

53. **Gibbs, G. W. and Lachance, M.,** Dust — fibre relationships in the Quebec chrysolite industry, *Arch. Environ. Health,* 28, 69, 1974.

54. **Peto, J.,** Lung cancer mortality in relation to measured dust levels in an asbestos textile factory, in *Biological Effects of Mineral Fibres,* Wagner, J. C., Ed., IARC Scientific Publication 30, Lyon, 1980.

55. **Dement, J. M., Harris, R. L., Symons, M. J., and Sly, C.,** Estimates of dose response for respiratory cancer among chrysotile asbestos textile workers. Inhaled Particles V., *Ann. Occup. Hyg.,* 26, 869, 1982.

56. **Finkelstein, M.,** Mortality among long-term employees of an Ontario asbestos-cement factory, *Br. J. Ind. Med.,* in press.

57. **Henderson, V. L. and Enterline, P. E.,** Asbestos exposure — factors associated with excess cancer and respiratory disease mortality, *Ann. N.Y. Acad. Sci.,* 330, 117, 1979.

58. **Peto, J.,** The hygiene standard for chrysotile asbestos, *Lancet,* i, 484, 1978.

59. **Borow, M., Conston, A., Livornese, L., and Schalet, N.,** Mesothelioma following exposure to asbestos: a review of 72 cases, *Chest,* 64, 641, 1973.

60. **Hughes, J. and Weill, H.,** Lung cancer risk associated with manufacture of asbestos cement products, in *Biological Effects of Mineral Fibres,* Wagner, J. C., Ed., IARC Scientific Publication 30, Lyon, 1980.

61. **Anderson, H., Lilis, R., Daum, S., Fischbein, A., and Selikoff, I. J.,** Household exposure to asbestos and risk of subsequent disease, in *Dusts and Disease,* Lemen, R. and Dement, J. M., Ed., Pathotox Publishers, Park Forest South, 1979.

62. **Anderson, H. A., Lilis, R., Daum, S. M., Fischbein, A. S., and Selikoff, I. J.,** Household contact asbestos neoplastic risk, *Ann. N.Y. Acad. Sci.,* 271, 311, 1976.

63. **Vianna, N. J. and Polan, A. K.,** Non-occupational exposure to asbestos and malignant mesothelioma in females, *Lancet,* 1, 1061, 1978.

64. **McDonald, A. D.,** Malignant mesothelioma in Quebec, in *Biological Effects of Mineral Fibres,* Wagner, J. C., Ed., IARC Scientific Publication 30, Lyon, 1980.

65. **Navratil, M. and Trippe, F.,** Prevalence of pleural calcification in persons exposed to asbestos dust, and in the general population in the same district, *Environ. Res.,* 5, 210, 1972.

66. **Kilviluoto, R.,** Pleural calcification as a roentgenologic sign of non-occupational endemic anthopyhllite-asbestosis, *Acta Radiol. Suppl.,* 194, 1, 1960.

67. **Yazicioglu, S.,** Pleural calcification associated with exposure to chrysotile asbestos in southeast Turkey, *Chest,* 70, 43, 1976.

68. **Zolov, C., Bourilkov, T., and Babadjov, L.,** Pleural asbestosis in agricultural workers, *Environ. Res.,* 1, 287, 1967.

69. **Newhouse, M. L. and Thompson, H.,** Mesothelioma of pleura and peritoneum following exposure to asbestos in the London area, *Br. J. Ind. Med.,* 22, 261, 1965.

70. **Bohlig, H. and Hain, E.,** Cancer in relation to environmental exposure, in *Biological Effects of Asbestos,* Bogovski, P., Gilson, J. C., Timbrell, V. and Wagner, J. C., Eds., IARC Scientific Publication 8, Lyon, 1973.

71. **Greenburg, M. and Lloyd Davies, T. A.,** Mesothelioma register 1967—68, *Br. J. Ind. Med.,* 31, 91, 1974.

72. **Pampalon, R., Siemiatycki, J., and Blanchet, M.,** Pollution environnementale par l'amiante et santé publique au Québec. Une analyse comparative de la mortalité dans les agglomérations du Québec, 1966—1977, *Union Med. Can.,* 111, 475, 1982.

73. **Siemiatycki, J.,** Mortality in the general population in asbestos mining areas, presented at the World Symp. Asbestos, Montreal, May 24—27, 1982.

74. **Hammond, E.C., Garfinkel, L., Selikoff, I. J., and Nicholson, W. J.,** Mortality experience of residents in the neighbourhood of an asbestos factory, *Ann. N.Y. Acad. Sci.,* 330, 417, 1979.

75. **Levy, B. S., Sigurdson, E., Mandel, J., Laudon, E., and Pearson, J.,** Investigating possible effects of asbestos on city water: surveillance of gastrointestinal cancer in Duluth, Minnesota, *Am. J. Epidemiol.,* 103, 362, 1976.

76. **Mason, T. J., McKay, E. W., and Miller, R. W.**, Asbestos-like fibres in Duluth water supply: relation to cancer mortality, *J. Am. Med. Assoc.*, 228, 1019, 1974.
77. **Wigle, D. T.**, Cancer mortality in relation to asbestos in municipal water supplies, *Arch. Environ. Health*, 32, 185, 1977.
78. **Toft, P., Wigle, D., Meranger, J. C., and Mao, Y.**, Asbestos and drinking water in Canada, *Sci. Total Environ.*, 18, 77, 1981.
79. **Kararek, M. S., Conforti, P. M., Jackson, L. A., Cooper, R. C., and Murchio, J. C.**, Asbestos in drinking water and cancer incidence in the San Francisco Bay area. *Am. J. Epidemiol.*, 112, 54, 1980.
80. **Conforti, P. M., Kanarek, M. S., Jackson, L. A., Cooper, R. C., and Murchio, J. C.**, Asbestos in drinking water and cancer in the San Francisco Bay Area: 1969—1974 incidence, *J. Chron Dis.*, 34, 211, 1981.
81. **Harrington, J. M., Craun, G. F., Meigs, J. M., Landrigan, P. J., Flannery, J. T., and Woodhull, P. S.**, An investigation of the use of asbestos cement pipe for public water supply and the incidence of gastrointestinal cancer in Connecticut 1935—1973, *Am. J. Epidemiol.*, 107, 96, 1978.
82. **Meigs, J. W., Walter, S. D., Heston, J. F., Millette, J. R., Craun, G. E., Woodhull, R. S., and Flannery, J. T.**, Asbestos cement pipe and cancer in Connecticut 1955—1974, *J. Environ Health*, 42, 187, 1980.
83. **Polissar, L., and Hutchinson, F.**, Case-control study — Puget Sound area, presented at U.S. Environmental Protection Agency Summary Workshop on Ingested Asbestos, Cincinnati, October 13—14, 1982.
84. **Davis, J. M. G.**, The biological effects of mineral fibres, *Ann. Occup. Hyg.*, 24, 227, 1981.
85. **Ehrenreich, T., and Selikoff, I. J.**, Asbestos fibers in human lung, *Am. J. Forensic Med. Pathol.*, 2, 67, 1981.
86. **Harrington, J. S.**, The biological effects of mineral fibres, especially asbestos, as seen from *in vitro* and *in vivo* studies, *Annals d'Anatomie Pathologique*, 21, 155, 1976.
87. **Stanton, M. F. and Wrench, C.**, Mechanisms of mesothelioma induction with asbestos and fibrous glass, *J. Natl. Cancer Inst.*, 48, 797, 1972.
88. **Stanton, M. F.**, Fiber carcinogenesis: is asbestos the only hazard? *J. Natl. Cancer Inst.*, 52, 633, 1974.
89. **Pott, F.**, Animal experiments on biological effects of mineral fibres, in *Biological Effects of Mineral Fibres*, Wagner, J. C., Ed., IARC Scientific Publication 30, Lyon, 1980.
90. **Harrington, J. S.**, Fiber carcinogenesis: Epidemiological observation and the Stanton hypothesis, *J. Natl. Cancer Inst.*, 67, 977, 1981.
91. **Davis, J. M. G.**, The use of animal inhalation experiments in the study of asbestos bioeffects, *Staub.-Reinhalt*, 40, 453, 1980.
92. **Vorwald, A. J., Durkan, T., and Pratt, P. C.**, Experimental studies of asbestosis, *Arch. Ind. Hyg. Occup. Med.*, 3, 1, 1951.
93. **Wagner, J. C., Berry, G., Skidmore, J. W., and Timbrell, V.**, The effects of the inhalation of asbestos in rats, *Br. J. Cancer*, 29, 252, 1974.
94. **Davis, J. M. G.**, The use of animal models for studies on asbestos bioeffects, *Ann. N.Y. Acad. Sci.*, 330, 795, 1979.
95. **Davis, J. M. G., Beckett, S. T., Bolton, R. E., Collings, P., and Middleton, A. P.**, Mass and number of fibres in the pathogenesis of asbestos-related lung disease in rats, *Br. J. Cancer*, 37, 673, 1978.
96. **Davis, J. M. G.**, Current concepts in asbestos fiber pathogenicity, in *Dusts and Disease*, Lemen, R. and Dement, J. M., Ed., Pathotox Publishers, Park Forest South, 1979.
97. **Davis, J. M. G., Beckett, S. T., Bolton, R. E., and Donaldson, K.**, A comparison of the pathological effects in rats of the UICC reference samples of amosite and chrysotile collected from the factory environment, in *Biological Effects of Mineral Fibres*, Wagner, J. C., Ed., IARC Scientific Publication 30, 285, 1980.
98. **Walton, W. H.**, The nature, hazards and assessment of occupational exposure to airborne asbestos dust: a review, *Ann. Occup. Hyg.*, 25, 117, 1982.
99. **Toft, P., Meek, M. E., Wigle, D., and Méranger, J. C.**, Asbestos in drinking Water, *C.R.C. Critical Reviews in Environmental Control*, 14, 151, 1984.
100. **Bonser, G. M. and Clayson, D. B.**, Feeding of blue asbestos to rats, British Empire Cancer Campaign for Research, Abstr. 45, 242, 1967.
101. **Gross, P., Harley, R. A., Swinburne, L. M., Davis, J. M. G., and Greene, W. B.**, Ingested mineral fibres: do they penetrate tissue or cause cancer? *Arch. Environ. Health*, 29, 341, 1974.
102. **Webster, I.**, The ingestion of asbestos fibres, *Environ. Health Perspect.* 9, 199, 1974.
103. **Gibel, W., Lohs, K. L., Horn, K. H., Wildner, G. P., and Hoffman, F.**, Animal experimental investigations of the carcinogenic activity of asbestos filter material following oral administration, *Archiv. Geschwulstforsch.*, 46, 437, 1976.
104. **Cunningham, H. M., Moodie, C. A., Lawrence, G. A., and Pontefract, R. D.**, Chronic effects of ingested asbestos in rats, *Arch. Environ. Contam. Toxicol.*, 6, 507, 1977.

105. **Wagner, J. C., Berry G., Cooke, T. J., Hill, R. J., Pooley, F. D., and Skidmore, J. W.,** Animal experiments with talc, in *Inhaled Particles IV*, Pergamon Press, New York, 1977, 647.

106. **Smith, W. E., Hubert, D. D., Sobel, H. J., Peters, E. T., and Doerfler, T. E.,** Health of experimental animals drinking water with and without amosite asbestos and other mineral particles, *J. Environ. Pathol. Toxicol.* 3, 277, 1980.

107. **Bolton, R. E., Davis, J. M. G., and Lamb, D.,** The pathological effects of prolonged asbestos ingestion in rats, *Environ. Res., 29*, 134, 1982.

108. **Donham, K. J., Berg, J. W., Will, L. A., and Leininger, J. R.,** The effects of long-term ingestion of asbestos on the colon of F 344 rats, *Cancer, 45*, 1073, 1980.

109. **McConnell, E.,** NIEHS carcinogenesis bioassay of amosite asbestos in Syrian golden hamsters, presented at Environmental Protection Agency Summary Workshop on Ingested Asbestos, Cincinnati, October 13—14, 1982.

110. **McConnell, E.,** NIEHS carcinogenesis bioassay of chrysotile asbestos in Syrian golden hamsters, presented at Environmental Protection Agency Summary Workshop on Ingested Asbestos, Cincinnati, October 13—14, 1982.

111. **Preuss, P. W., White, P., Cohn, M., and Spitzer, H.,** The assessment and control of risks resulting from exposure to asbestos in consumer products, pres. World Symp. Asbestos in Montreal, May 24—27, 1982.

112. Ambient Water Quality Criteria For Asbestos, U.S. Environmental Protection Agency, Washington, D.C., 1980.

113. **Schneiderman, M. S., Nisbet, I. C. T., and Brett, S. M.,** Assessment of risks posed by exposure to low levels of asbestos in the general environment, *Bundesgesunheitsamt-Berichte, 4*, 3/1, 1981.

114. **Schirripa, J. T.,** Airborne levels of asbestos fibers in general occupancy buildings and an estimate of carcinogenic risk, *J. Am. Ind. Hyg. Assoc.,* in press.

115. Technical Support Document For Regulating Action Against Friable Asbestos-Containing Materials in School Buildings, Office of Pesticides and Toxic Substances, U.S. Environmental Protection Agency, Washington, D.C., 1980.

116. **Peto, J.,** An Alternative Approach For the Risk Assessment of Asbestos in Schools, Rep. to the U.S. Environmental Protection Agency, Washington, D.C., 1981.

117. A National Survey For Asbestos Fibres in Canadian Drinking Water Supplies, Environmental Health Directorate Report 79-EHD-34, Department of National Health and Welfare, 1979.

118. **Vincent, J. H., Johnston, W. B., Jones, R. O, and Johnston, A. M.,** Static electrification of airborne asbestos: a study of its causes, assessment and effects on deposition in the lungs of rats, *J. Am. Ind. Hyg. Assoc., 42*, 711, 1981.

119. **Wagner, J. C., Berry, G., and Timbrell, V.,** Mesothelioma in rats after inoculation with asbestos and other minerals, *Br. J. Cancer, 28*, 173, 1973.

120. **Pott, F.,** personal communication, 1982.

121. Report on the Health Hazards of Asbestos, Report of Asbestos Ad Hoc Subcommittee, National Health and Medical Research Council, Australia, 1981.

122. *Luftqualitätskriterien. Umweltbelastung durch Asbest und andere faserige Feinstaübe*, Umweltbundesamt, Erich Shmidt Verlag, Berlin, 1980.

123. Limit Values, National Board of Occupational Safety and Health, Sweden, 1978.

124. **Millette, J. R., Clark, P. J., and Pansing, M. F.,** Exposure to Asbestos from Drinking Water in the United States, U.S. Environmental Protection Agency, Cincinnati, 1979.

125. **Suta, B. E. and Levine, R. J.,** Non-occupational asbestos emissions and exposure, in *Asbestos. Properties, Applications and Hazards*. Vol. 1, Michaels, L. and Chissick, S. S., Ed., John Wiley & Sons, Chichester, 1979.

126. **Lynch, K. M., McIver, F. A., And Cain, J. R.,** Pulmonary tumours in mice exposed to asbestos dust, *Arch. Ind. Health, 15*, 207, 1957.

127. **Wagner, J. C.,** Asbestosis in experimental animals, *Br. J. Ind. Med., 20*, 1, 1963.

128. **Holt, P. F., Mills, J., and Young, D. K.,** The early effects of chrysotile asbestos dust on the rat lung, *J. Pathol. Bact., 87*, 15, 1964.

129. **Holt, P. F., Mills, J., and Young, D. K.,** Experimental asbestosis with four types of fibers: importance of small particles, *Ann. N.Y. Acad. Sci., 132*, 87, 1965.

130. **Holt, P. F., Mills, J., and Young, D. K.,** Experimental asbestosis in the guinea pig, *J. Pathol. Bact., 92*, 185, 1966.

131. **Davis, J. M. G.** An electron microscopy study of the effect of asbestos dust on the lung, *Br. J. Exp. Pathol., 44*, 454, 1963.

132. **Botham, S. K. and Holt, P. F.,** The mechanism of formation of asbestos bodies, *J. Pathol. Bact., 96*, 443, 1968.

133. **Botham, S. K. and Holt, P. F.,** Development of asbestos bodies on amosite, chrysotile and crocidolite fibres in guinea-pig lungs, *J. Pathol., 105*, 159, 1971.

134. **Davis, J. M. G.**, The ultrastructure of asbestos bodies from guinea-pig lungs, *Br. J. Exp. Pathol.*, 45, 634, 1964.
135. **Gross, P., deTrevill, R. T. P., Tolker, E. B., Kaschak, M., and Babyak, M. A.**, Experimental asbestosis, *Arch. Environ. Health*, 15, 343, 1967.
136. **Shabad, L. M., Pylev, L. N., Krivosheeva, L. V., Kulagina, T. F., and Nemenko, B. A.**, Experimental studies on asbestos carcinogenicity, *J. Natl. Cancer Inst.*, 52, 1175, 1974.
137. **Wehner, A. P., Busch, R. H., Olson, R. J., and Craig, D. K.**, Chronic inhalation of asbestos and cigarette smoke by hamsters, *Environ. Res.*, 10, 368, 1975.
138. **Wehner, A. P., Stuart, B. O., and Sanders, C. L.**, Inhalation studies with Syrian golden hamsters, *Prog. Exp. Tumor Res.*, 24, 177, 1979.
139. **Wehner, A. P.**, Effects of inhaled asbestos, asbestos plus cigarette smoke, asbestos-cement and talc baby powder in hamsters, in *Biological Effects of Mineral Fibres*, Wagner, J. C., Ed., IARC Scientific Publication 30, Lyon, 1980.
140. **Tetley, T. D., Hext, P. M. Richards, R. J., and McDermott, M.**, Chrysotile-induced asbestosis: changes in the free cell population, pulmonary surfactant and whole lung tissue of rats, *Br. J. Exp. Pathol.*, 57, 505, 1976.
141. **Tetley, T. D., Richards, R. J., and Harwood, J. L.**, Changes in pulmonary surfactant and phosphatidylcholine metabolism in rats exposed to chrysotile asbestos dust, *Biochem. J.*, 166, 323, 1977.
142. **Wehner, A. P., Dagle, G. E., Cannon, W. C., and Buschbom, R. L.**, Biological effects of inhaled asbestos cement dust in the Syrian golden hamster, in *Dusts and Disease*, Lemen, R. and Dement, J. M., Ed., Pathotox Publishers, Park Forest South, 1979.
143. **Davis, J. M. G., Beckett, S. T., Bolton, R. E., and Donaldson, K.** The effects of intermittent high asbestos exposure (peak dose levels) on the lungs of rats, *Br. J. Exp. Pathol.*, 61, 272, 1980.
144. **Wagner, J. C., Berry, G., Skidmore, J. W., and Pooley, F. D.**, The comparative effects of three chrysotiles by injection and inhalation in rats, in *Biological Effects of Mineral Fibres*, Wagner, J. C., Ed., IARC Scientific Scientific Publication 30, Lyon, 1980.
145. **Johnson, N. F. and Wagner, J. C.**, A study by electron microscopy of the effects of chrysotile and man made mineral fibres on rat lungs, in *Biological Effects of Mineral Fibres*, Wagner, J. C., Ed., IARC Scientific Publication 30, Lyon, 1980.
146. **Crapo, J. D., Barry, B. E., Brody, A. R., and O'Neill, J. J.**, Morphological, morphometric and microanalytical studies on lung tissue of rats exposed to chrysotile asbestos in inhalation chambers, in *Biological Effects of Mineral Fibres*, Wagner, J. C., Ed., IARC Scientific Publication 30, Lyon, 1980.
147. **Lee, K. P., Barras, C. E., Griffith, F. D., Waritz, R. S., and Lapin, C. A.**, Comparative pulmonary responses to inhaled inorganic fibers with asbestos and fiberglass, *Environ. Res.* 24, 167, 1981.
148. **Wagner, J. C.**, Experimental production of mesothelial tumours by implantation of dust in laboratory animals, *Nature (London)*, 196, 180, 1962.
149. **Smith, W. E., Miller, L., Churg. J., and Selikoff, I. J.**, Mesotheliomas in hamsters following intrapleural injection of asbestos, *J. Mt. Sinai Hosp.*, 32, 1, 1965.
150. **Jagatic, J., Rubnitz, M. E., Godwin, M. C., and Weiskopf, R. W.**, Tissue response to intraperitoneal asbestos with preliminary report of acute toxicity of heat-treated asbestos in mice, *Environ. Res.*, 1, 217, 1967.
151. **Roe, F. J. C., Carter, R. L., Walters, M. A., and Harington, J. S.**, The pathological effects of subcutaneous injections of asbestos fibres in mice: migration of fibres to submesothelial tissues and induction of mesotheliomata, *Int. J. Cancer*, 2, 628, 1967.
152. **Wagner, J. C. and Berry, G.**, Mesotheliomas in rats following inoculation with asbestos, *Br. J. Cancer*, 23, 567, 1969.
153. **Davis, J. M. G.**, The long term fibrogenic effects of chrysotile and crocidolite asbestos dust injected into the pleural cavity of experimental animals, *Br. J. Exp. Pathol.*, 51, 617, 1970.
154. **Davis, J. M. G.**, The calcification of fibrous pleural lesions produced in guinea pigs by the injection of chrysotile dust, *Br. J. Exp. Pathol.*, 52, 238, 1971.
155. **Burger, B. F. and Englebrecht, F. M.**, The biological effects of long and short fibres of crocidolite and chrysotile A after intrapleural injection into rats, *S. Afr. Med. J.*, 44, 1268, 1970.
156. **Davis, J. M. G.**, The fibrogenic effects of mineral dusts injected into the pleural cavity of mice, *Br. J. Exp. Pathol.*, 53, 190, 1972.
157. **Englebrecht, F. M., and Burger, B. F.**, Biological effects of asbestos dust on the peritoneal viscera of rats, *S. Afr. Med. J.*, 47, 1746, 1973.
158. **Davis, J. M. G. and Coniam, S. W.**, Experimental studies on the effects of heated chrysotile asbestos and automobile brake lining dust injected into the body cavities of mice, *Exp. Mol. Pathol.*, 19, 339, 1973.
159. **Shin, M. L. and Firminger, H. I.**, Acute and chronic effects of intraperitoneal injection of two types of asbestos in rats with a study of the histopathogenesis and ultrastructure of resulting mesotheliomas, *Am. J. Pathol.*, 70, 291, 1973.

160. **Davis, J. M. G.**, Histogenesis and fine structure of peritoneal tumors produced in animals by injections of asbestos, *J. Natl. Cancer Inst.*, 52, 1823, 1974.

161. **Englebrecht, F. M. and Burger, B. F.**, Mesothelial reaction to asbestos and other irritants after intra-peritoneal injection, *S. Afr. Med. J.*, 49, 87, 1975.

162. **Stanton, M. F., Layard, M., Tegeris, A., Miller, E., May, M., and Kent, E.**, Carcinogenicity of fibrous glass: pleural response in the rat in relation to fiber dimension, *J. Natl. Cancer Inst.*, 58, 587, 1977.

163. **Stanton, M. F. and Layard, M.**, The carcinogenicity of fibrous minerals, in *Proc. Workshop on Asbestos: Definitions and Measurement Methods*, U.S. National Bureau of Standards, Gaithersburg, Md., 1978.

164. **Davis, J. M. G.**, The histopathology and ultrastructure of pleural mesotheliomas produced in the rat by injections of crocidolite asbestos, *Br. J. Exp. Pathol.*, 60, 642, 1979.

165. **Pylev, L. N.**, Pretumorous lesions and lung and pleural tumours induced by asbestos in rats, Syrian golden hamster and *Macaca mulatta* (Rhesus) monkeys, in *Biological Effects of Mineral Fibres*, Wagner, J. C., Ed., IARC Scientific Publication 30, International Agency for Research on Cancer, Lyon, 1980.

166. **Lafuma, J., Morin, M., Poncy, J. C., Massé, R., Hirsch, A., Bignon, J., and Monchaux, G.**, Mesothelioma induced by intrapleural injection of different types of fibres in rats; synergistic effects of other carcinogens, in *Biological Effects of Mineral Fibres*, Wagner, J. C., Ed., IARC Scientific Publication 30, International Agency for Research on Cancer, Lyon, 1980.

167. **Pott, F., Huth, F., and Spurny, K.**, Tumor induction after intraperitoneal injection of fibrous dusts, in *Biological Effects of Mineral Fibres*, Wagner, J. C., Ed., IARC Scientific Publication 30, International Agency for Research on Cancer, Lyon, 1980.

168. **Wagner, J. C., Hill, R. J., Berry, G., and Wagner, M. M. F.**, Treatments affecting the rate of asbestos-induced mesotheliomas, *Br. J. Cancer*, 41, 918, 1980.

169. **Monchaux, G., Bignon, J., Jaurand, M. C., Lafuma, J., Sebastien, P., Masse, R., Hirsch, A., and Goni, J.**, Mesotheliomas in rats following inoculation with acid-leached chrysotile asbestos and other mineral fibres, *Carcinogenesis*, 2, 229, 1981.

170. **Warren, S., Brown, C. E., Chute, R. N., and Federman, M.**, Mesothelioma relative to asbestos, radiation and methycholanthrene, *Arch. Pathol. Lab. Med.*, 105, 305, 1981.

171. **Bonser, G. M., Faulds, J. S., and Stewart, M. J.**, Occupational cancer of the urinary bladder in dyestuffs operatives and of the lung in asbestos textile workers and iron-ore miners, *Am. J. Clin. Pathol.*, 25, 126, 1955.

172. **Seidman, H., Selikoff, I. J., and Hammond, E. C.**, Mortality of brain tumors among asbestos insulation workers in the United States and Canada, *Ann. N.Y. Acad. Sci.*, 381, 160, 1982.

173. **Rous, V. and Studeny, J.**, Aetiology of pleural plaques, *Thorax*, 25, 270, 1970.

174. **Fears, T. R.**, Cancer mortality and asbestos deposits, *Am. J. Epidemiol.*, 104, 853, 1976.

175. **Graham, S., Blanchet, M., and Rohrer, T.**, Cancer in asbestos-mining and other areas of Quebec, *J. Natl. Cancer Inst.*, 59, 1139, 1977.

176. **Polissar, L.**, The effect of migration on comparison of disease rates in geographic studies in the United States, *Am. J. Epidemiol.*, 111, 1975, 1980.

177. **Gormley, I. P., Brown, R. C., Chamberlain, M., and Davies, R.**, Report on an international workshop on the *in vitro* effects of mineral dusts, *Ann. Occup. Hyg.*, 23, 225, 1980.

178. **Davis, J. M. G.**, *Evidence for Variations in the Pathogenic Effects of Different Forms of Commercially Used Asbestos*, Institute of Occupational Medicine Report No. TM/80/4, Edinburgh, 1980.

179. **Beck, E. G.**, Experimental pathology — in vitro studies — related to asbestos and other mineral fibres, in *Biological Effects of Mineral Fibres*, Wagner, J. C., Ed., IARC Scientific Publication 30, International Agency for Research on Cancer, Lyon, 1980.

180. **Chamberlain, M., Brown, R. C., and Griffiths, D. M.**, The correlation between the carcinogenic activities *in vivo* and the cytopathic effects *in vitro* of mineral dusts, in *In Vitro Effects of Mineral Dusts*, Academic Press, New York, 1980.

181. **Gross, P. and Harley, R. A.**, Asbestos-induced intrathoracic tissue reactions, *Arch. Pathol.*, 96, 245, 1973.

182. **Pigott, G. H. and Ishmael, J.**, Toxicological assessment of potential hazards from new inorganic fibres, *J. Soc. Occup. Med.*, 29, 20, 1979.

183. **Kung Vösamäe, A. and Vinkmann, F.**, Combined carcinogenic action of chrysotile asbestos dust and *n*-nitrosodiethanolamine on the respiratory tract of Syrian golden hamsters, in *Biological Effects of Mineral Fibres*, Wagner, J. C., Ed., IARC Scientific Publication 30, Lyon, 1980.

184. **King, E. J., Clegg, J. W., and Rae, V. M.**, The effect of asbestos, and of asbestos and aluminum, on the lungs of rabbits, *Thorax*, 1, 188, 1946.

185. **Humphrey, E. W., Ewing, S. L., Wrigley, J. V., Northrup, W. F., Kersten, T. E., Mayer, J. E., and Varco, R. L.**, The production of malignant tumors of the lung and pleura in dogs from intratracheal asbestos instillation and cigarette smoking, *Cancer*, 47, 1994, 1981.

186. **Gardner, L. U.**, *First Progress Report on Asbestosis Experiments at the Saranac Lab.*, 1937, cited in King, E. J., Clegg, J. W., and Rae, V. M., *Thorax*, 1, 188, 1946.
187. **Acheson, E. D. and Gardner, M. J.**, Asbestos: scientific basis for environmental control of fibres, in *Biological Effects of Mineral Fibres*, Wagner, J. C., Ed., IARC Scientific Publication 30, Lyon, 1980.
188. **Gisser, S. D., Nayak, S., Kaneko, M., and Tchertkoff, V.**, Adenocarcinoma of the rete testis: a review of the literature and presentation of a case with associated asbestosis, *Hum. Pathol.*, 8, 219, 1977.
189. **Beck, B., Konetzke, G., Ludwig, V., Röthig, W., and Sturm, W.**, Malignant pericardial mesotheliomas and asbestos exposure: a case report, *Am. J. Ind. Med.*, 3, 149, 1982.
190. **Goldsmith, J. R.**, Asbestos as a systemic carcinogen: the evidence from eleven cohorts, *Am. J. Ind. Med.*, 3, 341, 1982.
191. **McDonald, A. D. and McDonald, J. C.**, Malignant mesothelioma in North America, *Cancer*, 46, 1650, 1980.
192. **Gloague, D.**, Asbestos — can it be used safely?, *Br. Med. J.*, 282, 551, 1981.
193. **Smith, W. E., Miller, L., and Churg, J.**, An experimental model for study of cocarcinogenesis in the respiratory tract, *U.S. A.E.C. Symposium Series*, 21, 299, 1970.
194. **Smith, J. M., Wootton, I. D. P., and King, E. J.**, Experimental asbestosis in rats. The effect of particle size and of added alumina, *Thorax*, 6, 127, 1951.
195. *Summary Proc. Biol. Effects of Man Made Mineral Fibres — Occup. Health Conf.*, World Health Organization Regional Office for Europe, 1982.
196. **Wright, G. W. and Kuschner, M.**, The influence of varying lengths of glass and asbestos fibres on tissue response in guinea pigs, in *Inhaled Particles IV*, Walton, W. H., Ed., Pergamon Press, Oxford, 1978.
197. **Pott, F.**, Some aspects on the dosimetry of the carcinogenic potency of asbestos and other fibrous dusts, *Staub.-Reinhalt Luft*, 38, 486, 1978.
198. **Wagner, J. C., Pooley, F. D., Berry, G., Seal, R. M. E., Munday, D. E., Morgan, J., and Clark, N. J.**, A pathological and mineralogical study of asbestos-related deaths in the United Kingdom in 1977, in *Inhaled Particles V*, Walton, W. H., Ed., Pergamon Press, Oxford, 1982.
199. **Jones, J. S. P., Pooley, F. D., Clark, N. J., Owen, W. G., Roberts, G. H., Smith, D. G., Wagner, J. C., Berry, G., and Pollock, D. J.**, The pathology and mineral content of lungs in cases of mesothelioma in the United Kingdom in 1976, in *Biological Effects of Mineral Fibres*, Wagner, J. C., Ed., IARC Scientific Publication 30, Lyon, 1980.
200. **McDonald, A. D., McDonald, J. C., and Pooley, F. D.**, Mineral fibre content of lung in mesothelial tumours in North America, in *Inhaled Particles V, Ann. Occup. Hyg.*, 26, 417, 1982.
201. **Wagner, J. C.**, Opening discussion in *Biological Effects of Mineral Fibres*, Wagner, J. C., Ed., IARC Scientific Publication 30, Lyon, 1980.
202. **Sigurdson, E. E., Levy, B. S., Mandel, J., McHugh, R., Michienzi, L. J., Jagger, H., and Pearson, J.**, Cancer morbidity investigations: lessons for the Duluth study of possible effects of asbestos in drinking water, *Environ. Res.*, 25, 50, 1981.
203. **Toft, P., Wigle, D. T., Méranger, J. C. and Mao, Y.**, Asbestos and drinking water in Canada, in *Water Supply and Health*, van Lelyveld, H., and Zoeteman, B. C. J., Eds., Elsevier, Amsterdam, 1981, 77.
204. **Wigle, D. T., Mao, Y., Toft, P., and Méranger, J. C.**, Cancer mortality and drinking water quality in selected Canadian municipalities, in *Proc. Workshop Compatibility of Great Lakes Basin Cancer Registries*, International Joint Commission, Windsor, 1981, 41.
205. **Polissar, L., Severson, R. K., Boatman, E. S., and Thomas, D. B.**, Cancer incidence in relation to asbestos in drinking water in the Puget Sound region, *Am. J. Epidemiol.*, 116, 314, 1982.
206. **Millette, J. R., Clark, P. J., Pansing, M. F., and Twyman, J. D.**, Concentration and size of asbestos in water supplies, *Environ. Health Perspect.*, 34, 13, 1980.
207. **Cook, P. M.**, Mineral fiber contamination of Western Lake Superior: health risk assessment, presented at American Association for the Advancement of Science, Toronto, January 6, 1981.
208. **Hilding, A. C., Hilding, D. A., Larson, D. M., and Aufdenheide, A. C.**, Biological effects of ingested amosite asbestos, taconite tailings, diatomaceous earth and Lake Superior water in rats, *Arch. Environ. Health*, 36, 298, 1981.
209. **Gross, P., deTreville, R. T. P., Cralley, L. J., and Davis, J. M. G.**, Pulmonary ferruginous bodies, *Arch. Pathol.*, 85, 539, 1968.

Chapter 8

VINYL CHLORIDE — A CANCER CASE STUDY

I. F. H. Purchase, J. Stafford, and G. M. Paddle

TABLE OF CONTENTS

I. INTRODUCTION

Vinyl chloride monomer (VCM), more properly named monochloroethene, is a colorless gas normally handled under pressure as a liquid which boils at $-14°C$ at atmospheric pressure. Discovered around 1835, VCM's commercialization did not begin until the 1930s and did not reach high volume until after 1945. Present manufacture is around 12×10^6 metric tonnes per annum (tpa), nearly all of which is used to make the polymer, polyvinylchloride (PVC).

Until the 1960s, VCM was regarded as a material of low human toxicity and the main concerns were related to VCM's narcotic effect. Indeed there are many reports of employees exposed to VCM in monomer and polymer plants becoming dizzy and unconscious. Because VCM was considered to be relatively innocuous, for many years it had a Threshold Limit Value of 500 ppm based on an 8 hr time weighted average.[1,36,58] Measurements of employee exposure were infrequent since most measurement and warning systems were designed to ensure plant atmospheres were kept below the explosive limits, fire and explosion being the main hazards of VCM. Retrospective estimates of typical time weighted average personal exposures for polymerization workers have been cited as:[3]

1945 to 1955	1,000 ppm
1955 to 1960	400 to 500
1960 to 1970	300 to 400
mid 1973	150
1975	5

However in some jobs, particularly in the cleaning of autoclaves in which VCM is polymerized to PVC, exposures in thousands of ppm were undoubtedly experienced for short periods since in some plants operators from time to time became faint and unconscious.

The first clear indication of chronic health problems associated with VCM arose in the 1960s in men who entered VCM polymerization autoclaves to remove build-up of polymer from the walls. Some of these men developed acro-osteolysis (AOL).[14,31,55] Modification of working practices led to the reduction in incidence of AOL cases in autoclave cleaners. Though AOL is occasionally seen in people not exposed to VCM,[44,71] it is a rare disease and its appearance engendered a great deal of concern in occupational physicians in Western Europe, particularly Professor P. L. Viola of the Solvay Company. In the late 1960s Viola tried to reproduce AOL in rats by exposing them to high concentrations of VCM for long periods.[63] He failed to produce AOL but he reported an increase in incidence of tumors at various sites. For the first time, it had been suggested that VCM was an animal carcinogen[64-67] Outside Western Europe, scant attention was paid to Viola's findings.

As a direct result of the Viola work, four West European VCM/PVC manufacturing companies, viz. Montedison, Rhone-Poulence, Solvay, and ICI decided to support a comprehensive study of the animal toxicology of VCM by Professor C. Maltoni, Director of the Institute of Oncology at Bologna. Maltoni's work which extended over 8 years has proved to be the most comprehensive study of VCM toxicology.[39-41] By the end of 1972 Maltoni had found the rare tumor angiosarcoma of the liver (ASL) in some of the exposed rats and confirmed that VCM is indeed an animal carcinogen. The early findings were reported at an international symposium early in 1973.[38] Maltoni recommended epidemiological investigations and medical controls of exposed workers. Early in 1974 B. F. Goodrich announced that they had found three ASL cases in employees at their Louisville PVC polymerisation plant.[16] It was now clear that VCM was a human carcinogen and gave rise to a rare tumor whose only other known etiological agents in man were thorium dioxide, arsenic and possibly anabolic steroids.[39]

The year 1974 became a watershed in the history of the VCM and PVC industry. All over the industrialised world, there was intense activity to assess whether or not the manufacture and use of 10^7 tpa of a human carcinogen could be controlled adequately. The industry was large and of clear socioeconomic importance. More or less independently in different parts of the world, the manufacturing processes were modified, hygiene was improved significantly and as a result the health of workers was safeguarded. Though superficially different hygiene standards were instituted in different countries, the technical outcome is much the same around the world — typical PVC plants now operate in the average region of 1 to 3 ppm and VCM plants around 0.5 ppm. The derivation of these standards was not based solely on an assessment of likely health effects but also on the basis of what could be achieved by the application of reasonably practical technology.

Since 1974, the health hazards of VCM have been the subject of many investigations, scientific papers, seminars, etc.[13,25,32,53,56,61] The plethora of information (and misinformation) now available makes the writing of an objective historical case study of VCM an extremely difficult task.

II. DATA AVAILABLE FOR RISK ASSESSMENT

A. Experimental Studies

There is an extensive literature on the acute and subacute effects of VCM. The principal effect seen is anesthesia which occurs at relatively high doses (7 to 10%) in both animals and man. The doses responsible for acute toxicity are so much higher than the minimum dose which can result in carcinogenicity (about 1000-fold) that there is frequently no sign of overt organ toxicity prior to the development of the carcinogenic response.

VCM is mutagenic in a variety of test systems including *Salmonella typhimurium*,[49] *Saccharomyces*,[37] and *Drosophila*[62] In these experiments it was usually necessary to use some form of mammalian microsomal metabolizing system to convert VCM into its active metabolites, chloroethylene oxide, and chloroacetaldehyde. The data on mutagenicity of VCM provides useful qualitative information on its mode of action and metabolism, but is not suitable for quantitative estimation of risk to man.

The most useful experimental data are derived from long-term animal carcinogenicity studies. An extensive series of 17 studies has been carried out by Maltoni and co-workers[41] which gives a useful data base for risk assessment. Other studies[20,35] tend to confirm the findings made by Maltoni and co-workers except that effects may have been observed with greater or lesser frequency.

In Maltoni's experiments, carcinogenic effects were observed in mice, rats and hamsters. A complication in the selection of this data for risk assessment is the variety of tumor types observed (Table 1). Some of these occurred at very high exposure levels but mammary adenocarcinoma in females and angiosarcoma of the liver (ASL) in both sexes of both rats and mice occurred at 50 ppm or below, exposures which are similar to those believed to have occurred on manufacturing plants.[3]

B. Epidemiological Studies

Several major epidemiological studies on workers exposed to VCM have been reported (Table 2). The main organs which have been associated with higher incidences of cancer in workers exposed to VCM are the liver, lung and brain. Increases in the standardized mortality ratios of cancers of the buccal cavity and pharynx, of lymphomas and the lymphatic and cardiovascular systems have been reported in one or two studies. The analysis of cancer of the respiratory system is often confounded by smoking, making quantitative analysis of the contribution of VCM difficult. The excess of liver cancers is due to an excess of ASL in many of the studies.

An analysis of the statistical power of various studies for the association between VCM

Table 1
THE LOWEST CONCENTRATIONS AT WHICH A
SIGNIFICANT EXCESS OF VARIOUS TUMOUR TYPES
WAS OBSERVED IN THE RAT CARCINOGENICITY
STUDIES[41]

Tumour type	(ppm)
Fore stomach papilloma	30,000
Zymbal gland carcinoma	10,000
Neuroblastoma	10,000
Nephroblastoma	250
	100
Liver angiosarcoma	200
	50
Mammary gland adenocarcinoma	5

exposure and cancer of the lung, liver, and brain[6] concluded that the results for liver were consistent with an etological role for VCM. For brain cancer, where 3 of 5 studies had statistically significant findings, these results were more variable, positive findings occurring in the studies with the greatest statistical power. The authors concluded that the most reasonable interpretation was that the data are consistent with a causal association between VCM exposure and an excess of brain cancer. Infante[33] in reaching the same conclusion, points out that the relative risk for brain cancer is much lower than that for liver cancer. Only 2 out of 8 studies examined by Beaumont and Breslow on lung cancer yielded statistically significant results and because studies with a high power were negative, a causal association was considered unlikely.[6]

Angiosarcoma of the liver is the most suitable endpoint for analysis of the risk of exposure to VCM for a number of reasons. It is a rare cancer in unexposed population groups making attribution to VCM exposure on the basis of work history a reasonable approach. ASL occurs in both animals and humans exposed to VCM and it is unlikely that any other carcinogenic effect of VCM will be found to occur at lower exposures than the lowest exposures which induce ASL. For these reasons the majority of work on quantitative risk assessment of chronic exposure to VCM has used ASL as the endpoint to study. For the same reasons this case study concentrates on ASL.

C. Case Register

The availability of data from a comprehensive case register of ASL cases with a history of occupational exposure to VCM provides an opportunity of identifying risk factors for the induction of ASL. This forms Section V of this case study.

III. PERSONS POTENTIALLY EXPOSED TO VINYL CHLORIDE

Vinyl chloride manufacture and its polymerization to polyvinylchloride is a large industry which employed at its peak some 50,000 workers. It has been recognized since 1974 that VCM is a human carcinogen, but because historical records were so incomplete the definition of the precise risk which is associated with exposure to a particular level of VCM has proven difficult. The result is that the hygiene standards applied in most industrialized countries (time weighted average of 1 to 3 ppm) represent an acceptable best practice rather than a judgement of the acceptability of a clearly defined risk.

Current manufacture and use of VCM and PVC results in potential exposure of four groups of the population. The highest exposure category are the workers involved in the manufacture of VCM, its polymerization to PVC and certain other industrial uses of VCM. Within this

Table 2
EPIDEMIOLOGICAL STUDIES OF CANCER ASSOCIATED WITH EXPOSURE TO VCM

Ref.	No in Study (% follow up)	Sites with changes in SMR		Comments
		Increase	**No increase**	
43	?	Brain Lung Liver including ASL		
57	8384 (85%)	Buccal cavity and pharynx Respiratory system Unknown site Lymphoma Angiosarcoma	Genital Digestive Organs Urinary Tract Leukaemia	Significant SMR not significant but increases with exposure and time
18	2120	None		Some criticism of conduct of study
46	257 (99%)	ASL		
48	594 (99%)	All tumours ?		Arsenicals involved
9	771 (97%)	Liver/pancreas cerebral? cardiovascular		Significant increase (2 ASL) Increase not significant
47	10173 (95%)	Digestive tract		Proportional mortality rate (PMR) Study
50 51 70	11,028 (90%)	Malignant liver Lymphatic system GI Tract	Brain	Related to duration of exposure
69	1151	Brain Respiratory tract Lymphatic system ASL		Mixed exposure, not VC related
23	7409 (99%)	Primary liver ASL	Stomach Brain Lymphatic and Hae-mopoetic system	Not significant Significant
24	1618 (95%)	Colon/stomach Prostatic hyperplasia		
7	5441 (86%)	All tumours		
8	464 (100%)	Respiratory system		
10 and 11	3847	Digestive system	Breast	PMR Study of Female and male Fabricators Increase in PMR Not confirmed by case-controlled study
6		Liver Brain	Respiratory	Review of 9 studies
15	10173 (95%)	Brain	Buccal cavity and pharynx Respiratory Digestive	Follow up of Tabershaw and Gaffey's study. No increase in incidence of brain cancer with increase in exposure
60	1310 (97%)	Liver	Brain Lung	
21	3232 (100%)	Lymphatic and haemopoetic systems	Liver	No ASL
59	1310 (97%)	Digestive (Liver)	Lung	Lung SMR = 42.6

group, certain occupations, particularly the autoclave cleaners, have higher potential exposure than others although all groups would now be expected to have exposures complying with hygiene standards of 1 to 3 ppm.

The next category is those exposed as a result of using the PVC. Workers in the compounding and fabrication of PVC products are exposed to residual VCM released from PVC on heating (on being heated PVC does not decompose to VCM). In general the exposure levels for these workers are very low in comparison to PVC polymerization workers (from 10 to 100 times lower).

Consumers who eat food and drink beverages which have been packed in PVC may ingest unreacted VCM which has migrated into the food or beverage. Since 1974, the amount of VCM in PVC has been reduced to less that 1 mg/kg with the result that maximum daily intake of VCM in food and drink is 0.1 μg/day.[42]

The fourth group with potential exposure to VCM are those who live in the vicinity of VCM or PVC manufacturing or fabricating factories. The levels in ambient air around a factory are very low (in the parts per 10^9 range) but much larger population groups, which include all age groups, are involved.

For the workers in VCM manufacture and PVC polymerization and fabrication the route of exposure is by inhalation. Much of the animal carcinogenicity data is based on inhalation exposure and the human epidemiology is predominantly of populations exposed occupationally by inhalation. Thus an assessment of the risk factors and the quantitative risk of inhalation exposure is the main objective. For the consumer exposed to VCM via food and beverages the route is by ingestion. There are relatively few experimental studies which use oral administration and only one study which uses a comparable exposure pattern.[20] Similarly there are no specific epidemiological data on oral ingestion. Risk assessment for exposure via the oral route must rely on the existing animal data and an extrapolation from epidemiological and experimental studies of inhalation exposure.

IV. RISK ASSESSMENT FROM EXPERIMENTAL ANIMAL DATA

A. Assumptions

In carrying out a risk assessment on the basis of animal data, a number of assumptions have to be made. The first of these relates to the overall dosimetry. Experimental animals are exposed to concentrations of vinyl chloride or dosed with amounts of vinyl chloride which allow an estimate of the amount to which they have been exposed. It is possible to calculate a correction factor for these quantities so that they are applicable to man. However, rats and mice live for relatively short periods of time (up to 2 years) during which they develop cancers of a similar type to those seen in man. The latent period for the same tumors in man may be between 20 and 40 years. It is therefore assumed that the lifetime of man is equivalent to the lifetime of an experimental animal species even though the chronological time is substantially different.

Strictly speaking, mathematical extrapolation of risk on the basis of experimental animal data provides an estimate of the risk at low doses to the experimental animal under consideration. There are a variety of factors, particularly inherent biological susceptibility and differences in metabolism, which render the extrapolation of the data from animals directly to man subject to numerous errors. It is at this point that scientific judgment is necessary to be able to decide whether this data is applicable to the human situation.

B. Metabolism

For many chemicals, toxicity and carcinogenicity are a function of the amount of an active metabolite produced rather than of the chemical itself. VCM is no exception as there is evidence that it requires metabolism to produce its toxic effects. In rats, VCM has been shown to be metabolised extensively, producing a range of excretion products. After administration by gavage or inhalation, part of the dose is exhaled unchanged and the remainder is excreted or retained in the carcass. A general scheme of its metabolism in rats is given in Figure 1.

On the basis of this scheme, the highly reactive intermediates in the metabolic process (particularly chlorethylene oxide) react with cellular macromolecules, including DNA, producing the critical lesions leading to mutation or the induction of cancer.

In a series of papers, Gehring and co-workers have described the quantitative aspects of metabolism of VCM and showed that there is a dose dependency in the rate of metabolism. After administration of ^{14}C labeled VCM by gavage at doses between 0.5 mg/kg and 100 mg/kg to Wistar rats, the amount excreted in urine and faeces and that retained in the carcass was estimated over 72 hours.[68] As the dose of VCM was increased, the proportion exhaled increased and that excreted in the urine and feces decreased (Figure 2). The proportion retained in the carcass also decreased.

The same general phenomenon occurred after administration by inhalation although the magnitude of the differences in retention and excretion was less.[68]

In a further series of experiments, Gehring and co-workers (1978) studied the amount of non-volatile material retained in the carcasses of rats exposed to various levels of ^{14}C labelled VCM for six hours. Analysis of the exposure and retention data demonstrated that the metabolism of VC appeared to be in accordance with Michaelis-Menten kinetics. They were able to calculate the constants for maximum velocity of metabolism (V_m in μg metabolised/ 6 hrs) and the Michaelis constant (K_m in μg VCM/litre air). Thus the formula:

$$V = \frac{V_m S}{K_m + S}$$

(where V = velocity of metabolism in μg/6 hrs; S = concentration of VCM being inhaled) describes the relationship between VCM inhaled and VCM metabolised. The values for the constants were estimated as V_m = 8558μg metabolised/6 hrs and K_m = 860μg VCM/litre air. The importance of this observation is that there was a considerable change in the ratio of administered dose to metabolised dose as the concentration of exposure increased (Table 4). At the higher doses a smaller proportion of VCM was metabolised than at the low doses.

C. Review of Earlier Calculations of Risk

There have been a number of attempts to calculate the risk of developing angiosarcoma of the liver on the basis of extrapolation from experimental data. These have been reviewed by Barr and an adaptation of his data is presented in Table 3.[4] All of the published quantitative risk assessments are based on Maltoni's data, combined sometimes with data from epidemiological observations. The early reports of Maltoni's data that were available were of incomplete experiments, and only in about 1980 were the complete results of his experimental series published. Thus the early risk calculations published in the 1970s are on the basis of incomplete data.

The observation by Gehring and co-workers that the quantitative metabolism of VCM was dose-dependent, suggested that earlier attempts to calculate risk on the basis of applied dose (which assumed a linear relationship between applied dose and effective tissue dose over the whole dose range studied) were subject to some error. The introduction of biotransformation data into the estimation of risk increased the exposure which was calculated to cause a 10^{-6} lifetime risk from parts per billion to in excess of one part per million. A

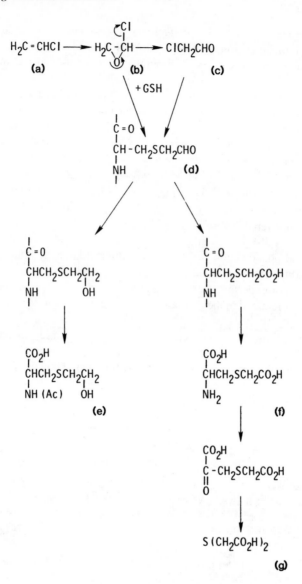

FIGURE 1.   Scheme showing the metabolism of VCM in rats to S-containing metabolites. VCM (a) is converted into chlorethylene oxide (b) which is transformed spontaneously to chloracetaldehyde (c). These two metabolites are mutagenic and hence considered as the proximate carcinogens. The urinary excretion products N-acetyl-s-(2 hydroxyethyl) cysteine (e) and thiodiglycollic acid (g) are derived from these mutagenic metabolites via (d) and in the case of S-(carboxymethyl)cysteine (f). After Green and Hathaway (1977).

further refinement of the technique which used DNA binding as the measure of dosimetry provided a similar estimate of the exposure.[2]

There is a variety of mathematical models which can be used for extrapolating below the experimental dose range. It is not possible to select from amongst these mathematical models on the basis of goodness of fit to experimental data. Attempts to do so have shown that most of the models fit the data equally well.[28] It is equally difficult to select amongst the models on the basis of the assumed mechanism of action of VCM. This is well illustrated by a comparison of the lifetime risks from calculation by the Armitage-Doll multistage model by the Food Safety Council[22] and Gaylor and Kodell.[26] For the same 10^{-6} lifetime risk, the

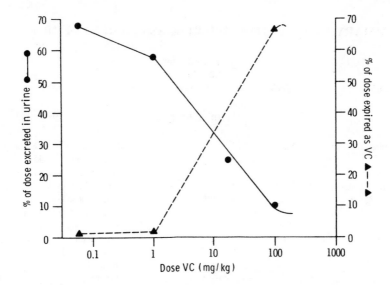

FIGURE 2. Summary of dose-dependent excretion of VCM ●————● urinary excretion. ▲————▲ pulmonary elimination. Urinary excretion represents metabolites of VCM while pulmonary elimination is unchanged VCM. (After Watanabe and Gehring, 1976.)

Food Safety Council estimated the dose as 2×10^{-2} ppm whereas Gaylor and Kodell estimated the dose as 5.2×10^{-4} ppm. The difference between these two estimates is due to alternative assumptions on the value of the expansion of the exponential term used.

Some of the studies, particularly the early ones, have been criticized on a variety of grounds. The use of selected data from Maltoni's experiment, the assumptions about the proportion of cancers represented by ASL and the use of selected data from epidemiological studies are all criticisms that have been made.

In general, calculations based on the amount of material metabolized or using human data have produced exposure values of about 1 ppm for a 10^{-6} lifetime risk. All the other studies have produced exposure values in the ppb range. A large variable appears to be the selection of the mathematical model applied to the experimental data. In the following section two models are used to calculate the exposure for a 10^{-6} risk from a variety of experimental animal data applying the correction for metabolism used by Gehring.

D. Calculation of Exposure for 10^{-6} Risk

Maltoni and co-workers have carried out 15 experiments in rats including 14 dose levels administered to Sprague Dawley rats 4 hours per day for 5 days per week for 52 weeks.[41] A summary of the crude incidence rates is given in Table 4. Similar data on Wistar rats exposed by inhalation (Table 5), rats administered VCM by ingestion (Table 6), and mice administered vinyl chloride by inhalation (Table 7) are also presented. Data from experiments with various exposure periods of short duration are given in Table 8.

For the calculation of dose metabolized in rats in the inhalation experiments, the formula described by Gehring and the constants used by him have been applied.[27] In the case of the experiment with Wistar rats, the K_m and V_m values derived for Sprague Dawley rats have been used. These estimates of metabolized dose have been included in the tables.

For the experiment in which VCM was given by gavage, the data from Figure 2 were used to estimate the amount of VCM exhaled unchanged. As even for the high dose animals the $t^{1/2}$ for exhalation of VCM was 14 minutes, these data based on a 72 hour period give a good estimate of the fraction of VCM exhaled in the 24 hours between doses. It has been

Table 3
SUMMARY OF QUANTITATIVE RISK ASSESSMENTS FOR VCM

Author	Species	Exposure for 10^{-6} lifetime risk (ppb)	Comments
By Inhalation			
Schneiderman et al. (1975)	Rat	73	Probit (slope = 1, Mantel)
		119	Logit (slope = 3.45)
		2	Logit (slope = 2.3,1 hit)
Kuzmack and McGaughy	Rat	14	Linear through zero
(1975)	Man	140 to 1400	Log — probit
Gehring (1975)	Rat	>1000	Biotransformation data included
	Man		Linear or log — probit
Food Safety Council (1978,	Rat	20	One — hit
1980)		20	Armitage Doll
		2.1×10^{-6}	Weibull
		3.9×10^{-7}	Multi-hit
Anderson (1980)	Rat	>1000	DNA binding used for dosimetry
	Man		
Gaylor and Kodell (1980)	Rat	0.7	Upper 97.5% confidence limit of linear model
		0.5	Armitage Doll
Carlborg (1981)	Rat	2.5×10^{-5}	Weibull
Barr (1982)	Man	>1000	Derived from Barr's negative epidemiology
This paper (Table 9)	Rat	0.025 to 9.16	Log — probit
	Mouse	2×10^{-15}	
	Man	0.63 to 90	Log — probit including biotransformation data for man
	Rat	$2 \times 10^{-3} - 2 \times 10^{-8}$	Weibull
	Mouse	6×10^{-43}	
	Man	0.067 to 8.14	Weibull including biotransformation for man
By Ingestion			
EPA (1980)	Rat	4μg/day	Food or water
NAS (1980)	Rat	3×10^{-5}mg/kg/day	Water
Crump and Guess (1980)	Man	0.7μg/day	Applying worker data to water
	Rat	0.5μg/day	Upper 95% confidence limits

From Barr, J. T., Presented at the 75th annual meeting of the Air Pollution Control Association, 1982. With permission.

assumed that the VCM not exhaled was metabolized, an assumption similar to the one used for estimating metabolized dose in the inhalation experiments. There is little comparable data on the metabolism of VCM by Wistar rats. Green and Hathway showed that VCM administered by gavage was exhaled and metabolized in a similar manner to the Sprague Dawley rats used by Gehring.[30] The V_m and K_m values derived from Sprague Dawley rats have been used. In the experiments using Wistar rats, the same assumptions about V_m and K_m have been made.[20] The quantity of VCM administered has been dealt with as if it was administered by gavage.

For mice, the data from Maltoni's series and from the experiments by Lee et al. have been combined in Table 7 in spite of the fact that the latter experiments were substantially shorter than those described by Maltoni.[35] The estimation of the dose metabolized in mice

Table 4
VINYL CHLORIDE DOSE AND INCIDENCE OF HEPATIC ANGIOSARCOMA IN SPRAGUE DAWLEY RATS EXPOSED 5 DAYS/ WEEK FOR 52 WEEKS

Concentration (ppm)	Amount metabolized		Angiosarcoma incidence (%)			Experiment No.[41]
	µg/4hr	µg total	Male	Female	Mean	
30,000	5647	1.47×10^6	16.6	43.3	30.0	BT 6
10,000	5521	1.44×10^6	10.0	13.3	11.7	BT 1
6,000	5403	1.41×10^6	10.3	33.3	22.0	BT 1
2,500	5030	1.3×10^6	20.0	23.3	21.7	BT 1
500	3413	8.8×10^5	0	20.0	10.0	BT 1
250	2435	6.3×10^5	3.4	6.7	5.1	BT 1
200	2129	5.5×10^5	11.7	8.3	10.0	BT 2
150	1761	4.6×10^5	1.7	8.3	5.0	BT 2
100	1309	3.4×10^5	0	1.7	0.8	BT 2
50	739	1.9×10^5	1.1	7.2	4.2	BT1,9
25	395	1.0×10^5	1.7	6.7	4.2	BT 15
10	169	4.4×10^4	0	1.7	0.8	BT15
5	84	2.2×10^4	0	0	0	BT15
1	17	4.4×10^3	0	0	0	BT15
0	0	0	0	0	0	BT1,2,9,15

Note: Experiment BT6 ended after only 68 weeks, the rest were all approximately 140 weeks; therefore % tumours in BT6 probably low relative to the rest because of short latency period available.

has been calculated using values for V_m that have been adjusted on the basis that, for a chemical which requires metabolism to its active form, the quantity metabolized will be proportional to the body surface area and must be expressed in terms of metabolized dose/ kg body mass. This technique has also been used by Gehring et al. for estimating the dose metabolized by man.[27]
Thus

$$V_m(\text{mouse}) = V_m(\text{rat}) \times \frac{0.011\text{m}^2}{0.045\text{m}^2}$$

$$= 5706\mu\text{g/4hr} \times \frac{0.011}{0.045}$$

$$= 1395\mu\text{g/4hr}$$

The values of 0.045 m^2 and 0.011 m^2 are the body surface area of a rat and a mouse respectively. Since toxicity is a function of the concentration of the toxic metabolite in tissue, the amount transformed must be normalized for mass to estimate an equivalent response. Thus V_m must be adjusted on the basis of the body weights of a rat (0.25 kg) and a mouse (0.03 kg) by dividing by $0.03/.25 = .12$. Thus the V_m for the mouse on a mass equivalent basis is

$$\frac{1395}{.12} = 11625\text{mg/4hr}$$

This value of V_m has been used in calculating the total amount of VCM metabolized (Table 7).

Table 5
VINYL CHLORIDE DOSE AND INCIDENCE OF HEPATIC ANGIOSARCOMA IN WISTAR RATS EXPOSED 5 DAYS/WEEK FOR 52 WEEKS

Concentration (ppm)	Amount metabolized μg/4hr	Amount metabolized μg total	Hepatic angiosarcoma (%) Male	Hepatic angiosarcoma (%) Female	Hepatic angiosarcoma (%) Mean	Experiment No.[41]
10,000	5521	1.4×10^6	29.6	—	—	BT 7
6,000	5403	1.4×10^6	11.5	—	—	BT 7
2,500	5030	1.3×10^6	12.0	—	—	BT 7
500	3413	8.8×10^5	10.7	—	—	BT 7
250	2435	6.3×10^5	3.7	—	—	BT 7
50	739	1.9×10^5	0	—	—	BT 7
1	17	4.4×10^3	0	—	—	BT 17
0	0	0	0	—	—	BT 17

Table 6
VINYL CHLORIDE DOSE AND INCIDENCE IN RATS ADMINISTERED VINYL CHLORIDE BY GAVAGE OR INGESTION

Dose admin. (mg/kg)	% exhaled[a]	Amount metabolized μg dose[b]	Amount metabolized μg total	Angiosarcoma incidence (%) Male	Angiosarcoma incidence (%) Female	Angiosarcoma incidence (%) Mean	Exp. No.[41]
50[c]	50	6,250	1.6×10^6	20	22.5	21.2	BT 11
16.65	35	2,705	7.0×10^5	10	15.1	12.5	BT 11
3.33	10	750	2.0×10^5	0	0	0	BT 11
1.0	2	3,245	7.26×10^4	1.3	2.7	2.0	BT 27[f]
0.3	1.7	74	2.16×10^4	0	1.4	0.7	BT 27
0.03	1.4	7.4	2.16×10^3	0	0	0	BT 27
0	—	0	0	0	0	0	BT 11,27
300[d]	80	15,000	6.2×10^6	49	53	51	
14.1[e]	32	2,390	1.65×10^6	49	16	32	Feron
5.0	16.5	1,040	7.25×10^5	10	4	7	et al.[20]
1.7	2	420	2.9×10^5	0	0	0	
0	69	0	0	0	0	0	

[a] Calculated from data derived from Watanabe and Gehring presented in Figure 1.
[b] Assuming a 250g rat.
[c] Sprague Dawley rats dosed by gavage in corn oil 5x weekly for 52 weeks.
[d] Wistar rats used as controls by Feron et al. and dosed for 83 weeks.
[e] Wistar rats receiving a diet containing VCM dissolved in PVC.
[f] BT27 dosed for 59 weeks.

From the variety of models used for low-dose risk extrapolation (Table 3) an arbitrary choice of models has been made to test the robustness of the extrapolation from the different animal studies. A log-probit analysis of the dose which would be expected to produce a lifetime risk of 10^{-6} of ASL is presented in Table 9a. This calculation can be carried out on the basis of the concentration inhaled, the daily dose metabolized or the total quantity metabolized during the whole experiment. As can be seen from the figures in Table 9a, there is a wide variation in the estimated dose depending on the data base used for calculation. The largest variation between doses derived from the rat experiments is 360-fold (0.025 ppb vs. 9.1 ppb) when exposure in ppb is considered, but this decreases to 100-fold for other estimates of dose. The results from mice are substantially lower when expressed in ppb (2 $\times 10^{-15}$ ppb) but less so for other expressions of dose.

Table 7
VINYL CHLORIDE DOSE AND INCIDENCE OF HEPATIC ANGIOSARCOMA IN MICE

Concentration (ppm)	Amount metabolized		Hepatic angiosarcoma (%)			Experiment No.[41]
	µg/4hr	µg total	Male	Female	Mean	
10,000	11245	1.7×10^6	3.8	30	17.8	BT 4[a]
6,000	11007	1.7×10^6	6.7	36.7	21.7	BT 4
2,500	10246	1.5×10^6	20.7	33.3	27.1	BT 4
1,000	8699	3.4×10^6	39.4	50.0	44.7	Lee et al.[b]
500	6952	1.0×10^6	20.0	26.7	23.3	BT 4
250	4959	7.4×10^6	30.0	30.0	30.0	BT 4
250	4959	7.4×10^6	24.0	47.0	36.5	Lee et al.[b]
50	1506	2.2×10^5	3.3	0	1.7	BT 4
1	1506	5.9×10^5	10.3	0	5.2	Lee et al.[b]
0	0	0	0	0	0	BT 4 and Lee

[a] Swiss mice, 81 week experiment, dosed for 30 weeks.
[b] CD_1, 52 week experiment, exposed 6 hr/day (Lee et al.[35]). These results have not been included in the calculations for Table 9 because the experimental design incorporated interim sacrifices.

Table 8
VINYL CHLORIDE DOSE AND HEPATIC INCIDENCE IN SPRAGUE-DAWLEY RATS ADMINISTERED VINYL CHLORIDE BY INHALATION

Conc. (ppm)	Sched.[a]	No. of doses	Amount metabolized[b]		Angiosarcoma incidence (%)			
			µg/4hr	µg total	Male	Female	Mean	Exp. no.[41]
10,000	I	260	5521	1.4×10^6	10	13.3	11.7	BT 1
10,000	II	85	5521	4.7×10^5	0	0	0	BT 3
10,000	III	25	5521	1.4×10^5	1.7	0	0.8	BT 10
10,000	IV	100	1379	1.4×10^5	1.7	0	0.8	BT 10
10,000	V	25	5521	1.4×10^5	0	1.7	0.8	BT 10
6,000	I	260	5403	1.4×10^6	10.3	33.3	22.0	BT 1
6,000	II	85	5403	4.6×10^5	0	3.3	1.7	BT 3
6,000	III	25	5403	1.4×10^5	0	0	0	BT 10
6,000	IV	100	1350	1.4×10^5	3.4	1.7	2.5	BT 10
6,000	V	25	5403	1.4×10^5	0	1.7	0.8	BT 10

[a] Schedules: I, 4 hr/day, 5 days/week, 52 weeks. II, 4 hr/day, 5 days/week, 17 weeks. III, 4 hr/day, 5 days/week, 5 weeks. IV, 1 hr/day, 4 days/week, 25 weeks. V, 4 hr/day, 1 day/week, 25 weeks.
[b] Amount metabolized (v) in 4 hr derived from the formula: $V(\mu g/hr) = \dfrac{V_m \times S}{K_m + S}$ where Vm is 4/6 of the 6 hr value.

Similar calculations based on a Weibull analysis have been carried out and the doses calculated to give a 10^{-6} lifetime risk are presented in Table 9b. This is a more conservative mathematical model and the estimates of doses are accordingly lower. The variation in estimates of dose is, if anything, larger than that observed with the log probit analysis (for example 10^{-5} difference between S derived from Wistar and Sprague Dawley rats). The doses for mice are so much lower than those calculated for rats or man that the assumptions used in their calculation must be suspect.

A further calculation which uses the transformation derived by Gehring to derive the human dose likely to produce a risk of 10^{-6} is given in Table 9 (S calculated for man).[27] These calculations are based on a V_m for man of 1675 mg/8 hr based on corrections for body surface area and mass. The values are substantially higher than those calculated for the rat and mouse and there is still a range of over 100-fold in the estimates derived from the different rodent experiments. When this amount of variability occurs in extrapolation of the risk of low dose exposure of VCM based solely on different experiments in the same species, the reliability and hence utility of these procedures is open to question.

In Maltoni's series of experiments variations in the frequency and duration of doses administered by inhalation to rats are also reported. It can be seen from Table 8 that the general relationship between dose administered and the incidence of angiosarcomas which has been derived from the 52 weeks exposure does not apply to exposures of shorter duration. It is however, noteworthy that in all the experiments, a total metabolized dose in excess of 5×10^5 µg was required to produce an incidence of angiosarcoma in excess of 1 to 2%. This relationship was seen in rats and mice and in experiments where VCM was administered by gavage or by inhalation. In these long-term inhalation studies, a total metabolized dose of 5×10^5 µg is equivalent to about 200 ppm administered over 52 weeks, and represents a practical threshold for this series of experiments.

In conclusion, it can be seen that there is a wide variation in the estimates of dose for a 10^{-6} lifetime risk. This variation is due both to the type of mathematical model which is applied, the assumptions which are made and the particular experiment which is used to provide data for the extrapolation. A high level of confidence cannot be placed on such low dose extrapolations when variables which would not be expected to alter the expression of risk have a profound effect on the estimated risk. In addition to this, the inter species extrapolation from experimental animals to man is largely intuitive. It is clear that estimates of risk should take into account all available data including epidemiology to provide a degree of reliability.

V. RISK ASSESSMENT FROM HUMAN STUDIES

A. Register of Cases of Angiosarcoma of the Liver

Since 1974, lists of reported ASL cases attributable to VCM exposure in the VCM/PVC industry have been kept by NIOSH,[54] IARC and by the VCM committee of the Association of Plastics Manufacturers in Europe (APME). It is believed that the most exhaustive compilation is that of the APME which since 1976 has been maintained by Dr. J. Stafford of ICI, who has published and circulated it to interested parties in universities, government, trade unions, research organizations, and industry. A paper based on this register of ASL cases is being prepared for publication.

Details of 99 cases in the register at the end of 1982 have been tabulated from information provided by the individual VCM/PVC manufacturing companies, by trade associations like the Japanese PVC Association, by organizations like IARC, NIOSH, and the Berufsgenossenschaft (German Industrial Injuries Insurance Institute) and by individual research workers in the occupational health field. The 99 cases have been analyzed by country and by manufacturing company and plant. Analysis of these data reveals several interesting features. The cases have been recorded from all major VCM/PVC manufacturing countries (Table 10) but not necessarily in proportion to the PVC production capacity now or prior to 1962. In the absence of data on the number of workers employed, production capacity is the only available indication of the numbers of people potentially exposed. (It is worth noting that there is a paucity of data from USSR and some other Eastern Bloc countries).

The majority of the ASL cases are PVC autoclave cleaners or men who have worked in or around autoclaves. There are three ASL cases among men who manufactured VCM and a few cases were involved both with monomer and with polymer production. It is interesting to note that only one case suffered both from acro-osteolysis and ASL, which may be considered surprising in the light of the view that both conditions result from high exposures.

Table 9a

QUANTITATIVE RISK ESTIMATIONS DERIVED FROM AVAILABLE ANIMAL CARCINOGENICITY DATA AND EXPRESSED AS AMOUNT OR CONCENTRATION OF VCM CALCULATED TO GIVE A LIFETIME RISK OF 10^{-6} BASED ON LOG PROBIT ANALYSIS, ESTIMATED USING MAXIMUM LIKELIHOOD.

Experimental data	Exposure for rodents (S ppb)	Amount metabolized in 6 hrs (V µg/6 hrs)	Total amount metabolized (TM mg)	Exposure in ppb calculated from V (S calculated for man)[a]
Sprague Dawley Rats inhalation (Table 4)	0.025	1.23	0.305	0.63
Wistar rats, male only inhalation (Table 5)	9.16	159	39.3	90
Rats, ingestion Wistar (Table 6) S-D Wistar and S-D combined	3×10^{-5} 9×10^{-4} mg/kg 6×10^{-4}	0.69 2.19 mg/dose 1.70	2.27 0.2 0.88	—
Mice, inhalation (Table 7 BT4 only)	2×10^{-15}	0.60	0.0063	0.03
Wistar and Sprague Dawley Rats combined, inhalation (Tables 4 and 5)	0.038	1.41	0.35	0.72
Sprague Dawley Rats Short-term inhalation (Table 8)	—	0.004	2.86	—

[a] Exposure calculated from V (in column 3) using the formula $S = V \times \dfrac{860_m}{1675 - V} \times S$ where V_m for man is 1675 µg/8 hrs.

Table 9b
QUANTITATIVE RISK ESTIMATIONS DERIVED FROM AVAILABLE ANIMAL CARCINOGENICITY DATA AND EXPRESSED AS AMOUNT OR CONCENTRATION OF VCM CALCULATED TO GIVE A LIFETIME RISK OF 10^{-6} USING A WEIBULL DISTRIBUTION[a]

Experimental data	Exposure (S ppb)	Amount metabolized in 6 hrs (V µg/6 hrs)	Total amount metabolized (TM mg)	Exposure in ppb calculated from V (S calculated for man)[b]
Sprague Dawley Rats inhalation (Table 4)	2×10^{-8}	0.013	0.0032	0.067
Wistar rats, male only inhalation (Table 5)	2×10^{-3}	15.7	3.68	8.14
Rats, ingestion (Table 6)				
Wistar	9×10^{-14} mg/kg	3×10^{-6} mg/dose	0.0002	
Sprague Dawley	4×10^{-6} mg/kg	0.33 mg/dose	0.003	
Wistar and Sprague Dawley combined	2×10^{-8} mg/kg	0.005 mg/dose	0.0015	
Mice, inhalation (Table 7 BT4 only)	6×10^{-43}	2×10^{-5}	2×10^{-9}	1×10^{-5}
Wistar and Sprague Dawley Rats combined, inhalation (Tables 4 and 5)	6×10^{-8}	0.0172	0.0042	0.009
Sprague Dawley Rats Short-term inhalation (Table 8)	—	3×10^{-6}	0.19	

a   Estimated by maximum likelihood as a generalized linear model.

b   Exposure calculated from V (in column 3) using the formula $S = V \times \dfrac{860 \times S}{1675 - V}$ where V_m for man is 1675µg/8 hrs.

Table 10
DISTRIBUTION OF ASL CASES BY COUNTRY

Country	No. of ASL cases	PVC production nameplate capacity (in kilotonnes per annum)		
		1952	1962	1972
USA	29	193	704	2,090
W. Germany	21	22	260	1,155
France	14	11	176	627
Canada	10	5	22	88
UK	7	27	177	502
Sweden	5	3	20	105
Yugoslavia	4	3	8	60
Italy	3	9	212	778
Czechoslovakia	2	1	25	48
Japan	2	12	384	1,699
Belgium	1	3	25	195
Norway	1	2	20	65
Total	99			
Western Europe	52	82	951	3,950
North America	39	198	726	2,178
Rest of World	8	51	709	3,334
Total	99	331	2,386	9,462

A larger proportion of the PVC plants in Western Europe recorded cases of ASL than was the case for North America. There are 31 pre-1962 PVC plants in Western Europe and 18 of these recorded cases of ASL. Of the 29 pre-1962 PVC plants in North America, only nine recorded cases of ASL. The ASL cases tended to occur in larger numbers in some plants than in others (Table 11). Of the total of 39 ASL cases recorded in North America, 34 have occurred at four PVC plants; over 40 North American PVC plants have not so far recorded an ASL case.

The average latent period between starting work in an occupation with VCM exposure and death from ASL for the 99 cases is 21.9 years. It is noteworthy that the average latent period in France, Sweden and USA is high (between 24 and 25 years) whereas in Germany it is low (about 18 years).

If the Maltoni studies had not alerted epidemiologists in 1973 to search for ASL cases, the accumulation of human ASL cases would probably have become apparent in the USA (and perhaps Canada) in the 1974 to 1976 period as is indicated in Table 12.

The ASL cases known to date were all first exposed to VCM before Viola and Maltoni began their animal experimental research (see Table 13). The latest dates of first exposure occur in the mid-1960s; a date which coincides with the discovery of acro-osteolysis and the beginnings of changes in procedures for cleaning autoclaves. There is as yet therefore no information to indicate whether these improvements have had any effect on the incidence of ASL in exposed populations.

It is still too early to predict whether the annual number of ASL cases amongst VCM workers has reached a peak or not. Taken at its face value, Table 14 indicates that ASL cases appeared earlier in North America than in Western Europe and that the annual number of ASL cases in North America is perhaps tending to decrease while the number in Western Europe is still high. This is a curious observation in that VCM/PVC production started earlier in Europe, particularly in Germany, but on the other hand production capacity was

Table 11
CLUSTERING OF ASL CASES IN
INDIVIDUAL PVC PLANTS

			No. of ASL cases
		West Europe	
Plant[a]	1	West Germany	10
Plant	2	West Germany	4
Plant	3	West Germany	2
Plant	4	West Germany	2
Plant	1	France	5
Plant	2	France	5
Plant	3	France	2
Plant	1	UK	5
Plant	2	UK	2
Plant	1	Sweden	5
Total			42
		North America	
Plant	1	Canada	10
Plant	1	USA	11
Plant	2	USA	9
Plant	3	USA	4
Total			34
		Rest of World	
Plant	1	Japan	2
Plant	1	Yugoslavia	4
Plant	1	Czechoslovakia	2
Total			8

[a] For the purposes of this case study, it is not necessary to identify the precise ownership and location of these plants.

built up much more quickly in the USA than Europe after World War II ended (see Table 10). It is possible that some early cases of ASL have been overlooked or misdiagnosed; however it is encouraging to note that 33 out of the 99 cases were diagnosed before the connection between ASL and VCM exposure had been discovered.

On the basis of the data in this case register, it is possible to draw certain conclusions about risk factors associated with ASL. The large number of ASL cases in some factories and the absence of ASL cases in others of similar age indicates that variations in manufacturing practices between factories may be the cause of this notable difference. These variations may reflect both differences in the types of job which individual workers carried out and differences in engineering practices. It is clear that at the time when VCM was believed to be of low mammalian toxicity, workers were undoubtedly exposed to levels which are now recognized to be extremely high. It was these extremely high exposure levels that appear to have resulted in the ASL cases. Unfortunately precise hygiene measurements were not made at the time and it is therefore difficult to quantify the dose which gives a high risk of ASL and the dose which is of low risk. The bulk of the cases have however occurred in highly exposed autoclave cleaners with relatively few in other PVC or VCM production jobs and so far no well authenticated cases in PVC compounding or fabrication where many more people were exposed but at a much lower dose.

Table 12
ASL CASE NUMBERS BY YEAR OF DEATH AND GEOGRAPHICAL LOCATIONS (EXCLUDING IT 01)[a]

Key events		Western Europe	North America	Rest of world
	1955		C1	
	6			
	7		C2	
	8			
	9			
	1960			
	1		US8	
	2		C3	
	3			
	4		US5	Cz2
	5			
	6			
	7	F1		
	8		C4,C5,US4,US7,US10	
	9	G1	US12,US16	
Viola	1970	Sw1	US11	
	1	G2	C6,US2	
	2	N1,Sw2,UK1,It2	C7	
Maltoni	3	G3[b]	C8,US1,US3,US23	Y1,Y2,Cz1
Goodrich	4	G4,G5,UK3	C9,US13	
	5	F2,F3,G6, G7,G8,It3	US6,US9,US18,US26	Jap1
	6	B1,F4,F5,F6,F7,Sw3	US19,US20,US22	Jap2,Y3
	7	F8,F9,G10,G11,G12,Sw4	C10,US21,US24[c]	
	8	F10,F11,G9,G13,G15,G16,G17	US27,US28	Y4
	9	F12,F13,UK4,UK5,G18		
	1980	UK6,UK7,G19,Sw5,G20,G21	US17,US29,US30,US32	
	1	It4,F14,UK8,G22		
	2			
Totals		52	38[d]	8

[a] Italian case 01 was not a typical ASL; his primary tumour was probably of the pericardium. This man was engaged in extrusion of PVC sacks.

[b] Aerosol can filler.

[c] Cholangiosaracoma.

[d] Does not include US31 and Alive.

Key: B = Belgium   G = W Germany   Sw = Sweden
C = Canada   It = Italy   UK = United Kingdom
Cz = Czechoslovakia   Ja = Japan   Y = Yugoslavia
F = France   N = Norway   US = USA

Thus G9 = Case No 9 in West Germany.
Cases UK2, G14, US14, 15 and 25 were shown not to be associated with VCM exposure and hence withdrawn from the list.

B. Prediction of Future ASL Cases as a Consequence of Pre-1974 Exposure

The causal relationship between VCM and ASL is proved beyond doubt by the specificity of the tumor, the high relative incidence of that tumor in highly exposed workers, the consistency of the excess in different parts of the world, the time relationship between exposure and diagnosis and the dose response relationship. As a consequence, it should be

Table 13
ASL CASE NUMBERS BY YEAR OF FIRST EXPOSURE AND GEOGRAPHICAL LOCATIONS (EXCLUDING IT 01)[a]

Key events		Western Europe	North America	Rest of World
	1939		US24[b]	
	40			
	1	Fr11	C3,US27	
	2		US13,US29	
	3	Fr14	C2,US19	
	4	UK1	C1,C5,US5,US7,US28	
	5	Sw2	C4,US3,US9	
	6	Fr1,Fr3,Sw4	C7,C9,US8,US11,US21,US31[c]	
	7	Sw3	C6,US22,US26	
	8	Fr9	US1	
	9	Fr12,Fr4	US12	
	1950	Fr7,N1,UK8	US16	Y2,Cz2
	1	Sw1,UK5	US10,US32	
	2	G3[d]	US4	
	3	G15,It3	C10	Jap1,Y1
	4	G7,G8,UK4,G19	US18	Y3
	5	G11,G16,G18	US2,US17,US20	
	6	Fr10,G1		
	7	Fr8,G4,It2,G2		Cz1
	8	Fr6,B1	US23	Jap2,Y4
	9	Fr2,It4		
	1960	G5,G13		
	1	G9,G10,G12,G17,G20,G22	C8	
	2	G6,UK6,G21	US6	
	3	Fr13,UK7		
	4	Sw5	US30	
	5	Fr5		
	6	UK3		
	7			
	8			
	9			
Viola	1970			
	1			
	2			
Maltoni	3			
Totals		52	39	8

a   It01 is not consistent with other ASL cases; the primary tumor may have been of the pericardium. The man extruded PVC sacks.

b   Cholangiosarcoma.

c   US31 is still Alive.

d   Aerosol can filler.

possible to carry out an intensive analysis of the pre-1974 cohorts to establish as far as possible the dose response curve for ASL after VCM exposure and predict the likely outcome for the future.

It will be impossible to collect a complete data set on which to calculate risks of angiosarcoma for the whole world. Within a single company however there may be closer definition of the cohort, the number of cases, and the pattern of exposure. Using these data as a base and averaging across the worldwide population exposed to VCM, it is possible to calculate

Table 14
ANNUAL INCIDENCE OF ASL CASES (DATE OF DEATH) BY GEOGRAPHICAL AREA

	Western Europe	North America	Rest of World	Annual Total	Cumulative Total	Key Events
1955		1		1	1	
7		1		1	2	
1961		1		1	3	
2		1		1	4	
4		1	1	2	6	
7	1			1	7	
8		5		5	12	
9	1	2		3	15	
1970	1	1		2	17	Viola
1	1	2		3	20	
2	4	1		5	25	
3	1	4	3	8	33	Maltoni
4	3	2		5	38	Goodrich
5	6	4	1	11	49	
6	6	3	2	11	60	
7	6	3		9	69	
8	7	2	1	10	79	
9	5			5	84	
1980	6	4		10	94	
1	4	[a]		4	98	
2	Nil to date	Nil to date	Nil to date	Nil to date		
Total	52[b]	38[a]	8	98[a]	98[a]	

[a] Does not include US31 still alive.
[b] Includes G03 aerosol can filler but omits It01 bag extruder.

the future incidence of ASL. In order to do this, some relatively crude assumptions have to be made which can only be tested in the fullness of time when the predictions can be judged against the final outcome.

Despite all the reservations about using historical exposure and epidemiological records, it is appropriate to use assessments of VCM exposures and relate them to ASL incidence, because they are among the best sets of data ever likely to be available for such purposes. To conduct a calculation of the expected number of ASL cases worldwide due to VCM exposure prior to 1974 requires very little information over and above that already in hand. The data required are

1. Annual populations of employees classified by age
2. Annual exposure estimates for each person in 1
3. An exposure response latency model for ASL induced by VCM

The data under item 1 are the raw material of any cohort study and are available in most countries and particularly in the UK as a result of the Fox and Collier study.[23] In the course of any epidemiological study, data of this type are extracted from the relevant occupational records. Exposure data for item 2 are more difficult to obtain. Measurement of atmospheric levels of VCM in the workplace have been taken for many years to ensure compliance with hygiene standards, but they have not been collected sufficiently and rigorously to permit assessment for epidemiological purposes. All that is known is that atmospheric levels in monomer plants and in drying, bagging, and packing areas of polymerization plants have always been relatively low, while those in the autoclaving area were much higher. The highest exposures occurred when men entered the autoclaves to clean them out when the autoclaves were opened for evacuation. Information on duration of exposure can be gleaned from the same records as are used to define the occupational population. The problem of changing occupation, which occurred frequently has been dealt with by using the principal employment category or the highest exposed employment category. It is not likely that the combination of exposure data and duration of exposure will provide an accurate estimate of dose such as would be obtained from personal monitoring devices now in use. Nevertheless, the estimation of time-weighted average exposures for the lowest exposed employees is straightforward as the exposures were essentially continuous and constant. For autoclave cleaners, maintenance workers, and laboratory workers exposures in the course of a single shift could vary from zero to near narcotic levels. In spite of these problems, estimates of exposure were obtained by interviewing the longest serving members of the cohort under study. In individual studies, the occupational records and exposure estimates can then be combined to derive annual exposure estimates for each individual. Although it is easy to be skeptical about these estimates, discussions between manufacturing companies in other countries have disclosed that there is a surprisingly close agreement between the figures found in spite of the different methods used to derive them. In the calculations described below, it has been possible to avoid using the exposure data directly by relying on the similarity in exposure levels in differing locations. The exposure/response/latency data indicated under item 3 can be derived from established cases.

The key data for these procedures are the set of cases worldwide together with the descriptive data contained in Table 15. As this table has built up, it has been possible to calculate an incidence rate for each latency period for each exposure level for each age group. The patterns observed in these rates can then be used to derive a simple model of dose response latency which can be applied to the population data. Certain of the broad conclusions which may be drawn from the case register are relevant. Most of the cases reported have a latency of about 20 years, but exposure levels did not fall dramatically until the 1970s. It is likely therefore that cases will continue to occur for the next 10 years. As cases with an extremely long latency in excess of 40 years have not yet had a chance to occur, the shape of the dose response latency curve for these cases is as yet unknown. This is unlikely to be an important problem because the population at risk after 50 years latency is greatly diminished due to mortality from other causes. The great majority of cases have had extensive duration of exposure at high exposure levels.

The form of calculation used to estimate the future number of ASL cases is set out in Table 15. An assumption has been made that when exposures were reduced to low levels, the future risk of ASL was negligible. Two dates at which the negligible risk levels were attained have been selected; 1964 when levels were reduced to 100s of ppm and 1974 when the levels were reduced to below 10ppm following the discovery of the association between ASL and VCM exposure. The results from calculations using both dates are given in Table 15. A hypothetical exposed population of 100,000 has been used because this may not be dissimilar from the total numbers employed in VCM manufacture and polymerization to PVC; it should be noted, however, that the number used in the calculation is unimportant (see (1) below). The component of workers employed by ICI within the Fox and Collier Study has been used to estimate the age distribution within the hypothetical total exposed population of 100,000.[23] The persons already exposed during the whole of the various latent periods are shown in the third and seventh columns, where it can be seen that the numbers with a latency of 30 years or more form only a small proportion of the total. The total number of cases seen world wide for each latency are listed in column 2 and the 5 year incidences (columns 4 and 8) are obtained by division. The numbers of persons at risk in future (columns 5 and 9) are calculated by advancing time in 5 year periods taking account

Table 15
HYPOTHETICAL CALCULATION OF FUTURE ASL CASES USING TWO DIFFERENT ASSUMPTIONS ABOUT THE DATE AT WHICH LEVELS BECAME FREE OF RISK

		Calculations assuming no risk after 1964				Calculations assuming no risk after 1974			
Latency (years)	Cases to date	Persons at risk to date	5 year incidence	Future persons at risk	Future cases	Persons at risk to date	5 year incidence	Future persons at risk	Future cases
1 to 5	0	100,000	0.00	0	0	100,000	0.00	0	0
6 to 10	1	98,250	0.01	0	0	94,500	0.01	3,750	0
11 to 15	11	95,500	0.12	0	0	78,000	0.14	17,500	2
16 to 20	28	84,750	0.33	6,750	2	46,500	0.60	45,000	27
21 to 25	28	61,400	0.46	24,550	11	28,750	0.97	57,200	57
26 to 30	18	36,750	0.49	41,750	20	18,750	0.96	59,750	57
31 to 35	6	21,250	0.28	48,100	13	10,850	0.55	58,500	32
36 to 40	6	6,750	0.89	51,750	46	3,500	1.71	55,000	94
41 to 45	0	600	?	45,750	?	310	?	46,350	?
46 to 50	0	0	?	34,500	?	0	?	34,500	?
51 —	0	0	?	47,600	?	0	?	47,600	?
16 —			0.50	300,750	150		0.80	403,900	323

For details of the assumptions and methods see text Section V.B.

of the age dependent death rates in the population at large. Death rates for an intermediate year for the male population of England and Wales have been used in this calculation and the numbers have been rounded because of the approximate nature of the model. Future cases (column 10) are then obtained by multiplication. The incidence figures for long latent periods (> 25 years) are unreliable or nonexistent. However, the incidence estimates (columns 4 and 8) are fairly constant for latencies of 15 years or more and therefore values of 0.5 and 0.8 cases/1000 persons have been used for all latency periods over 15 years to calculate the expected number of cases. The calculation displayed in Table 15 is clearly unrealistic in many respects but most if not all of the simplifications are unlikely to affect the estimate of future cases by more than a small factor. For example:

1. The population size used for the calculation is probably larger than the exposed population. This is of little importance because the calculation depends on the ratio of "person-years to come" and "person-years experienced" and this ratio is the same for any population size.
2. Exposure level has been ignored. The calculations are based on the overall risk to the cohort. Sub-cohorts may well have higher risks (as is suggested in Section V) and the calculations could be repeated for these high-risk, highly-exposed sub-cohorts. The incidence figures for the sub-cohort would be higher but, as is explained in (1.) above, the estimate of future cases will change very little. Similarly duration of exposure has been ignored even though it is doubtless an important variable. The incidence data are insufficient to justify any sub-division by both latent period and duration of exposure.
3. ICI is not typical of the world-wide growth in the exposed population. This explains the high incidence figure for the latency of 35 to 40 years which is derived from the total number of cases (most of which occurred in the USA) divided by a population estimated from UK exposed persons. Undoubtedly the pattern of cases will vary from one location to another, but as it is not possible to obtain age and exposure data for

workers at each location, the calculation of total future cases must be subject to some error.

4.   No account has been taken of plant improvements occurring prior to 1964 and hence fewer cases may occur in, for example, the 1980 to 2000 period than are estimated from 1940 to 1980 experience.

An assumption that the risk of ASL ceased in 1964 rather than 1974 results in a considerable reduction in the estimate of future cases. For either assumption, the number of new cases observed annually should soon begin to decline and the rate of decline will indicate which assumption is nearer to the truth.

C. Risk Assessment on the Basis of Negative Epidemiology

One of the most frequent limitations of occupational epidemiological studies is that the limited number of people exposed occupationally limits the statistical power of the study. In the case of ASL, however, which has such a low natural incidence, it is possible to calculate the upper bounds of risk on the basis of estimates of exposure of large population groups in the vicinity of VCM and PVC manufacturing sites. Barr[4] has made such an estimate. In his paper he states:

One further evaluation of human risk can be made from the experience of persons residing near VCM/PVC plants. The EPA estimated (Kuzmack and McGaughy, 1975) that five million persons lived within five miles of these plants and were exposed to an annual average concentration of 17 ppb. The present distribution of plants was generally well established by 1959, thus we have 22 years of history or about 110 million person years. About five or six of these plants, with 1-2 million neighbours, go back another 20 years but these data are not firm enough for inclusion. The fact that no case of ASL has been confirmed as arising from these ambient exposures places the upper bound* of risk at less than 2.7×10^{-7}/ppm/year. It is believed that the exposure data were over estimated by EPA, and thus this result may be too low but it is in the same general range as that arrived at by Gehring (1979) and Anderson (1980) after making corrections for pharmacokinetics. Extending this crude calculation, these five million persons are now supposed by EPA to be exposed to 0.2 ppb (probably a high figure), which would predict no more than 0.0003 deaths per year, or one per 3,700 years in that whole population due to VCM exposure. But it also must be recognized that with approximately 20 cases per year of ASL in the general population, there can be expected from a purely statistical basis that there should be one case every two years or so among this group of five million plant neighbours.

The absence of cases of ASL in populations residing in the vicinity of VCM/PVC production facilities has been confirmed in the UK, where an extensive study of ASL cases failed to demonstrate any nonoccupational association with ASL.[5] These observations suggest that man is less susceptible to low doses of VCM than are rodents or, expressed differently, that the slope of the dose-response curve in man is much steeper than in rodents. Thus on the basis of the observations of ASL incidence in the USA, the estimated exposure for a lifetime risk (40 years exposure) of 10^{-6} was greater than 100 ppb.

VI. SUMMARY AND CONCLUSIONS

There is little doubt that exposure to high levels of VCM as a consequence of occupation can result in increased incidence of angiosarcoma of the liver. A review of 20 epidemiological studies involving about 45,000 workers occupationally exposed to VCM showed that neoplasms of the liver were increased in incidence in the majority of studies. For brain cancer the association between exposure to VCM and increased incidence was less clear because of the lower relative risk. Neoplasms of the respiratory tract, digestive system, lymphatic and hemopoietic system, buccal cavity and pharynx, cardiovascular system, colon/stomach were reported to have an increased incidence in one or more studies, but were not increased, or in some cases were decreased, in incidence in other studies. In view of the increased

* Upper bound calculated by assuming that the reported zero cases were at most 0.5 cases.

incidence of breast neoplasms in rodents exposed to VCM, the studies of Chiazze et al.[11] which did not confirm these findings in man are of crucial importance.

The register of ASL cases now contains records of 99 persons with confirmed ASL and occupational exposure to VCM. The average latent period between first exposure and death from ASL is 21.9 years. The majority of cases occurred in autoclave workers, who are recognized to have been exposed to extremely high levels. Although precise estimates of exposure are not available for the periods of most interest, the pattern of cases roughly suggests that extremely high exposures by modern standards were necessary for the induction of ASL. For example, ASL cases tended to occur in larger numbers in some plants than in others which can be explained most easily by differences in exposure patterns.

There is an extensive series of animal studies on the carcinogenicity of vinyl chloride some of which precede the epidemiological studies confirming the association between VCM exposure and ASL in man. ASL and neoplasms of a number of other organs have been induced by VCM in laboratory rodents. Estimation of the exposure levels likely to cause a lifetime risk of ASL of 10^{-6} on the basis of these data give extremely low levels (down to 3.9×10^{-7} ppb) which appear to be unrealistic estimates for man. Part of the reason for this is that laboratory studies have shown that VCM is metabolized in the liver (and elsewhere in the body) to the reactive metabolites chloroethylene oxide and chloroacetaldehyde. The rate of conversion is limited at high levels of exposure giving inaccurate estimates of the slope of the dose/response relationship. It has not been possible to estimate the rate of conversion in man, and hence extrapolation of these low risk dose estimates is conjectural. The second part of the problem of extrapolation to low risk is the selection of the most suitable mathematical model for extrapolation. Using Maltoni's data from rats, there is a substantial range (up to 10^8) of low-risk dose estimates which depend on the mathematical model and the assumptions used in applying the models. Using the same (probit and log dose) model and different sub-sets of the experimental data, again a large range of estimates is obtained even after correcting for the non-linear kinetics of metabolism at high dose (which reduces this range to about 10^2). Larger differences are obtained when the calculations use Weibull analysis as a basis of low dose estimation, suggesting that this is a problem with the use of mathematical models rather than being associated only with the log probit analysis. Although there is considerable variability in the dose response relationship in the different experiments reported, in all a total metabolized dose of $5 \times 10^5 \mu g$ (equivalent to lifetime inhalation of 200 ppm) was required to produce an elevation in ASL incidence; this dose represents a practical threshold in rodents. At this stage in their development, mathematical models for low-risk dose estimates are not sufficently reliable or reproducible to engender confidence in their use.

Using negative epidemiological studies of populations living in the vicinity of VCM production facilities, an estimate of the dose for a 10^{-6} lifetime risk in man may be made. The value (100 ppb) is similar to the highest estimates based on animal data and including biotransformation but is substantially larger than the lowest estimates which are up to 10^{10} lower (3.9×10^{-7} ppb using a multi-hit model). These higher estimates are compatible with occupational experience and suggest that the current hygiene standard of around 1ppm is sufficiently low for protecting the health of VCM/PVC workers and the estimates give a considerable safety factor for the general public consuming PVC packed food and drink or living near VCM/PVC facilities.

It has been possible to provide a crude estimate of the number of cases of ASL which may occur in the future from exposure to VCM prior to 1974. Using the age structure of employees in one company the total number of cases of ASL reported to date and the mortality pattern expected from a normal population, the possible future number of ASL cases has been estimated.

ACKNOWLEDGMENT

We thank Dr. M. Thomas for his help with the calculations.

REFERENCES

1. Threshold limit values for chemical substances and physical agents in the workroom environment with intended changes for 1974. American Conference of Governmental Industrial Hygienists, Cincinnati, Ohio, 1974.
2. **Anderson, M. W., Hoel, D. G., and Kaplan, N. L.,** A general scheme for the incorporation of pharmacokinetics in low-dose risk estimation for chemical carcinogenesis: example — vinyl chloride. *Toxicol. Appl. Pharmacol.,* 55, 154, 1980.
3. **Barnes, A. W.,** Vinyl chloride and the production of PVC., *Proc. Royal Soc. Med.,* 69, 277, 1980.
4. **Barr, J. T.,** Risk assessment for vinyl chloride in perspective. Presented at the 75th annual meeting of the Air Pollution Control Association, New Orleans, June 1982.
5. **Baxter, P. J., Anthony, P. P., MacSween, R. N. M., and Scheuer, P. J.** Angiosarcoma of the liver: incidence and etiology in Great Britain. *Brit. J. Ind. Med.,* 37, 213, 1980.
6. **Beaumont, J. J. and Breslow, N. E.** Power considerations in epidemiologic studies of vinyl chloride exposed workers. *Amer. J. Epid.,* 114, 725, 1981.
7. **Bertazzi, P. A., Villa, A., Foa, V., Saia, B., Febri, L., Mapp. C., Marcer, C., Manno, M., Marchi, M., Mariani, F., and Bottasso** An epidemiological study of vinyl chloride exposed workers in Italy., *Arch. Higrade Toxicol.,* 30, 379, 1979.
8. **Buffler, P. A., Wood, S., Eifler, C., Suarez, L., and Kilian, D. J.** Mortality experience of workers in a vinyl chloride monomer production plant., *J. Occup. Med.,* 21, 195, 1979.
9. **Byren, D., Engholm, G., Englund, A., and Westerholm, P.** Mortality and cancer morbidity in a group of Swedish VCM and PVC production workers. *Environ. Health Perspect.,* 17, 167, 1976.
10. **Carlborg, F. W.,** Dose-response functions in carcinogenesis and the Weibull model. *Fd. Cosmet. Toxicol.,* 19, 255, 1981.
11. **Chiazze, L., Warg, O., Nichols, W. E., and Ference, L. D.,** Breast Cancer Mortality Among PVC Fabricators. *J. Occup. Med.,* 22, 677, 1980.
12. **Chiazze, L. and Ference, L. D.,** Mortality among PVC-fabricating employees. *Environ. Health Perspect.* 41, 137, 1981.
13. Conference to re-evaluate toxicity of VCM and PVC *Environ. Health Perspect.,* 41, 1, 1981.
14. **Cook, W. A., Greve, P. M., Dinman, B. D., and Magnuson, H. J.,** Occupational acro-osteolysis II. An industrial hygiene study. *Arch. Env. Health,* 22, 74, 1981.
15. **Cooper, W. C.,** Epidemiological study of vinyl chloride workers: mortality through December 31, 1972. *Environ. Health Perspect.,* 41, 101, 1981.
16. **Creech, J. L. and Johnson, M. N.,** Angiosarcoma of the liver in the manufacture of PVC. *J. Occup. Med.,* 16, 150, 1974.
17. **Crump, K. S. and Guess, H. A.,** Drinking water and cancer. PB 81-128167, December 1980.
18. **Duck, B. W., Carter, J. T., and Coombes, E. J.,** Mortality study of workers in a polyvinyl chloride production plant. *Lancet* 2, 1197, 1975.
19. The cost of clean air and clean water. Annual report to the Congress. Environmental Protection Agency, December 1979. Senate Document No. 96-38, U.S. Government Printing Office, Washington, D.C., 1980.
20. **Feron, V. J., Hendriksen, C. F. M., Speek, A. J., Til, H. P., and Spit, B. J.** Lifespan oral toxicity study of vinyl chloride in rats., *Fd. Cosmet, Toxicol,* 19, 317, 1981.
21. **Filatova, V. S., et al.,** The Blastomogenic Hazard of Vinyl Chloride (a clinico-hygiene and epidemiological study)., *Gig. Tr. Prof. Sabol.,* 1, 28, 1982.
22. Food Safety Council Final Report. Proposed system for food safety assessment, Washington, D.C., 1980, also in *Fd. Cosmet. Toxicol,* 16, Suppl. 2, Dec. 1978.
23. **Fox, A. J. and Collier, P. F.,** Mortality experience of workers exposed to vinyl chloride monomer in the manufacture of polyvinylchloride in Great Britain., *Brit. J. Industr. Med., 34,* 1, 1977.
24. **Fretzel-Beyme, R., Schmitz, T., and Thiess, A. M.,** Mortalitatsstudie bei VC/PVC arbieterne der BASF aktiengesellschaft, Ludwigshafen am Rhein. Arbeitsmed. Sozialmed., *Preventivmed,* 13, 218, 1978.
25. **Gauvain, S.,** Vinyl chloride. *Proc. Royal Soc. Med.,* 69, 275, 1976.

26. **Gaylor, D. W. and Kodell, R. L.,** Linear interpolation algorithm for low-dose risk assessment of toxic substances. *J. Environ. Pathol. Tox.,* 4, 305, 1980.
27. **Gehring, P. J., Watanabe, P. G., and Park, C. N.,** Resolution of dose-response toxicity data for chemicals requiring metabolic activation: example — vinyl chloride., *Tox. Appl, Pharmacol,* 44, 581, 1978.
28. **Gehring, P. J., Watanabe, P. G., and Park, C. N.,** Risk of angiosarcoma in workers exposed to vinyl chloride as predicted from studies in rats. *Tox. Appl. Pharmacol,* 49, 15, 1979.
29. **Green, T. and Hathway, D. E.,** The biological fate in rats of vinyl chloride in relation to its oncogenicity. *Chem-Biol. Interactions,* 11, 545, 1975.
30. **Green, T. and Hathway, D. E.,** The chemistry and biogenesis of the S-containing metabolites of vinyl chloride in rats. *Chem-Biol. Interactions,* 17, 137, 1977.
31. **Harris, D. K. and Adams, W. G. F.,** Acro-osteolysis occurring in men engaged in the polymerisation of vinyl chloride. *Brit. Med. J.,* 3, 712, 1967.
32. **IARC** (International Agency for Research into Cancer), Evaluation of the carcinogenic risk of chemicals to humans. Monograph No. 19, 377, 1979.
33. **Infante, P. F.,** Observations of the site specific carcinogenicity of vinyl chloride to humans. *Environ. Health Perspect.,* 41, 89, 1981.
34. **Kuzmack, A. M. and McGaughy, R. E.,** Quantitative risk assessment for community exposure to vinyl chloride. U.S. EPA, Washington, D.C., Dec. 5, 1975.
35. **Lee, C. C., Bhandari, J. C., Winston, J. M., House, W. B., Dixon, R. C., and Woods, J. S.,** Carcinogenicity of vinyl chloride and vinylidene chloride. *J. Toxicol. Environ. Health,* 4, 15, 1978.
36. **Lester, D., Greenberg, L. A., and Adams, W. R.,** Effects of single and repeated exposures of humans and rats to vinyl chloride. *Amer. Ind. Hyg. Assoc. J.,* 24, 265, 1963.
37. **Loprieno, N., Bavale, R., Baroncelli, S., Bartsch, H., Brouzetti, G., Gammelini, A., Corsi, C., Freza, D., Nieri, R., Leporini, C., Rosellini, D., and Rossi, A. M.,** Induction of gene mutagens and gene conversions by vinyl chloride metabolites in yeast. *Cancer Res.,* 36, 253, 1977.
38. **Maltoni, C.,** Occupational carcinogenesis. 2nd International Symposium on Cancer Detection and Prevention, Bologna, in *Advances in Tumour Prevention, Detection and Characterization,* Excerpta Medica, 2, 26, 1977.
39. **Maltoni, C. and Rondinella, R.,** Hepatic angiosarcoma in workers exposed to vinyl chloride in Italy. *Acta Oncologia,* 1, 35, 1980.
40. **Maltoni, C., Lefemine, G., Ciliberti, A., Cotti, G., and Carretti, D.,** Epidemiologie animale et epidemiologie humane: le cas de chlorure de vinyl monomere. In XXe reunion de Club de Cancerogenese Chimique, ISBN 2.86315.007.3, Publications Essentielles, Paris, 15, 1980.
41. **Maltoni, C., Lefemine, G., Ciliberti, A., Cotti, G., and Carretti, D.,** Carcinogenicity bioassays of vinyl chloride monomer: a model of risk assessment on an experimental basis. *Environ. Health Perspect.,* 41, 3, 1981.
42. Ministry of Agriculture, Fisheries and Food, Survey of vinyl chloride content of polyvinyl chloride for food contact and of foods, Her Majesty's Stationery Office, 1978.
43. **Monson, R. R., Peters, J. M., and Johnson, M. N.,** Proportional mortality among vinyl chloride workers. *Lancet,* 2, 397, 1974.
44. **Meyerson, L. B. and Meier, G. C.,** Cutaneous lesions in acro-osteolysis. *Arch. Dermatol.,* 106, 224, 1972.
45. **NAS** (1980). Drinking Water and Health. Vol. 3, p. 38, National Academy Press,
46. **Nicholson, W. J., Hammond, E. C., Seidman, H., and Selikoff, I. J.,** Mortality experience of a cohort of vinyl chloride/polyvinyl chloride workers., *Lancet,* 2, 1197, 1975.
47. **ORC** (1976). Mortality data collected by ORC concerning the effects of vinyl chloride exposure in PVC Fabrication.
48. **Ott, M. G., Langner, R. R., and Holder, B. H.,** Vinyl chloride exposure in a controlled industrial environment. *Arch. Environ. Health,* 30, 333, 1975.
49. **Rannug, V., Gothe, R., and Wachtmeister, C. A.,** The mutagenicity of chloroethylene oxide, chloroacetaldehyde, 2-chloroethanol and chloracetic acid, conceivable metabolites of vinyl chloride. *Chem. Biol. Interact.,* 12, 251, 1976.
50. **Reinl, W. and Weber, H.,** Stand der epidemiologischen Forschung uber die vinylchlorid-krankheit. *Zetralbl Arbeitsmed,* 26, 97, 1976.
51. **Reinl, W., Weber, H., and Greiser, E.,** Epidemiology study of the mortality of workers exposed to vinyl chloride in FRG. Paper presented at the 19th Internation Conference for Occupational Health, Dubrovnic, September 1978.
52. **Schneiderman, M. A., Mantel, N., and Brown, C. C.,** From mouse to man — or how to get from the laboratory to Park Avenue and 59th Street. *Ann. N.Y. Acad. Sci.,* 246, 237, 1975.
53. **Selikoff, I. J.,** Toxicity of vinyl chloride/polyvinyl chloride. *Anns. N.Y. Acad. Sci.,* 246, 1975.
54. **Spirtas, R. and Kaminski, R.,** Angiosarcoma of the liver in vinyl chloride/polyvinyl chloride workers. *J. Occup. Med.,* 20, 427, 1978.

55. **Suciu, J., Drejman, I., and Valaskii, M.,** Contributions to the study of disease by vinyl chloride. *Med. Interna.,* 15, 967, 1963.
56. **Szadkowski, D. and Lehnert, G.,** Vinylchlorid als krankheitsursache. Eine Bibliographie VKE, Frankfurt. 1982.
57. **Tabershaw, J. R. and Gaffey, W. R.,** Mortality study of workers in the manufacture of vinyl chloride and its polymers. *J. Occup. Med.,* 16, 509, 1974.
58. **Torkelson, T. R., Ogen, F., and Rowe, V. K.,** The toxicity of vinyl chloride as determined by repeated exposure of laboratory animals. *Am. Ind. Hyg. Assoc. J.,* 22, 354, 1961.
59. **Theriault, G.,** Cancer mortality of Canadian workers exposed to VCM: a three year follow-up. *J. Occup. Med.,* 24, 730, 1982.
60. **Theriault, G. and Allard, P.,** Cancer Mortality of a Group of Canadian Workers Exposed to Vinyl Chloride Monomer. *J. Occup. Med.,* 23, 671, 1981.
61. Selected abstracts on the carcinogenicity of VCM. United States Department of Health Education and Welfare. 1980.
62. **Verburgt, F. G. and Vogel, E.,** Vinyl chloride mutagenesis in *Drosophila melanogaster. Mut. Res.,* 48, 327, 1977.
63. **Viola, P. L.,** Pathology of vinyl chloride. Proceedings of the 16th Internation Congress on Occupational Health, Tokyo, 1969.
64. **Viola, P. L.,** Cancerogenic effect of vinyl chloride. Abstracts 10th International Cancer Conference, Houston, Texas, Abstract No. 29, 1970.
65. **Viola, P. L.,** Pathology of vinyl chloride. *Med. Lavoro,* 61, 174, 1970.
66. **Viola, P. L., Bigotti, A., and Caputo, A.,** Oncogenic response of rat skin, lungs and bones to vinyl chloride. *Cancer Res.,* 31, 516, 1971.
67. **Viola, P. L.,** La malattia da cloruro di vinile. *Med. Lavoro,* 65, 81, 1974.
68. **Watanabe, P. G. and Gehring, P. J.,** Dose-dependent fate of vinyl chloride and its possible relationship to oncogenicity in rats. *Environ. Health Perspect.,* 17, 145, 1976.
69. **Waxweiller, R. J., Stringer, W., Wagoner, J. K., Jones, J., Falk, H., and Carter, C.,** Neoplastic risk among workers exposed to vinyl chloride. *Ann. N.Y. Acad. Sci.,* 271, 40, 1976.
70. **Weber, H., Reinl, W., and Greiser, E.,** German investigations on morbidity and mortality of workers exposed to vinyl chloride. *Environ. Health Perspect.,* 41, 95, 1981.
71. **Wilson, R. H., McCormick, W. E., Tatum, C. F., and Creech, J. L.,** Occupational acro-osteolysis. *JAMA.,* 201, 577, 1967.

Chapter 9

AN INTEGRATED APPROACH TO THE STUDY OF FORMALDEHYDE CARCINOGENICITY IN RATS AND MICE

Thomas B. Starr, James E. Gibson, and James A. Swenberg

TABLE OF CONTENTS

I. ABSTRACT

A chronic inhalation toxicology study of formaldehyde in rats and mice demonstrated the ability of inhaled formaldehyde to cause nasal irritation and tumors. Effects were dose-related, with prominent effects induced at 14.3 ppm in comparison to 2.0, or 5.6 ppm of formaldehyde in inspired air. Delivered dose, as contrasted with administered dose, appears to be an important determinant of these effects. At low concentrations, a series of potentially protective barriers and processes may serve to limit the amount of formaldehyde that reaches the DNA of target cells in the nasal mucosa. These include the mucus blanket and mucociliary clearance apparatus, removal by intercellular fluids, the cell membrane, and intracellular metabolism and binding to macromolecules other than DNA. At high concentrations, saturation, in some cases, damage to these barriers and processes, as evidenced in part by enhanced cell death and proliferation, may allow exogenous formaldehyde more direct access to the DNA of proliferating cells, thus enhancing the likelihood of genotoxic events. Also, a significant difference in intensity of the respiratory depression reflex in rats and mice exposed to the same ambient air concentration results in mice receiving a smaller delivered dose. This difference accounts for the observed disparity in rat and mouse carcinogenic responses to the same administered dose. Results of using delivered dose as the independent variable in quantitative risk assessment models indicate the importance of using the amount of toxic agent at the site of injury, rather than the concentration in inspired air, for the prediction of human response to low ambient air concentrations of formaldehyde.

II. LONG-TERM INHALATION STUDIES IN RATS AND MICE

The first evidence that formaldehyde is a rodent carcinogen was obtained in a chronic inhalation study sponsored by the Chemical Industry Institute of Toxicology (CIIT).[1,2] The study utilized 960 Fisher-344 rats and 960 B6C3F1 mice, with 120 animals of each species and sex exposed to either 0, 2.0, 5.6, or 14.3 ppm of formaldehyde vapor 6 hr/day, 5 days/week for up to 24 months. The principal finding was that exposed rats and mice both developed squamous cell carcinomas of the nasal cavity. Nearly half (103) of the rats exposed to 14.3 ppm formaldehyde exhibited this neoplastic lesion upon necropsy, while only two mice (rats) exposed to 14.3 (5.6) ppm formaldehyde developed the lesion by 24 months. More recently, exposure to 14.1 ppm formaldehyde gas has been shown to induce squamous cell carcinomas in the nasal cavities of Sprague-Dawley rats.[3]

Table 1 provides a summary of the neoplastic lesions found in the nasal cavities of rats and mice during the course of the CIIT bioassay.[4] Several features are noteworthy. First, the incidence of squamous cell carcinomas in rats appears to be highly nonlinear, with nearly a 50-fold increase from 5.6 to 14.3 ppm formaldehyde. Second, mice appear to be far less responsive than rats to the same ambient air concentration, since only two nasal tumors of any type were found among exposed mice, and these were squamous cell carcinomas in the group exposed to 14.3 ppm formaldehyde for 24 months. Third, although the incidence of polypoid adenomas in rats was not statistically significant with an adjusted pairwise analysis, a significant adjusted trend test indicated that the incidence of this benign tumor may have been enhanced by exposure to formaldehyde.[2,5] These exophytic growths were not found in association with epithelial dysplasia or squamous metaplasia as were many of the squamous cell carcinomas, and these lesions did not appear to progress to squamous cell carcinomas.[4]

Figure 1 depicts the nonparametric Kaplan-Meier[6] estimate of survival for rats in the 5.6 and 14.3 ppm dose groups, censored at the time of death for those animals that did not exhibit this tumor. The complement of this function is an estimate of the cumulative distribution of time-to-squamous cell carcinoma. The high dose group's probability of survival to 24 months was reduced nearly sevenfold relative to that for the group exposed to 5.6

Table 1
SUMMARY OF NEOPLASTIC LESIONS IN THE NASAL CAVITY OF FISCHER-344 RATS AND B6C3F1 MICE EXPOSED TO FORMALDEHYDE GAS

Diagnosis	0 ppm				2.0 ppm				5.6 ppm				14.3 ppm			
	Mouse		Rat		Mouse		Rat		Mouse		Rat		Mouse		Rat	
	M	F	M	F	M	F	M	F	M	F	M	F	M	F	M	F
	(109)	(114)	(118)	(114)	(100)	(114)	(118)	(118)	(106)	(112)	(119)	(116)	(106)	(109)	(117)	(115)
Squamous cell carcinoma	0	0	0	0	0	0	0	0	0	0	1	0	2	0	51	52
Nasal carcinoma	0	0	0	0	0	0	0	0	0	0	0	0	0	0	1[a]	1
Undifferentiated carcinoma or sarcoma	0	0	0	0	0	0	0	0	0	0	0	0	0	0	2[a]	0
Carcinosarcoma	0	0	0	0	0	0	0	0	0	0	0	0	0	0	1	0
Polypoid adenoma	0	0	1	0	0	0	4	0	0	0	6	0	0	0	4	1
Osteochondroma	0	0	1	0	0	0	0	0	0	0	0	0	0	0	0	0

Note: Numbers in parentheses are number of nasal cavities evaluated.

[a] One rat in this group also has a squamous cell carcinoma.

Reprinted from Swenberg, J. A., Barrow, C. S., Boreiko, C. J., et al., *Carcinogenesis*, 1983, in press. With permission.

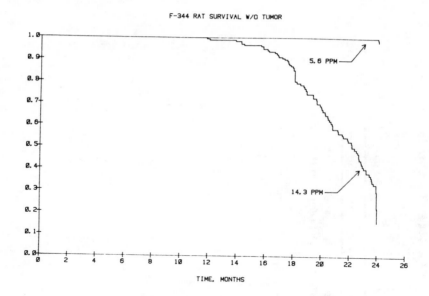

FIGURE 1.   The nonparametric Kaplan-Meier estimate for Fischer-344 rat survival to
death with squamous cell carcinomas.

ppm formaldehyde. Also notable is the strong apparent dependence of time-to-tumor on
dose. Approximately the same proportion of rats survived through 14 months of exposure
to 14.3 ppm as survived through 24 months of exposure to 5.6 ppm.

III. ANALYSES OF RESPONSE VS. ADMINISTERED DOSE

Since a nonzero response was observed only in the high dose group of mice, the parameters
of quantal and time-to-tumor response models for this species cannot be properly estimated
by maximum likelihood methods. Attention in this section will therefore be restricted to
analyses of the squamous cell carcinoma data for rats. The dramatic difference in species
responsiveness to identical ambient air concentrations of formaldehyde will be addressed in
Section IV.

A. Quantal Response
A number of mathematical dose-response models have been proposed for the purpose of
low dose extrapolation. These include the probit, logit or logistic, Weibull or extreme value,
gamma multihit, and multistage. The properties of these models are well established and
have been described elsewhere. An excellent review is provided by Krewski and Van Ryzin.[7]
Typically, these models describe experimental data for a given time point, such as the
final sacrifice, about equally well, and they cannot be distinguished from one another on
the basis of objective criteria such as goodness of fit in the experimental dose range. This
is most definitely the case for the formaldehyde data. Typically as well, however, the
maximum likelihood estimates of excess risk associated with low doses, or, alternatively,
of the ''virtually safe doses'' (VSDs) associated with given levels of excess risk differ by
many orders of magnitude across models, and some researchers have argued that these
model-based estimates of excess risk are therefore meaningless.[8,9]
In the case of formaldehyde, however, the VSD estimates provided by these models lie
within a factor of 70 of each other for excess risks ranging over three orders of magnitude,
as is illustrated in Table 2 (VSD and lower confidence limit estimates have been calculated
using the adjusted incidence figures reported by Gibson[10]).

Table 2
VSD PREDICTED USING THE PROBIT, LOGIT, WEIBULL, GAMMA MULTIHIT, AND MULTISTAGE MODELS

	Formaldehyde VSD (ppm)			
Excess risk	1/100,000	1/1,000,000	1/10,000,000	1/100,000,000
Probit	2.60	2.14	1.79	1.52
Multihit	2.05	1.52	1.14	0.86
Logit	1.41	0.89	0.56	0.35
Weibull	1.26	0.76	0.46	0.28
Multistage	0.38	0.17	0.08	0.04

Adapted from Gibson, J. E., *Formaldehyde Toxicity*, Gibson, J. E., Ed., Hemisphere Publishing, Washington, 1983, 295.

Table 3
LCL95 FOR VSDS PREDICTED USING THE PROBIT, LOGIT, WEIBULL, GAMMA MULTIHIT, AND MULTISTAGE MODELS

	Formaldehyde LCL95 (ppm)			
Excess risk	1/100,000	1/1,000,000	1/10,000,000	1/100,000,000
Probit	1.84	1.44	1.15	0.94
Multihit	1.24	0.82	0.55	0.37
Logit	0.77	0.42	0.23	0.13
Weibull	0.66	0.34	0.18	0.09
Multistage	0.007	0.0008	0.00006	0.00001

Adapted from Gibson, J. E., *Formaldehyde Toxicity*, Gibson, J. E., Ed., Hemisphere Publishing, Washington, 1983, 295.

Note, as is typical, that the probit model is the least conservative, while the multistage, which has been restricted to three or fewer stages in order to avoid an identifiability problem, is the most conservative. (Removal of the restriction on number of stages under the assumption that the coefficients of powers of dose lower than cubic are identically zero leads to multistage VSD and lower confidence limit estimates that are much closer to, but still lower than, those for the other models[10]). For a given excess risk level, the largest and smallest VSD estimates differ by as little as a factor of 7 (for a risk of 1/100,000) and by at most a factor of 40 (for a risk of 1/100,000,000). Furthermore, all of the VSD estimates are less than a factor of 150 below 5.6 ppm, the lowest dose at which a positive response was observed. Thus, excess risks as low as 1/100,000,000 are predicted at concentrations little more than two orders of magnitude below the experimental dose range.

Of course, VSD estimates are uncertain, and a reasonable numerical estimate of that uncertainty is provided by the lower 95% confidence limit (LCL95) on the VSD for a given excess risk, or, alternatively, by the upper 95% confidence limit on the excess risk for a given low dose. Table 3 presents the LCL95s for the VSDs presented in Table 2. For the probit, logit, Weibull, and gamma multihit models these estimates are based on the asymptotic normality of the logarithm of dose. For the multistage, asymptotic normality of this quantity obtains only when all of the multistage parameters are nonzero,[11] which does not appear to be the case for the formaldehyde data set. Consequently, the LCL95s for this model were estimated by Monte Carlo simulation. Briefly, 1000 simulated data sets were constructed by random sampling from a population whose true dose-response function was taken to be the multistage with parameters equal to the maximum likelihood estimates obtained with the original data. The multistage model parameters were then reestimated for each simulated data set, and corresponding VSDs were calculated and ranked in ascending order. The 51st VSD was taken to be the desired LCL95.

Comparison of Table 2 and 3 reveals that, with the exception of the multistage model, the model-specific LCL95s are all less than a factor of three below their corresponding VSDs. Thus, the uncertainty in the VSD estimates for models other than the multistage is quite small. The multistage model, however, behaves very differently. The simulation process generates some data sets for which a term linear in dose is estimated to be nonzero. For these cases, the excess risk is predicted to decrease only linearly with dose, and consequently, the estimated VSD associated with a given excess risk is quite small. As a result, the ratio of VSD to LCL95 estimates for the multistage model is 54 for an excess risk of 1/100,000, while it is 4000 for an excess risk of 1/100,000,000.

Parenthetically, it should be noted that on April 2, 1982, the U.S. Consumer Product Safety Commission (CPSC) announced a decision to ban the sale of urea formaldehyde foam insulation (UFFI) in the United States.[12] This decision was based on a quantitative risk assessment which concluded that living in a UFFI-equipped residence for 9 years would increase an individual's lifetime risk of developing cancer by as much as 51 per million.[12] This estimate of excess human risk was constructed by first fitting the multistage model (limited to three stages as described above) to the bioassay data for rats. The fitted model was then used to estimate the risk associated with the exposure concentrations expected in UFFI-equipped residences (0.07 ppm, time-weighted average).[12]

The CPSC defended its selection of the multistage model over other quantal response models by citing evidence that formaldehyde interacted directly with genetic material in certain test systems. In addition, they assumed that different exposure regimens that yielded the same concentration × time product elicited the same response in order to convert risk estimates based on the bioassay protocol to equivalent risk estimates for exposure of 16 hr/day, 7 days/week. The commission also assumed that there was no difference between human and Fischer-344 rat sensitivities to the toxicity, and particularly, the carcinogenicity of formaldehyde. Furthermore, it was assumed, albeit implicitly, that strict linear proportionality holds between the concentration of formaldehyde in ambient air and in the target tissue (replicating cells of the respiratory epithelium). Since each of these assumptions can be legitimately called into question, it is not surprising that the ban was challenged in a U.S. Court of Appeals. In March 1983, the Fifth Circuit Appeals Court in New Orleans overturned the ban, ruling that the CPSC had failed to provide "substantial evidence" that formaldehyde posed an unreasonable health risk.[13]

B. Time-to-Tumor Response

The product form of the hazard rate is a relatively general model for carcinogenic risk that encompasses both the dose level and the time required for a response to develop. Special cases have been studied by a number of authors.[14-17] In the Armitage and Doll version of the multistage model, the largest exponent of time in the cumulative hazard function is equal to the number of stages in the carcinogenic process.[14] Furthermore, the number of stages, and hence the number of multistage parameters that are estimable, is constrained to be no more than the number of bioassay dose groups (other than controls). In the more general polynomial forms considered by Hartley et al.[16,17] there is no practical constraint on the order of the time polynomial, although it is a common practice to keep it relatively small. Nonpolynomial but still factorable forms are also possible, although such forms have received only limited attention.[18]

The comparatively early onset of responses in the 14.3-ppm group (Figure 1) and the absence of a response until the 24-month sacrifice in the 5.6-ppm group (Table 1) are two features of the rat bioassay that cannot simultaneously be described adequately by an Armitage-Doll type multistage model that is restricted to three stages. Fitting of this model to the rat data leads to an unsatisfactory compromise that rather severely overestimates (underestimates) the response at nearly all times in the 5.6-(14.3) ppm dose group (data not shown).

IV. DELIVERED VS. ADMINISTERED DOSE

Quantitative risk assessments have for the most part been forced to rely on the administered dose from bioassays as the independent variable in estimated dose-response functions. However, the biological arguments used to defend the choice of a particular mathematical form have clearly used the term dose to represent the quantity or concentration of the final active form of a compound in target tissue.[14,19,20] The implicit assumption that administered dose is the relevant measure of exposure might be justifiable if assessments were required only for an animal of the same species and sex employed in a bioassay and for exposure levels within the experimental dose range. It is much harder to justify if risk assessments are required for an animal of an entirely different species and for exposure levels that are orders of magnitude below the experimental dose range. Dose-dependent differences across species in the distribution and metabolism of compounds can play a crucial role in the success or failure of the interspecies conversion of risk estimates. Even within the same species and sex, qualitative differences in distribution, pharmacokinetics, and lesion repair processes between low and high doses can invalidate low dose extrapolations based on the assumption of a purely linear relationship between administered and delivered doses.[21-24] We now turn our attention to some of the critical factors which determine this relationship for the case of formaldehyde exposure by inhalation.

A. Respiratory Depression Reflex: Rat vs. Mouse

Formaldehyde is a potent sensory irritant, capable of stimulating a variety of respiratory tract receptors, particularly those in the nasal cavity associated with the trigeminal nerve.[25,26] One of the most important responses to stimulation by an irritant is a concentration-dependent decrease in minute ventilation which serves to minimize uptake and deposition of potentially toxic agents.[27,28]

Studies of the sensory irritant effects of formaldehyde indicate that mice are far more sensitive to these effects than rats. For example, while the ambient air concentration required to elicit a 50% decrease in respiratory rate is 31.7 ppm in Fischer-344 rats, it is only 4.9 ppm in B6C3F1 mice.[28] Since tidal volume remains essentially unchanged, minute volume also decreases by approximately 50% at these concentrations. As a result, mice inhale less formaldehyde per unit time than rats when both are exposed to the same ambient air concentration of 6 (15) ppm formaldehyde.[29] Fischer-344 rats had a 10% (20%) reduction in minute volume compared to a 40% (70%) reduction in B6C3F1 mice, as is illustrated in Figure 2a.

Almost 100% of the inspired formaldehyde is deposited in the nose.[30] An estimate of the average rate of formaldehyde deposition per unit surface area of the nasal mucosa can be computed by dividing the amount of formaldehyde inhaled per unit time by the surface area of the nasal cavity. During a 6 hr exposure to 15 ppm formaldehyde the time-weighted average deposition rate for mice was approximately half that for rats (Figure 2b). At 6 ppm both species were predicted to receive similar doses. Furthermore, there was no significant difference either in the dose available for deposition or tumor incidence between rats exposed to 6 ppm and mice exposed to 6 or 15 ppm. Additional studies using whole body autoradiography have qualitatively confirmed the difference in deposited dose observed in rats and

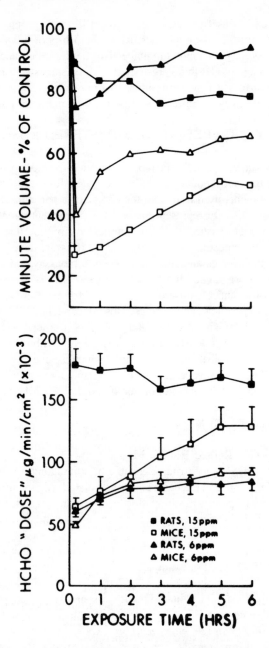

FIGURE 2.   Time-response curves for minute volume from rats and mice exposed to 6 or 15 ppm formaldehyde for 6 hr. (B) Time-weighted averages of the theoretical formaldehyde dose available for deposition in the nasal cavity of rats and mice during a 6-hr exposure to 6 or 15 ppm. (From Swenberg, J. A., Barrow, C. S., Boreiko, C. J., et al., *Carcinogenesis,* 4(8), 945, 1983. With permission.)

mice exposed to 15-ppm formaldehyde.[29] It is thus essential to consider interspecies differences in pulmonary ventilation and nasal cavity volume to surface area relationships when estimating risks from exposure to airborne irritants.

B. Mucociliary Function

The nasal squamous cell carcinomas induced in rats by formaldehyde occurred in areas normally lined by respiratory epithelium and covered by a continuous layer of mucus that

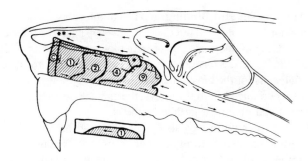

FIGURE 3. Diagram of a rat head opened adjacent to the midline with the nasal septum removed to demonstrate the outline of the turbinates, part of the lateral wall, and the nasopharynx. The inset represents the left nasoturbinate which is placed to reveal its lateral aspect. The general direction of mucus flow is indicated by the arrows. Progressive extension of areas of formaldehyde induced mucostasis and partial to complete ciliastasis with increasing days of exposure (15 ppm; 1, 2, 4, or 9 days, 6 hr/day) is indicated by the cross-hatching. "C" indicates the anterior areas of respiratory epithelium in which mucus flow and ciliary activity are generally not apparent in control rats. The medial nasoturbinate (**) clears via the nasal septum, and flow from the lateral wall (*) clears via the dorsal or anterior aspect of the maxilloturbinates. (From Swenberg, J. A., Barrow, C. S., Boreiko, C. J., *Carcinogenesis*, 4(8), 945, 1983. With permission.)

flows over the surface.[1,5,31] The mucociliary apparatus protects the nasal passages and lower airways from particulate matter deposited upon the mucus surface, but the ability of this system to protect the underlying epithelium from toxic gases has yet to be established.[32] However, in order for chemicals or particles to affect directly the underlying epithelium, they must first penetrate the superficial mucus layer, which consists primarily of water ($\sim$ 95%), mucus glycoproteins (0.5 to 1%), free proteins and salts, and other materials in much smaller amounts.[33] Formaldehyde binds readily to proteins and also reacts with polysaccharides at room temperature.[34,35] It has also been shown recently that G.I. tract mucus provides a diffusion barrier to nutrients and hydrogen ions.[36,37] When combined with constant mucus removal and replacement, factors such as these might be expected to retard and attenuate the penetration of formaldehyde to underlying epithelial cells, and thus provide these cells with some measure of protection against formaldehyde toxicity.[31,38]

However, formaldehyde can also impair mucociliary function in many species, with the sequential induction of reduced mucus flow rate, mucostasis, and ciliastasis.[39-43] A concentration effect has been established recently for the inhibition of the rat nasal mucociliary apparatus following in vivo formaldehyde exposure, with mucostasis being more extensive than ciliastasis.[31,38] Impaired mucociliary clearance occurred at specific locations in the rat nasal passages following 6-hr exposures to 15-ppm formaldehyde for 1 to 9 days (Figure 3).[31] Exposure to this concentration produced distinct areas of ciliastasis on the ventral and lateral aspects of the nasoturbinate and the medial aspect of the maxilloturbinate, with progression of this effect more deeply into the nasal cavity with increasing days of exposure. By 9 days, ciliastasis had reached the lateral wall posterior to the maxillo- and nasoturbinates. Exposure to 6-ppm formaldehyde had only focal effects on mucociliary activity in anterior portions of naso- and maxilloturbinate. With 2 ppm only small areas of ciliastasis were observed in some animals. These areas were confined primarily to the anterioventral margin of the nasoturbinate. Animals exposed to 0.5-ppm formaldehyde showed no evidence of impaired mucociliary function.[38]

Thus, it appears that formaldehyde can incapacitate the nasal mucociliary apparatus at specific locations in the nose. If this system does modulate the amount of inspired formaldehyde that reaches the underlying nasal epithelium, then localized disruption of its function could account partly for the subsequent appearance of epithelial lesions in these locations.

C. Intracellular Metabolism and Binding to Macromolecules

Formaldehyde is a normal biological intermediate that appears to be present in all biological tissues. A detailed review of its biochemical toxicology has been provided by Heck and Casanova-Schmitz.[44] In its active form (N^5, N^{10}-methylene-tetrahydrofolate), formaldehyde is used by mammalian cells for the biosynthesis of purines, thymine, methionine, and serine. Active formaldehyde is produced from serine (via the enzyme L-serine:tetrahydrofolate-5,10-methylene transferase) and glycine (via the glycine cleavage reaction).[45] Other compounds containing N-, O-, and S-methyl groups also yield formaldehyde as a primary metabolic or catabolic product. Although active formaldehyde is used for many biosynthetic reactions that require one-carbon units, it does not appear to be the predominant form in mammalian tissues. Rather, formaldehyde appears to be bound primarily to other nucleophiles, such as water, sulfhydryls including glutathione, or amines.[46]

Formaldehyde is reactive and cytotoxic, and cells have developed specific enzymatic pathways for its removal. Formaldehyde dehydrogenase catalyzes the nicotinamide adenine dinucleotide (NAD+)-dependent oxidation of the reversible formaldehyde adduct, S-hydroxymethylglutathione, to S-formylglutathione.[47] Some isozymes of aldehyde dehydrogenase, which can act upon a broader range of substrates, catalyze the (NAD+)-dependent oxidation of formaldehyde to formate. These enzymes are widely distributed in tissues. Thus, rapid metabolism of formaldehyde is typical, and may have been partly responsible for the fact that exposure of rats to 6- or 15-ppm formaldehyde did not appear to increase the concentration of free and reversibly bound formaldehyde in the nasal mucosa.[48]

The reactivity of formaldehyde toward biological macromolecules has been well documented.[34,49] Extensive cross-linking of DNA and protein has been observed in a variety of mammalian cells in culture following treatment with formaldehyde.[50-53] These cross-links are rapidly repaired (repair half-time < 4 hr) in mouse L1210 and human bronchial cells. Repair probably proceeds via excision, which may account for the small numbers of single-strand DNA breaks observed in formaldehyde-treated cells.[50,51,54] A portion of the DNA of rat respiratory mucosa was rendered nonextractable from proteins except after enzymatic proteolysis following exposure to formaldehyde at concentrations of 6 ppm and higher.[55] The inability to extract DNA from proteins under denaturing conditions is presumptive evidence, but not proof, of the formation of DNA-protein cross-links in vivo.[56-58] Thus it appears that at least some formaldehyde directly penetrated to the DNA of nasal cavity epithelial cells at concentrations of 6 ppm and higher.

D. Cytotoxicity and Cell Turnover

Acute cytotoxicity was apparent in the nasal turbinates of rats following a single 6-hr exposure to 15-ppm formaldehyde.[59] Additional exposures produced a series of toxic responses ranging from mild rhinitis to severe ulceration. Lesions in rats were of much greater severity than those in mice after as little as 5 days exposure to 15-ppm formaldehyde.[29,59] A prominent response to this cytotoxicity was restorative cell proliferation and hyperplasia. When the nasal passages of rats and mice were examined after 3 days of exposure (6 hr/day) to 0, 0.5, 2, 6, or 15 ppm formaldehyde, a dramatic concentration-related gradient in the percent of labeled respiratory epithelial cells was evident.[60] A 20- to 10-fold increase in the labeling index was found in mice and rats exposed to 6 or 15 ppm, respectively. No increase was detectable in animals exposed to 0.5 or 2 ppm. The respiratory epithelium of control rodent nasal passages showed very little cell replication, and only 0.1 to 0.5% of the cells were labeled.

Cell proliferation was also examined in groups of rats exposed to the same daily dose (in ppm-hr), but to different ambient air concentrations and corresponding durations of exposure; i.e., 12 ppm for 3 hr, 6 ppm for 6 hr, and 3 ppm for 12 hr.[60] Increases in cell proliferation depended both on concentration and location within the nose. In the most anterior portion of the nasal cavity, where mucociliary clearance is minimal, cell replication increased fivefold in all exposure groups. This effect is in marked contrast to the effects in the main portion of the respiratory epithelium, the primary site of the squamous cell carcinomas found in the CIIT bioassay, and where mucociliary clearance is normally prominent. There, cell proliferation was strongly and nonlinearly concentration dependent, with no significant increase at a concentration of 3 ppm.

Such concentration × time effects are not confined to acute studies. Rats exposed for 6 months to 14.3 ppm formaldehyde 6 hr/day 5 days/week (429 ppm-hr/week) had much more severe inflammatory, hyperplastic, and metaplastic lesions than did similar animals exposed for 6 months to 3-ppm formaldehyde 22 hr/day, 7 days/week (462 ppm-hr/week).[2,61]

V. EFFECTS OF DELIVERED DOSE ON TEST ANIMAL RISK ASSESSMENT

As noted earlier, quantitative risk assessments that are based exclusively on the dose administered in a bioassay implicitly assume a purely linear relationship between this dose measure and the amount of the final active form of a compound that reaches a specific target tissue. Yet, as we have illustrated in the preceding section, inspired formaldehyde must travel a highly complex path through a number of potential barriers, escaping removal and metabolic detoxification processes along the way, before it can reach the DNA of respiratory epithelial cells that are capable of replication.

Specifically, it must first be deposited on the outer surface of the mucus blanket. It must then diffuse downward through this layer into the underlying periciliary fluid before it is advected away from the target tissue, via mucociliary clearance, ultimately to be swallowed. It must then escape removal via intercellular fluids, including the blood, and penetrate the membrane of a target cell. Once inside a cell, it must not be detoxified by the normal metabolic processes that very effectively regulate endogenous formaldehyde levels. It must also not be bound to the many extranuclear macromolecules for which it has a high affinity. Finally, it must penetrate the nuclear membrane and bind to single-strand nuclear DNA. In order to produce a mutation the target cell must then undergo replication before repair of the damaged DNA can occur.

It is thus reasonable to represent the amount of formaldehyde fixed in DNA lesions that escape repair before replication as the difference between the amount deposited on the surface of the mucus blanket and the amount that is removed, detoxified, or otherwise accommodated by the respiratory epithelium of the nasal cavity. Furthermore, it is reasonable to presume that the removal, detoxification, binding (to macromolecules other then DNA), and DNA repair processes are saturable in the sense that exposure to sufficiently high levels of formaldehyde can lead to their impairment. At high exposure levels, proportionately more of the deposited formaldehyde is thus likely to reach DNA than at low exposure levels. In other words, the relationship between administered and delivered doses is likely to be nonlinear.

An hypothetical relationship between these two dose measures that embodies the above considerations is illustrated in Figure 4. The aggregate rate of removal, detoxification, binding to macromolecules other than DNA, and DNA repair is presumed to exhibit Michaelis-Menten kinetics, with a saturation constant of 1 ppm formaldehyde in ambient air, and with a low dose-limiting rate equal to the deposition rate.[62] Note that the relationship is nearly linear in the observable response range of administered doses, since the above-mentioned processes are taken to be saturated in this range. Below 1 ppm, however, the relationship departs significantly from linearity, becoming essentially quadratic below about 0.05 ppm. Indeed, at 0.007 ppm, the LCL95 for the multistage model at which the risk was estimated to be 1/100,000 (Table 3), the delivered dose predicted by this relationship is approximately 0.00005 ppm. Now the LCL95 for the multistage model is essentially linear

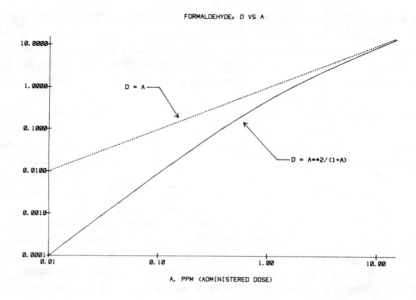

FIGURE 4.   An hypothetical nonlinear relationship between delivered and administered doses of formaldehyde assuming Michaelis-Menten kinetics for removal, metabolic detoxification, and binding to macromolecules other than DNA. (From Starr, T. B., *Formaldehyde: Toxicology, Epidemiology, and Mechanisms,* Clary, J. J., Gibson, J. E., and Waritz, R. S., Eds., Marcel Dekker, New York, 1983, 253. With permission.)

in the excess risk at such low doses. Thus, if the multistage model with parameters as estimated from the original data were actually to represent the relationship between tumor incidence and delivered rather than administered dose, the true excess risk would be 144 times smaller than the nominal value of 1/100,000. Alternatively, a true excess risk of 1/100,000 would be associated with a LCL95 of 0.0872 ppm for administered dose, 12 times larger than the corresponding LCL95 of .007 ppm for delivered dose.

Thus, the mild Michaelis-Menten nonlinearity produces a dramatic effect on the estimates of excess risk associated with exposure to low ambient air concentrations of formaldehyde, while leaving estimates in the experimental dose range essentially unchanged. Of course, the nonlinear form selected here is hypothetical, and is very likely to represent a gross oversimplification of the actual relationship between administered and delivered dose. Determination of the precise form of that relationship should be a high priority for future mechanistically oriented research.

VI. IMPLICATIONS FOR HUMAN RISK ASSESSMENT

A number of issues have been raised in preceding sections that appear to have a direct bearing on the assessment of risk to humans from exposure to formaldehyde. We have seen how interspecies differences in responsiveness to the sensory irritant properties of formaldehyde gas can create significant differences in the rate of formaldehyde deposition in the nasal cavity, even when the different species are exposed to the same ambient air concentration. However, for human risk, the exact implications of the findings regarding the respiratory depression reflex in rodents are still unclear, since rodents are obligatory nose breathers, while humans breathe through both the nose and mouth.[63] Relative to nasal breathing, though, an oral-nasal pattern of respiration would be expected to diminish the dose of formaldehyde deposited in the nasal cavity and simultaneously increase that delivered to the oral cavity, trachea, and lungs.

We must also be concerned about possible differences between humans and rodents in sensitivity, efficiency, and saturability of the other processes discussed in Section IV that may serve to protect target cells from the toxic effects of formaldehyde. For example, human exposure to comparatively low concentrations of formaldehyde (0.38 to 1.63 ppm, for up to 5 hr) reduced mucus flow rate in the nasal passages by 10% to 50% while only relatively local effects were observed at comparable concentrations in rats.[64] Resolution of such concerns must await the results of further comparative mechanistic studies in different species.

Furthermore, since laboratory study protocols rarely, if ever, coincide with conditions of human exposure, adjustments must be made to account for the differences in laboratory animal and human exposure regimens. The CPSC assessment for urea formaldehyde foam insulation assumed that excess cancer risk for rats was the same for different regimens that yielded the same daily concentration × time product, i.e., exposure to 6 ppm for 6 hr/day was assumed to produce the same effect as exposure to 2.25 ppm for 16 hr/day.[12] However, if the excess cancer risk in rats from exposure to formaldehyde were better correlated with increased cell proliferation than with ppm-hr, as appears to be true, then the adjustment based on ppm-hr yields excessively high estimates of risk at low ambient air concentrations.

All of these issues will obviously require further study before their implications for human exposure to formaldehyde can be fully elucidated. In the meantime, prudence dictates that formaldehyde levels in the workplace, in homes, and, indeed, in the ambient air in general be kept as low as is practical.

ACKNOWLEDGMENTS

In preparing this case study, we have liberally drawn from several recent reviews of formaldehyde toxicity.[4,44] Most notable are the contributions from Craig Barrow, Kevin Morgan, and Henry Heck. We also thank Raymond Buck for constructing the Kaplan-Meier survival function estimates and for performing the multistage model Monte Carlo simulations and parameter, VSD, and LCL95 calculations.

REFERENCES

1. **Swenberg, J. A., Kerns, W. D., Gralla, E. J., and Pavkov, J. L.,** Induction of squamous cell carcinomas of the rat nasal cavity by inhalation exposure to formaldehyde vapor, *Cancer Res.*, 40, 3398, 1980.
2. A Chronic Inhalation Toxicology Study in Rats and Mice Exposed to Formaldehyde, Docket #10922, Chemical Industry Institute of Toxicology, 1981.
3. **Albert, R. E., Sellakumar, A. R., Laskin, S., et al.,** Nasal cancer in the rat induced by gaseous formaldehyde and hydrogen chloride, *J. Natl. Cancer Inst.*, 68, 597, 1982.
4. **Swenberg, J. A., Barrow, C. S., Boreiko, C. J., et al.,** Nonlinear biological responses to formaldehyde and their implications for carcinogenic risk assessment, *Carcinogenesis*, 4(8), 945, 1983.
5. **Kerns, W. D., Pavkov, K. L., Donofrio, D. J., et al.,** Carcinogenicity of formaldehyde in rats and mice after long-germ inhalation exposure, *Cancer Res.*, 43, 4382, 1983.
6. **Kaplan, E. L. and Meier, P.,** Nonparametric estimation from incomplete observations, *J. Am. Stat. Assoc.*, 53, 457, 1958.
7. **Krewski, D. and Van Ryzin, J.,** Dose response models for quantal response toxicity data, in *Statistics and Related Topics*, Csorgo, M., Dawson, D. A., Rao, J. N. K., and Saleh, A. K. Md. E., Eds., North-Holland, New York, 1981, 201.
8. **Mantel, N.,** Limited usefulness of mathematical models for assessing the carcinogenic risk of minute doses, *Arch. Toxicol. Suppl. 3*, 305, 1980.
9. **Gaylor, D. W.,** Mathematical approaches to risk assessment: squamous cell carcinoma in rats exposed to formaldehyde vapor, in *Formaldehyde Toxicity*, Gibson, J. E., Ed., Hemisphere Publishing, Washington, 1983, 279.

10. **Gibson, J. E.,** Risk assessment using a combination of research and testing results, in *Formaldehyde Toxicity,* Gibson, J. E., Ed., Hemisphere Publishing, Washington, 1983, 295.
11. **Crump, K. S., Guess, H. A., and Deal, K. L.,** Confidence intervals and test of hypotheses concerning dose response relations inferred from animal carcinogenicity data, *Biometrics,* 33, 437, 1977.
12. Consumer Product Safety Commission, Part IV: Consumer Product Safety Commission ban of urea formaldehyde foam insulation, withdrawal of proposed labeling rule, and denial of petition to issue a standard, *Federal Register,* 47, 14366—14419, 1982.
13. **Sun, J.,** Formaldehyde ban is overturned, *Science,* 220, 699, 1983.
14. **Armitage, P. and Doll, R.,** The age distribution of cancer and a multistage theory of carcinogenesis, *Br. J. Cancer,* 8, 1, 1954.
15. **Crump, K. S.,** Dose response problems in carcinogenesis, *Biometrics,* 35, 157, 1979.
16. **Hartley, H. O. and Sielken, R. L.,** Estimation of safe doses in carcinogenic experiments, *Biometrics,* 33, 1, 1977.
17. **Hartley, H. O., Tolley, H. D., and Sielken, R. L.,** The product form of the hazard rate model in carcinogenic testing, in *Statistics and Related Topics,* Csorgo, M., Dawson, D. A., Rao, J. N. K., and Saleh, A. K. Md. E., Eds., North-Holland, New York, 1981, 185.
18. **Kalbfleisch, J. D., Krewski, D., and Van Ryzin, J.,** Dose response models for time to response toxicity data, *Can. J. Stat.,* 11(1), 25, 1983.
19. **Sugimura, T., Wakabayashi, K., Yamada, M., et al.,** Activation of chemicals to proximal carcinogens, in *Mechanisms of Toxicity and Hazard Evaluation,* Holmstedt, B., Lauwerys, R., Mercier, M., et al., Eds., Elsevier/North Holland Biomedical Press, New York, 1980, 205.
20. **Guess, H. A. and Crump, K. S.,** Low-dose-rate extrapolation of data from animal carcinogenicity experiments — analysis of a new statistical technique, *Math. Biosci.,* 32, 15, 1976.
21. **Cornfield, J.,** Carcinogenic Risk assessment, *Science,* 198, 693, 1977.
22. **Gehring, P. J., Watanabe, P. G., and Park, C. N.,** Resolution of dose-response toxicity data for chemicals requiring metabolic activation: example — vinyl chloride, *Toxicol. Appl. Pharmocol.,* 44, 581, 1978.
23. **Anderson, M. W., Hoel, D. G., and Kaplan, N. L.,** A general scheme for the incorporation of pharmacokinetics in low-dose risk estimation for chemical carcinogenesis, *Br. J. Cancer,* 8, 1, 1980.
24. **Hoel, D. G., Kaplan, M. L., and Anderson, M. W.,** The implication of nonlinear kinetics on risk estimation in carcinogenesis, *Science,* 219, 1032, 1983.
25. **Kane, L. E. and Alarie, Y.,** Sensory irritation to formaldehyde and acrolein during single and repeated exposures in mice, *Am. Ind. Hyg. Assoc. J.,* 38(10), 509, 1977.
26. **Kulle, T. J. and Cooper, G. P.,** Effects of formaldehyde and ozone on the trigeminal nasal sensory system, *Arch. Environ. Health,* 30, 237, 1975.
27. **Barrow, C. S., Steinhagen, W. H., and Chang, J. C. F.,** Formaldehyde sensory irritation, in *Formaldehyde Toxicity,* Gibson, J. E., Ed., Hemisphere Publishing, Washington, 1983, 16.
28. **Chang, J. C. F., Steinhagen, W. H., and Barrow, C. S.,** Effects of single or repeated formaldehyde exposure on minute volume of B6C3F1 mice and F-344 rats, *Toxicol. Appl. Pharmacol.,* 61, 451, 1981.
29. **Chang, J. C. F., Gross, E. A., Swenberg, J. A., et al.,** Nasal cavity deposition, histopathology, and cell proliferation after single or repeated formaldehyde exposures in B6C3F1 mice and F-344 rats, *Toxicol. Appl. Pharmacol.,* 68, 161, 1983.
30. **Egle Jr., J. L.,** Retention of inhaled formaldehyde, propionaldehyde, and acrolein in the dog, *Arch. Environ. Health,* 26, 119, 1982.
31. **Morgan, K. T., Patterson, D. L., and Gross, E. A.,** Formaldehyde and the nasal mucociliary apparatus, in *Formaldehyde: Toxicology, Epidemiology, and Mechanisms,* Clary, J. J., Gibson, J. E., and Waritz, R. S., Eds., Marcel Dekker, New York, 1983, 193.
32. **Proctor, D. F.,** The mucociliary system, in *The Nose, Upper Airway Physiology and the Atmospheric Environment,* Proctor, D. F. and Anderson, I., Eds., Elsevier/North Holland Biomedical Press, New York, 1982, 245.
33. **Creeth, J. M.,** Constituents of mucus and their separation, *Br. Med. Bull.,* 34, 17, 1978.
34. **French, D. and Edsall, J. T.,** The reactions of formaldehyde with amino acids and proteins, *Adv. Prot. Chem.,* 2, 277, 1945.
35. **Kihara, Y., Kasuya, M., and Tanaka, K.,** Reaction of aldehydes on starch, *Denpun Kogyo Gakkaishi,* 10, 1, 1982.
36. **Smithson, K. W., Millar, D. B., Jacobs, L. R., et al.,** Intestinal diffusion barrier: unstirred water layer or membrane surface mucous coat?, *Science,* 214, 1241, 1981.
37. **Williams, S. E. and Turnberg, L. A.,** Retardation of acid diffusion by pig gastric mucus: a potential role in mucosal protection, *Gastroenterology,* 79, 299, 1980.
38. **Morgan, K. T.,** Localization of areas of inhibition of nasal mucociliary function in rats following in vivo exposure to formaldehyde, *Am. Rev. Resp. Dis.,* 27, 166, 1983.
39. **Cralley, L. V.,** The effect of irritant gases upon the rate of ciliary activity, *J. Ind. Hyg. Toxicol.,* 24, 193, 1942.

40. **Dalhamn, T.,** Mucous flow and ciliary activity in the trachea of healthy rats and rats exposed to respiratory irritant gasses (SO$_2$, H$_3$N, HCHO), *Acta Physiol. Scand.,* 36(123), 5, 1956.
41. **Falk, H. L., Kotin, P., and Rowlette, W.,** The response of mucus-secreting epithelium and mucus to irritants, *Ann N.Y. Acad. Sci.,* 106, 583, 1963.
42. **Carson, S., Goldhamer, R., and Carpenter, R.,** Responses of ciliated epithelium to irritants, *Am. Rev. Resp. Dis.,* 93, 86, 1966.
43. **Kensler, C. J. and Battista, S. P.,** Chemical and physical factors affecting mammalian ciliary activity, *Am. Rev. Resp. Disease,* 93, 93, 1966.
44. **Heck, H. d'A. and Casanova-Schmitz, M.,** Biochemical toxicology of formaldehyde, *Rev. Biochem. Toxicol.,* 6, 155, 1984.
45. **Neuberger, A.,** The metabolism of glycine and serine, in *Comprehensive Biochemistry,* Vol. 19A, Neuberger, A. and van Deenen, L. L. M., Eds., Elsevier, Amsterdam, 257.
46. **Heck, H. d'A., White, E. L., and Casanova-Schmitz, M.,** Determination of formaldehyde in biological tissues by gas chromatography/mass spectrometry, *Biomed. Mass Spectrom.,* 9, 347.
47. **Uotila, L. and Koivusalo, M.,** Formaldehyde dehydrogenase from human liver: purification, properties, and evidence for the formation of glutathione thiol esters by the enzyme, *J. Biol. Chem.,* 249, 7653, 1974.
48. **Heck, H. d'A. and Casanova-Schmitz, M.,** Reactions of formaldehyde in the rat nasal mucosa, in *Formaldehyde: Toxicology, Epidemiology, and Mechanisms,* Clary, J. J., Gibson, J. E., and Waritz, R. S., Eds., Marcel Dekker, New York, 1983, in press.
49. **Feldman, M. Y.,** Reactions of nucleic acids and nucleoproteins with formaldehyde, *Prog. Nucl. Acid Res. Mol. Biol.,* 13, 1, 1973.
50. **Ross, W. E. and Shipley, N.,** Relationship between DNA damage and survival in formaldehyde-treated mouse cells, *Mutat. Res.,* 79, 277, 1980.
51. **Bedford, P. and Fox, B. W.,** Role of formaldehyde in methylene dimethane sulphonate-induced DNA cross-links and its relevance to cytotoxicity, *Chem.-Biol. Interact.,* 38, 119, 1981.
52. **Ross, W. E., McMillan, D. R., and Ross, C. F.,** Comparison of DNA damage by methylmelamines and formaldehyde, *J. Natl. Cancer Inst.,* 67, 217, 1981.
53. **Fornace Jr., A. J., Lechner, J. F., Grafstrom, R. C., and Harris, C. C.,** DNA repair in human bronchial epithelial cells, *Carcinogenesis,* 3, 1373, 1982.
54. **Fornace, A. J., Jr.,** Detection of DNA single strand breaks produced during the repair of damage by DNA-protein crosslinking agents, *Cancer Res.,* 42, 145, 1982.
55. **Casanova-Schmitz, M., and Heck, H. d'A.,** Effects of formaldehyde exposure on the extractability of DNA from proteins in the rat nasal mucosa, *Toxicol. Appl. Pharmacol.,* 70, 121, 1983.
56. **Mangana-Schwencke, N. and Ekert, B.,** Biochemical analysis of damage induced in yeast by formaldehyde. II. Induction of cross-links between DNA and protein, *Mutat. Res.,* 51, 11, 1978.
57. **Klatt, O., Stehlin, J. S., Jr., McBride, C., et al.,** The effect of nitrogen mustard treatment on the deoxyribonucleic acid by sensitive and resistant Ehrlich tumor cells, *Cancer Res.,* 29, 286, 1969.
58. **Todd, P. and Han, A.,** UV-induced DNA to protein cross-linking in mammalian cells, in *Aging, Carcinogenesis and Radiation Biology,* Smith, K. C., Ed., Plenum Press, New York, 1976, 83.
59. **Swenberg, J. A., Gross, E. A., Martin, J., et al.,** Mechanisms of formaldehyde toxicity, in *Formaldehyde Toxicity,* Gibson, J. E., Ed., Hemisphere Publishing, Washington, 1983, 132.
60. **Swenberg, J. A., Gross, E. A., Randall, H. W., et al.,** The effect of formaldehyde exposure on cytotoxicity and cell proliferation, in *Formaldehyde: Toxicology, Epidemiology, and Mechanisms,* Clary, J. J., Gibson, J. E., and Waritz, R. S., Eds., Marcel Dekker, New York, 1983, 225.
61. **Rusch, G. M., Bolte, H. F., and Rinehart, W. E.,** A 26-week inhalation toxicity study with formaldehyde in the monkey, rat, and hamster, in *Formaldehyde Toxicity* Gibson, J. E., Ed., Hemisphere Publishing, Washington, 1983, 98.
62. **Starr, T. B.,** Mechanisms of formaldehyde toxicity and risk evaluation, in *Formaldehyde: Toxicology, Epidemiology, and Mechanisms,* Clary, J. J., Gibson, J. E., and Waritz, R. S., Eds., Marcel Dekker, New York, 1983, 237.
63. **Proctor, D. F. and Chang, J. C. F.,** Comparative anatomy and physiology of the nasal cavity, in *Nasal Tumors in Animals and Man,* Reznik, G. and Stinson, S. F., Eds., CRC Press, Boca Raton, Fla., 1983.
64. **Anderson, I. and Molhave, L.,** Controlled human studies with formaldehyde, in *Formaldehyde Toxicity,* Gibson, J. E., Ed., Hemisphere Publishing, Washington, 1983, 154.

Chapter 10

DETERMINATION OF HUMAN RISK IN REGULATING POLYCHLORINATED BIPHENYLS (PCBs) — A CASE STUDY

Frank Cordle, Raymond Locke, and Janet Springer

TABLE OF CONTENTS

I. ABSTRACT

The problem of polychlorinated biphenyls (PCBs) became a national concern in 1971 when several incidents of accidental contamination of foods were reported. Following these episodes of food contamination, extensive efforts were successfully undertaken by the FDA to reduce the residues of PCBs in food. However, continuing PCB levels in several species of fresh-water fish and the continuing finding of PCB residues in human breast milk have raised concern about the PCB residues from environmental contamination, and it is this concern which has prompted a reassessment of the human risk due to consumption of such fish. The best evidence that a chemical may produce adverse health effects in humans is provided by adequate epidemiologic data which is either confirmed or supplemented by data from valid animal tests.

Traditionally, where the regulatory agencies have used information obtained in animal toxicology experiments to evaluate hazard and predict hypothetical safety in humans, "safety factors" such as 1:10 or 1:100 have been used. The size of the safety factor and the potential exposure to a chemical are established by properly informed scientific judgment. More recently, efforts have been directed toward the use of a combination of human and animal data and a variety of mathematical models to determine risk.

II. INTRODUCTION

Of the estimated 4.3 million chemicals in existence, some 63,000 are believed to be in use in the U.S. and up to 1000 new ones are coming into use each year. Although the discovery and increasing utilization of synthetic chemicals and other substances have in many ways improved the quality of life in the U.S., many of these substances have proven to be hazardous to humans and animals.

One of the major challenges of regulatory agencies such as the U.S. Food and Drug Administration (FDA) is to determine how best to reconcile these potential hazards with the potential benefits to be obtained from the use of these substances. This problem has been recognized in numerous laws and regulations. Some statutes require the banning of a chemical shown to pose hazards to human and animal health; others require regulatory control of a chemical in order to reduce the hazards resulting from its use. Perhaps the most important issue in policy decisions relating to the control of toxic substances in the environment is an assessment of the human risk from exposure to such substances.

The best evidence that an agent is toxic to humans, or is a human carcinogen with subsequent human risk that can be quantified, is provided by adequate epidemiology data backed by confirmatory animal data. However, for practical purposes, most decisions on human toxicity and carcinogenicity are based on animal studies.

Traditionally, the regulatory agencies have used information obtained in animal toxicology experiments to evaluate hazard and predict hypothetical safety in humans by the application of so-called "safety factors," e.g., 1:10 or more often 1:100. The size of the safety factor and the potential exposure to a chemical are established by properly informed scientific judgement.

More recently efforts have been directed toward the use of a combination of data that might provide a reasonable assessment of the risk of a variety of adverse health effects resulting from exposure to environmental contaminants. To aid in determining a proper regulatory action regarding a toxic chemical in the environment, whether it be a feed additive, industrial pollutant, or natural toxicant, it is helpful to have some knowledge about the background level of the particular adverse health outcome that might be expected from exposure to a particular substance. Since this kind of information is usually difficult to obtain directly from human data, it then becomes necessary to use a combination of human and animal data either to set an acceptable level through the use of safety factors based on human exposure to the substance and extrapolation from animal data to human outcome, or to use one or more of a variety of mathematical models to predict risk.

In the case study of the regulation of PCBs to be described here, both methods are used to describe the problems involved in attempting to determine the risk of adverse human health from exposure to polychlorinated biphenyls (PCB) in some species of fish regulated by FDA.

III. HISTORY OF EXPOSURE

The term PCBs refers to a complex mixture of different chlorobiphenyls and isomers, i.e., two compounds which have the same number of chlorine substitutents on the biphenyl molecule but at different locations.

PCBs were reportedly first synthesized in 1881 but were not commercially available until 1930.[1] The only domestic producer was the Monsanto Company. The use of PCBs in a wide variety of industrial applications steadily increased from 1930 until 1971, when the manufacturer voluntarily restricted their distribution in the U.S. to closed systems, including electrical transformers, capacitors, and heat exchangers.

PCB products manufactured by Monsanto in the U.S. are identified by the trade name "Aroclor®", and the particular kind of Aroclor® is identified by a four-digit number, e.g., Aroclor® 1254 or Aroclor® 1260. The first two digits refer to the 12 carbon atoms that make up the biphenyl and the second two digits refer to the approximate percent by weight of chlorine in the mixture.

Under this numbering system, Aroclor® 1254 contains 12 carbon atoms and about 54% chlorine, while Aroclor® 1260 contains 12 carbon atoms and about 60% chlorine. A typical residue from fish resembles the Aroclor® 1254 mixture more than it does the other Aroclors®.[2,3]

PCBs became a national concern in 1971, when several incidents of accidental contamination of foods were reported. In addition, the extent of the environmental contamination and its persistence were not known. Subsequently, various regulatory actions were taken by the agencies involved, and with the cooperation of the only U.S. producer the situation was felt to be under control.

Then, in 1975, high levels of PCB contamination in Hudson River fish were reported; national attention was refocused on PCBs. It was soon apparent that the actions and control measures of the early 1970s had not succeeded in even substantially alleviating the problems associated with PCB contamination in the environment.

To determine human exposure to PCBs through dietary fish one must know the levels of PCB residues in the edible portions of fish and the amounts of the various kinds of fish consumed by the population at large and by special subgroups of the population. Information has been compiled by the National Marine Fisheries Service-NOAA[4] on the most important types of fish eaten in the U.S. today and the mean daily amount of each type eaten by the subpopulations of actual users. This study included a sample of 25,947 fish eaters selected to represent all fish eaters in the United States.

Information from the survey indicates that some 20 species comprise 95% of all the fish products eaten. Although 93% of the U.S. population (197 million) eat fish, the average annual per capita consumption is small: 15.0 lb/year, the major part of which is a large "unclassified" fish fraction that exists in the U.S. diet, ranking just below tuna in importance. This unclassified fraction of fish represents a variety of fish species, each of which, taken separately, would contribute only a minor proportion of the diet. However, when taken as a group, those species represent a major contribution to fish consumption in the U.S. The major portion of many of our most familiar types of seafood is imported, and fresh-water species, led by trout, bass, and catfish, comprise about 9% of our total fish diet.

IV. EPIDEMIOLOGY OF HUMAN EXPOSURE

Considerable scientific interest has centered on the Yusho incident in Japan. In 1968, human intoxication with Kanechlor® 400, a PCB manufactured in Japan, was noted when a heat exchanger leaked this PCB into rice oil (''Yusho'' oil) that was consumed by Japanese families.

The typical clinical findings included chloracne and increased pigmentation of the skin, increased eye discharge, transient visual disturbances, feeling of weakness, numbness in limbs, headaches, and disturbances in liver function. Most of the babies born to mothers with Yusho had skin discoloration which slowly regressed as the children grew. Adult Yusho patients had protracted clinical disease with a slow regression of symptoms and signs, suggesting a slow metabolism and excretion of the PCB in humans, probably due to a long biological half-life.

A review of the literature in 1972 revealed the following facts: The average PCB content of the rice oil in the dose-response epidemiologic study was 2500 ppm. In this study, the average cumulative intake of PCBs leading to overt symptomatology was 2000 mg, and the lowest dose leading to overt symptomatology was 500 mg. Originally, rice oil contaminated with a heat exchanger, Kanechlor® 400, a polychlorinated biphenyl, was associated with Yusho symptomatology. PCBs were identified in the contaminated rice oil consumed and in the blood and tissues of Yusho patients. Therefore, the effects seen were attributed to PCBs.

In the review by Kuratsune et al.[5] a new factor was introduced into the system; namely, the canned rice oil was also contaminated with chlorinated dibenzofurans (PCDFs) to the extent of 5 ppm. In addition, in this same paper Kuratsune presented data of Nagayama et al.[6] showing that polychlorinated dibenzofurans were present in the liver and adipose tissue of Yusho patients, while none was found in that of a control group. Nagayama et al.[6] reported that the ratios of PCBs to PCDFs in Kanechlor® 400, in a Yusho oil of February 5 or 6, 1968, in adipose tissue, and in liver from a Yusho patient were 50,000, 200, 144, and 4:1, respectively. Thus, relative to Yusho oil, the liver with a PCB to PCDF ratio of 4:1 appears to concentrate PCDFs selectively relative to the PCBs. If PCDF is 200 to 500 times more toxic than PCB, the contaminated rice oil would be 2 to 3.5 times more toxic than expected from the PCB content alone.

Additional work has indicated that the original estimate of PCBs in the rice oil was in error because the analysis was based on total organic chlorine present, and that the rice oil contained approximately 1000 ppm of chlorinated quaterphenyls in addition to the PCB residues. A discussion of the use of these data in attempting to set so-called safe levels of PCBs in the diet will be presented in a later section of this chapter.

The earliest reports of adverse health effects due to exposure of workers to PCBs in this country are probably those of Schwartz[7] who described skin lesions and symptoms of systemic poisoning among workers who were said to have inhaled chlorobiphenyls; their complaints included digestive disturbances, burning of the eyes, impotence, and hematuria. Patch tests with the chlorobiphenyls were negative, and Schwartz speculated that the cause of the skin lesions was mechanical plugging of the follicles of the skin as the fumes solidified on it; the chlorine present in the products then exerted an irritating effect on the plugged follicles and thus caused suppuration. No quantitative data were reported, but a number of preventive practices were recommended.

There have been numerous reports over the ensuing years[1] of cutaneous eruptions, referred to as chloracne, and of systemic manifestations as well, among marine electricians, machinists, capacitor and transformer manufacturing workers, and others occupationally exposed to PCBs. However, in many of these reports the exposures are described as having been to mixtures of chlorinated hydrocarbons, quite often of chlorinated naphthalenes and PCBs.[8]

As indicated previously, low levels of human exposure to PCBs in the U.S. population may occur from air and water.[9] The Great Lakes represent a significant area of the nation contaminated with PCBs. The Michigan Water Resources Commision[10] reported that many surface water samples contain PCBs at concentrations above the detection limit of 10 ppt. In this study, residues of PCBs were found at ten locations from rivers and streams discharging into the Great Lakes. Effluents from wastewater treatment plants servicing industrialized communities have been found to be highly contaminated.

The Michigan Department of Public Health has reported the results of a study[11] which attempted to assess some of the consequences of human exposure to PCBs from the consumption of sportsfish caught in different areas of Lake Michigan. The study included exposed and control subjects from five areas of Michigan bordering on Lake Michigan. Exposed study subjects were those individuals who consumed at least 24 to 26 lb of Great Lakes fish per year. Control subjects were those individuals who consumed less than 6 lb of Great Lakes fish per year.

An assessment of the findings in the study indicates that the most frequently recorded quantity of fish consumed by the study participants was in the 24 to 25 lb/year range. The highest recorded fish consumption over the two-year period of the study was 180 lb/year and the highest single-season consumption was 260 lb. Mean PCB levels are reported as 18.93 ppm in whole lake trout in 1973 and 22.91 ppm in 1974, and as 12.17 ppm in Coho salmon in 1973 and 10.45 ppm in 1974. However, comparisons of PCB levels in raw vs. cooked fish indicated that actual human exposure to PCBs from fish consumption is less than might be expected from the raw fish data. This is not unexpected, since preparation (trimming away fatty tissue) and cooking have resulted in a decrease in the amount of PCBs actually consumed from fish. For example, the PCB level in cooked lake trout consumed by the study participants ranged from 1.03 to 4.67 ppm; in cooked salmon from 0.48 to 5.38 ppm; and in other cooked fish from 0.36 ppm to 2.06 ppm. These levels are decidedly lower than the level of PCB contamination reported in raw trout and salmon.

PCBs were found in all blood specimens collected during the study period from the 182 study participants, including controls. The values ranged from a low of 0.007 ppm in blood in the control group to a high of 0.366 ppm in the exposed group. Although there was a wide range of blood values for each quantity of fish consumed, there was a positive correlation between the reported quantity of Lake Michigan fish consumed and the concentration of PCB in the blood of study participants. No annual variation in PCB blood levels in humans could be demonstrated. The mean PCB blood values for the control and exposed groups did not appear to change markedly from 1973 to 1974.

In addition, abstinence from consumption of Lake Michigan fish for a period of 90 days or more did not change the PCB blood levels significantly. PCB blood levels over the abstaining period show variation but no steady decline in PCB. In fact, more subjects showed no change or a rise than showed a decline in PCB blood levels over time.

The calculated quantity of PCBs ingested by eating Lake Michigan fish averaged 46.5 mg/year and ranged from 14.17 to 114.31 mg/year. PCB ingestion for each individual was determined by proportioning his/her reported annual fish consumption by frequency of species eaten and the cooked fish PCB levels for those fish. The community average for cooked fish was used in instances in which cooked fish determination was not available for study participants. Because fish consumption was found to vary from year to year, the average annual consumption for each individual for the two baseline years of study was used in each case.

Results from this study indicate that the calculated mean daily dose received by the exposed group is 1.7 μg/kg/day and ranges from 0.09 to 3.94 μg/kg/day. If the average annual rate of PCB ingestion from fish indicated by these study results were continued over the years, the average sports fisherman consuming contaminated fish could receive a total PCB dose equal to 200 mg in approximately 4.3 years. Under the same set of assumptions, individuals consuming greater than average amounts of contaminated fish would reach the total dose level sooner. No adverse health effects or groups of symptoms that were clearly related to PCB exposure could be identified in the exposed group.

This implies that exposure to PCBs from eating contaminated fish at the levels observed and the presence of PCBs in these exposed persons have not caused any observable adverse health effects similar to those observed in the Yusho population. However, this does not exclude the possibility that effects too subtle for detection are occurring, or the possibility of long-term health effects.

A recent report describing the residue levels of PCBs in human breast milk by Wickizer et al.[12] illustrates the potential long-term presence of such residues in humans who have been exposed in the past and the need to continue the reduction of such exposure in the future. Results of the study, which was carried out in the State of Michigan, indicate that of the 1075 breast milk samples collected from 68 of the state's 83 counties, all contained PCB residues ranging from trace amounts to 5 ppm on a fat basis.

The mean level was approximately 1.5 ppm; 49.5% of the samples had PCB levels of 1 to 2 ppm, 17.4% had levels of 2 to 3 ppm, and 6.14% had levels of 3 ppm or greater. The public health significance of PCB residues in human breast milk and its effects on breast-fed infants are unclear at the present time. Although public health officials and pediatricians have become increasingly concerned about PCB residues in human breast milk and its potential adverse effect on breast-fed infants, in view of the known benefits of breast feeding such authorities have been reluctant to recommend changes in current breast feeding practices.

Because of the lack of sufficient human data, risk assessments for potential toxic effects of chemicals must of necessity be estimated from animal experiments. In the absence of contradictory kinetic or metabolic data, the animal data are used to estimate potential human risks. Because the numbers of animals used in tests are limited, doses above the human exposure levels are used to increase the probability of detecting potentially toxic chemicals.

Thus, the risks at low doses must be estimated from higher experimental doses. Because of the inability to observe the low end of the dose response curve with precision, the linear (or when necessary one-hit) extrapolation from high to low doses is often used.[13-16] Because of the many uncertainties involved in risk estimation, this approach is conservative and is biased in favor of public health. Also, linear (one-hit) extrapolation is the limiting case for the multistage model of carcinogenesis at low doses. Since the shapes of dose-response curves at low doses are unknown, it must be remembered that actual estimates of risk are not being obtained. Based upon plausible assumptions, however, it generally is possible to place upper bounds on potential risk by use of linear (one-hit) extrapolation based on animal data.

Estimates of risk based on such an approach have increasingly been attempted: an example is the assessment by the National Research Council on halomethanes in drinking water.[17] The results, while crude, have some value in the decision-making, if only to place in perspective the hazards of pollutants whose toxic properties are known.

V. ANIMAL DATA

Several elements related to risk must be considered in any assessment of animal data, namely, (1) the similarity of exposure in animals compared to humans, e.g., types of Aroclors® used in animal experimentations (most human exposure is to Aroclor®s 1254, and a typical residue from fish resembles the Aroclor® 1254 mixture more closely than it does the other Aroclors®;[2,3] and (2) the kinds of outcome that might be comparable to those for humans. In this case risk assessment, or the estimation of the acceptable daily intake, will be based on general toxicity, carcinogenicity, and the effects on reproduction for which there is little or no previous experience in such risk assessment.

The acute toxicity (oral and dermal) of the PCBs is of relatively low order when the substances are administered as a single dose. In contrast, the subacute toxicity of both the PCBs and individual chlorinated biphenyls appears to be of far greater concern; species sensitivity and cumulative toxic effects appear after continuous exposure at low levels.[18-20]

The effects of various PCB compounds have been studied in a number of animal species, and the results have been compiled and evaluated in recent reviews.[1,21,22] In assessing the possible toxicological hazard posed by PCBs for humans, it is preferable to utilize animal feeding studies in which the PCBs are added at low levels to the diet and the treatment is continued essentially throughout the life span of the animals. For assessing the risk posed to humans as a result of exposure to PCBs, three long-term studies have been chosen which are cited and discussed in detail below: (1) The National Cancer Institutes bioassay of Aroclor® 1254 for possible carcinogenicity in male and female Fischer 344 rats, (2) the study of Kimbrough et al.[18] on the induction of liver tumors in Sherman strain female rats by Aroclor® 1260, and (3) the 11-month study by Kimbrough and Linder[19] of the toxic effects of Aroclor® 1254 in male BALB/cJ mice.

In addition, because of the known extreme sensitivity to PCB-related toxicity of the rhesus monkey compared to rodent species, the short-term study by Allen and Norback[23] of the pathological responses of these primates to PCB exposure as a basis for risk assessment is also described.

In the National Cancer Institute's bioassay of Aroclor® 1254,[24] groups of male and female Fischer 344 rats (24 of each sex per group) were administered the test compound in the diet at 25, 50, or 100 ppm for a period of 104 to 105 weeks. Matched controls consisted of groups of 24 untreated rats of each sex. All animals were observed daily for signs of toxicity and palpated for tissue masses at each weighing. Moribund animals were sacrificed and subjected to gross and microscopic pathological examination, as were the animals sacrificed at the end of the experimental period. It was concluded that, under the conditions of the bioassay, Aroclor® 1254 was not carcinogenic in Fischer 344 rats; however, it was suggested that the high incidences of hepatocellular proliferative lesions of the GI tract in the Aroclor®-treated males and females might be associated with the administration of the compound.

In the study by Kimbrough et al.,[18] 200 Sherman strain female rats were fed a diet containing 100 ppm of Aroclor® 1260 for approximately 21 months, and treatment was discontinued for 6 weeks before the animals were sacrificed at 23 months. A group of 200 untreated female rats served as controls. All animals were observed daily, and moribund animals were sacrificed and subjected to gross and microscopic pathological examination, as were the animals sacrificed at the end of the experimental period. The authors concluded that Aroclor® 1260, when fed in the diet, had a hepatocarcinogenic effect in these rats. No significant differences could be observed between experimental and control animals with regard to the incidence of tumors in other organs.

In another study by Kimbrough and Linder,[19] 50 male BALB/cJ mice were fed a diet containing 300 ppm of Aroclor® 1254 for a period of 11 months. Another group of 50 male mice received a diet containing 300 ppm of Aroclor® 1254 for a period of 6 months and a control rat chow diet for the next 5 months.

Control males were fed a plain rat chow diet throughout the 11-month study. Food consumption and body weight were monitored during the study. The animals were sacrificed at the end of 11 months and the organs were examined grossly for pathology. The liver and any other abnormal-appearing tissues were examined microscopically. Of the 22 surviving animals which had received a diet containing 300 ppm of Aroclor® 1254 for 11 months, 9 animals exhibited hepatomas. One of the 24 mice which had received a diet containing 300 ppm of Aroclor® 1254 for 6 months and the control diet for the next 5 months exhibited a hepatoma. None of the control animals developed hepatomas during the course of the experiment.

PCB mixtures also appear to enhance hepatocarcinogenesis when they are given coincidentally with some chemical carcinogens[25] and to inhibit hepatocarcinogenesis when they are given with other chemical carcinogens.[26,27] Preston et al.[28] have reported the promoting effects of Aroclor® 1254 and polychlorinated dibenzofuran-free Aroclor® 1254 on diethylnitrosamine-induced tumors in rats. The rats were treated with diethylnitrosamine (DENA) in drinking water for 5 weeks and subsequently given either a controlled diet or a diet supplemented with either Aroclor® 1254 or Aroclor® 1254 from which the polychlorinated dibenzofurans (PCDF) were removed. Of those animals receiving DENA alone, 16% exhibited hepatocellular carcinomas. Of the rats treated with DENA followed by the diet with Aroclor® 1254, 64% had hepatocellular carcinoma. Of the rats treated with DENA followed by the PCDF-free Aroclor® 1254, 84% developed hepatocellular carcinoma.

Allen and Norback[23] have investigated PCB-related reproductive dysfunctions in the rhesus monkey, an animal species known to be more susceptible than rodents to the toxic effects of PCBs. In one series of experiments, eight female monkeys were fed a diet containing 2.5 ppm of Aroclor® 1248 for 6 months, eight other females were fed a diet containing 5.0 ppm of Aroclor® 1248, and twelve females served as controls. At the end of 6 months, all experimental and control animals were bred to control males.

Six of the eight animals receiving the test compound at 5.0 ppm in the diet conceived. The remaining two were bred on five separate occasions without conceiving. Four of the six animals which did conceive experienced abortions early in gestation, and one gave birth to a stillborn infant. Eight females receiving diets containing 2.5 ppm of Aroclor® 1248 were able to conceive, but only five females in this group were able to carry their infants to term, while the remaining three animals experienced abortions. The infants born to all experimental mothers receiving the test compound at either 2.5 ppm of 5.0 ppm in the diet were small at birth and exhibited detectable concentrations of PCBs in the skin. All of the 12 control females conceived and had normal births. Each of the three toxicological studies of PCBs described above were used to assess the carcinogenic or reproductive risks posed by these compounds.

In the risk assessment described below, the average and an upper limit for PCB intake per day are estimated. The nationwide survey of fish consumption conducted for NMFS-NOAA during 1973 to 1974 has been used to estimate consumption of fish. This survey included 25,947 persons representative of the U.S. population who recorded their fish and seafood consumption for each family member for a 1-month period. Of these 25,947 persons, 3939 ate the species of fish which had levels of PCB above 1 ppm.

Since the effect of instituting a tolerance of PCB levels down to 1 ppm would change the PCB contributed to the diet by fish only in those fish which showed PCB levels above 1 ppm, risks corresponding to tolerances of 5 ppm, 2 ppm, 1 ppm, or 0 were calculated for those 3939 persons who ate the species of interest. Because analytical methods for regulatory purposes are not presently available for levels less than 1 ppm, no risks were calculated for lower levels. The calculated risks could then be extrapolated to that proportion of the total U.S. population which is expected to eat these species, that is, 3939/25,947, or 15.2%.

For these persons, the consumption per day of each type of fish was multiplied by a mean PCB level estimated for each tolerance level to give a total PCB intake from fish per person per day. The 50th and 90th percentiles of PCB intake from fish for those eaters of the species of interest were then used to calculate risks.

The mean PCB level estimated when a given tolerance is in effect is perhaps the most difficult part of the risk estimation. The effect of a tolerance on the distribution of PCB levels depends to a large degree on the actual distribution of PCBs before a tolerance is instituted. The most recent data available on PCB levels in fish were the 1978 and 1979 FDA survey data, consisting of 713 samples for 1978 and 179 samples for 1979 collected from all of the FDA districts. This sampling is not representative or extensive enough to

Table 1
ANIMAL DATA USED FOR RISK EXTRAPOLATION TO HUMANS

Animal studies	Dose of Aroclor® fed (ppm)						
	0	2.5	5.0	25	50	100	300
NCI bioassay — Fischer rats fed Aroclor® 1254							
Total malignancies							
Males	5/24			2/24	9/24	12/24	
Females	4/24			13/24	8/24	9/24	
Combined	9/48			15/48	17/48	21/48	
Liver carcinoma and adenomas							
Males	0/24			0/24	1/24	2/24	
Females	0/24			0/24	1/24	2/24	
Combined	0/48			0/48	2/48	4/48	
Hematopoietic system							
Males	3/24			2/24	5/24	9/24	
Females	4/24			6/24	6/24	6/24	
Combined	7/48			8/48	11/48	15/48	
Kimbrough — female Sherman rats fed Aroclor® 1260							
Hepatocellular carcinomas	1/173					26/184	
Kimbrough — BALB/cJ male mice fed Aroclor® 1254							
Hepatomas, neoplastic nodules	0/5						9/22
Allen — female monkeys fed Aroclor® 1248[a]							
Problems of reproduction	0/12	3/8	7/8				

[a] For the monkeys fed Aroclor® 1248, assuming a body weight of 5 kg and daily food consumption of 250 g: 2.5 ppm = 125 µg/kg body weight per day; 5.0 ppm = 250 µg/kg body weight per day.

estimate an underlying nationwide distribution by species. As a rough approximation of the effect of a tolerance on the distribution, the values of PCB above the assumed tolerance were eliminated from the sample distribution and the mean was recalculated for each species. It should be noted that assuming a zero tolerance is not equivalent to using all values, inasmuch as the 1975 to 1979 survey was carried out when a tolerance of 5 ppm was in effect. Thus, the effect of going from zero tolerance to a tolerance of 5 ppm would be greater than shown here. Tuna and shellfish were assumed to have 0.0 ppm mean levels of PCB. The limited data available on tuna shows mean levels of less than 0.01 ppm. The mean values were then multiplied by consumption figures as described above to obtain intake of PCBs per day, with the parts per million of the diet calculated by assuming 1500 g total food intake per person per day.

Data from the NCI bioassay program in which Aroclor® 1254 was fed to Fischer rats are presented in Table 1, which shows the numbers for total malignancies, liver carcinomas plus adenomas, and malignancies of the hematopoietic system in males, females, and males and females combined at various feeding levels. Similar data are also presented in Table 1 for the feeding studies of Kimbrough in which female Sherman rats were fed 100 ppm Aroclor® 1260 and BALB female mice were fed 300 ppm Aroclor® 1254, and for the Allen reproductive data in which Aroclor® 1248 was fed at 2.5 or 5.0 ppm.

Using the data for Table 1, the upper confidence limits (99%) on lifetime risks for cancer and problems of reproduction in eaters of the 12 fish species of interest at the 50th percentile are presented in Table 2. Upper limits on estimated human risks have been computed from the NCI data for total malignancies for males plus females, liver carcinomas plus adenomas in males plus females, and malignancies of the hematopoietic system in males plus females. Risks computed from the Kimbrough data and from the Allen data are also presented in Table 2. The various risks shown are based on PCB values in fish, assuming zero tolerance, or a tolerance of 5, 2, or 1 ppm.

In describing the potential risk to human health from exposure to PCBs it seems appropriate also to review the general scheme of establishing safe regulatory levels. In the Yusho incident individuals consumed an average of 15,000 mg/day of the contaminated oil. The oil itself was contaminated at levels of 2000 to 3000 ppm PCBs and other contaminants (such as polychlorinated quaterphenyls); the average level of the contamination in the oil was 2500 ppm. The levels of contamination of the rice oil were calculated at the time of the incident by comparing the known organic chlorine content of the rice oil with the known organic chlorine content of Kanechlor® 400.

Based on the two average levels (consumption of rice oil and residue levels in the rice oil), the average daily intake of the combination of contaminants was 37.5 mg/day. The average cumulative dose of the contaminants causing an overt effect in the Japanese victims was reported to be 2000 mg. Thus 53 days of exposure was required to consume this amount. The period of exposure no doubt varies around this figure. However, it was estimated that the maximum exposure time was 100 days. It must also be assumed that the adverse health effects result from the combination of contaminants and that these effects are reasonably similar for the levels of PCBs as well as for the levels of chlorinated quaterphenyls or other contaminants.

Humans in the U.S. have not been exposed to PCBs at the high residue levels that occurred in the Yusho incident. PCB exposure in the United States has been assumed to be sporadic and self-limiting in nature, as far as the general public was concerned. Accordingly, in developing temporary tolerances based on the data from the Yusho incident, a time period of 1000 days of exposure was used. As previously stated, this was not an analysis based on lifetime exposure. Rather, it was postulated that PCB levels in food in the U.S. would steadily decrease over the 1000-day time period used in the calculation. This has in fact taken place for most food. Jelinek and Corneliussen[29] reported that from the 1969 to 1975 period, there were significant decreases in PCB levels in all foods with the exception of fish, where no particular trend had been noted.

In calculating a total allowable exposure from the average overt dose in the Yusho incident, a safety factor of 1:10 was used for those effects observed in the Yusho population, resulting in a total allowable exposure of 200 mg. Because of the sporadic and self-limiting nature of PCB exposure in the U.S., the total exposure (200 mg) was spread out over the 1000-day time period, providing a tolerable daily exposure of no more than 200 μg/day. Transforming this figure and using an average body weight of 70 kg for an adult produced a value of 3 μg/kg body weight per day.

Infants and young children may be more susceptible than adults to toxicants such as PCBs. They also consume a greater amount of food per kilogram of body weight and therefore have a proportionately greater exposure to PCBs than adults. Thus, in calculating the temporary tolerances, it seems appropriate to use an additional safety factor for infants and young children. The acceptable daily exposure for children is therefore calculated by using the lowest total dose producing an adverse health effect in the Yusho incident, which was determined to be 500 mg of the contaminants. Using the 1:10 safety factor spread over 1000 days, the tolerable daily exposure is 50 μg/day. Infants and young children should, therefore, not be exposed to PCBs at a level greater than 1 μg/kg body weight per day. An adult who

Table 2
UPPER CONFIDENCE LIMITS (99%) ON LIFETIME RISKS OF CANCER AND PROBLEMS OF REPRODUCTION IN EATERS OF FISH SPECIES OF INTEREST

Animal studies on which risks are based	50th Percentile eaters				90th Percentile eaters			
	Assuming no tolerance	Assuming tolerance = 5 ppm[a]	Assuming tolerance = 2 ppm	Assuming tolerance = 1 ppm	Assuming no tolerance	Assuming tolerance = 5 ppm	Assuming tolerance = 2 ppm	Assuming tolerance = 1 ppm
NCI bioassay, total malignancies for male and female	4.1	3.7	2.7	1.6	10.6	9.8	7.2	4.4
NCI bioassay, liver carcinoma and adenomas for male and female	0.9	0.9	0.6	0.4	2.5	2.3	1.7	1.0
NCI bioassay, hematopoietic for male and female	2.7	2.4	1.8	1.1	7.0	6.5	4.7	2.9
Kimbrough, rats liver carcinoma	1.3	1.2	0.8	0.5	3.4	3.1	2.3	1.4
Kimbrough, mice liver hepatomas	2.0	1.8	1.2	0.8	5.2	4.8	3.5	2.2
Allen, female—male reproduction[b]	337	307	222	132	883	811	595	367

Note: All risks are lifetime risks computed as rates per 100,000 of the population at risk.

[a] For each assumed tolerance, PCB values below the tolerance were eliminated.
[b] Inasmuch as the monkeys were fed Aroclor® 1248 for only 6 months, the risk computed for problems of reproduction are not true lifetime risks.

consumes a balanced and varied diet should not be expected to ingest more than the tolerable daily exposure of 200 μg/day. Similarly, infants or young children consuming a balanced and varied diet should not be expected to ingest more than their tolerable daily exposure.

VI. DISCUSSION

There is considerable disparity in reproductive results of monkeys compared to other species. Yet the consuming public tends to consider that monkeys are more like humans than rodents are. Thus it is necessary to consider why such disparities exist. It has been pointed out by McNulty[30] and by Allen et al.[23,31] that rhesus monkeys are very sensitive to PCBs, not only in the subacute effects, but in problems of reproduction as well. These monkeys received 2.5 ppm or 5.0 ppm in the diet over a period of 6 months; total intake ranged from 250 to 400 mg.

In contrast, Keplinger et al.[32] reported low mating indices and decreased survival of pups in rodents receiving Aroclor® 1242 at 100 ppm. No reproduction effects were found with Aroclor® 1242 or 1254 at 1 or 10 ppm although other studies in rodents[33] have demonstrated reproduction effect; e.g., in Sherman rats exposed to Aroclor® 1254 or Aroclor® 1260, the exposures producing such effects have been at levels considerably higher than the 2.5 or 5.0 ppm used in the monkey studies (20, 100, or 500 ppm).

Additional differences between the monkey and rodent sensitivities to PCBs are illustrated by the fact that the Sherman rats in the Kimbrough study that were fed 100 ppm of Aroclor® 1260 for 23 months apparently exhibited no subacute effects, which is also true of the NCI study in which Fischer rats were fed 25, 50, or 100 ppm for 104 to 105 weeks. In contrast, the monkeys fed 2.5 or 5.0 ppm in the Allen study displayed some subacute effects similar to Yusho as early as 2 months into the study.

The data upon which the various risk projections of carcinogenicity are made also illustrate the difficulties in such exercises. For example, the data used for projection of risk for Aroclor® 1254, the most common Aroclor® of human exposure, are taken from an NCI bioassay project in which the Aroclor® 1254 was found under the conditions of the test to be negative for carcinogenicity. In addition, an examination of Table 2 shows that the upper limit of lifetime risk of cancer for the 50th percentile of eaters of the species of interest is generally of the same magnitude in any of the studies shown.

Let us assume that what appears to be increased risks in reproduction from exposure to PCBs based on monkey studies is directly associated with the levels fed, the particular Aroclor® fed, and the differences in species sensitivity, i.e., the monkeys are considerably more sensitive to the effect of PCBs than are rodents or humans. It then becomes difficult to explain the differences between the Kimbrough study, which was positive for carcinogenicity in female Sherman rats fed Aroclor® 1260 at 100 ppm, and the NCI feeding studies which were negative for Aroclor® 1254 fed to Fischer rats at 25, 50 or 100 ppm.

These differences in carcinogenic outcome could be attributed to any of several causes: (1) the Kimbrough study used Sherman rats, whereas the NCI study used Fischer rats; (2) three levels of the Aroclor® were fed in the NCI study but only one level in the Kimbrough study; and (3) Kimbrough used 184 test animals at the 100 ppm level, whereas NCI used only 24 animals in each sex group at the 100 ppm level. On the other hand the difference may be purely statistical in which outcome could be changed in either direction by using comparable protocols and a similar number of animals. Obviously the other elements of unknown quantity in these or any other similar assessments involve the adequacy of the fish consumption data as well as the PCB residue data.

VII. SUMMARY

In any estimates of human risk derived from the extrapolation of animal data, close attention should be given not only to the levels of exposure to the various Aroclors® in a variety of animal studies but also to the way in which the exposure relates to human experience. For example, studies in monkeys have reported signs and symptoms similar to those of Yusho after 2 months of exposure to Aroclor® 1248 at levels of 2.5 and 5.0 ppm in the diet or 125 and 250 μg/kg body weight per day, respectively. Reproduction problems were reported in these monkey studies at each of these levels after 6 months of exposure.

In contrast, problems of reproduction have been observed in rodents only at considerably higher levels of exposure, e.g., in rats fed Aroclor® 1254 at 7.2 and 37.0 mg/kg body weight per day or 100 and 500 ppm in the diet. Rats fed Aroclor® 1260 at 500 ppm or 35.4 mg/kg body weight per day also exhibited reproduction problems. No problems were observed with Aroclor® 1260 at levels of 5 ppm (0.39 mg/kg body weight per day), 20 ppm (1.5 mg/kg body weight per day), or 100 ppm (7.4 mg/kg body weight per day).

In one study, the carcinogenicity of PCBs in rats (Fischer strain) fed Aroclor® 1254 at 25 ppm (1.9 mg/kg body weight per day), 50 ppm (3.8 mg/kg body weight per day), or 100 ppm (7.16 mg/kg body weight per day) was reported to be negative under the test conditions. Although some malignancies were observed, there was no statistical difference between test animals and controls. In another study, female Sherman rats fed Aroclor® 1260 at 100 ppm (7.4 mg/kg body weight per day) exhibited a statistically significant difference between test animals and controls for hepatocellular carcinomas.

In contrast, there appears to be little evidence of human exposure to these levels in the U.S. especially for the consumption of fish. Even in the Yusho experience in Japan where clinical signs and symptoms were observed, the average level of consumption of PCB residues in the rice oil is estimated at 0.75 mg/kg body weight per day.

In Michigan sportfishermen, who are presumed to be among the higher consumers of fish with PCB residues, the average intake has been reported at 1.7 μg/kg body weight per day with a range of 0.09 to 3.94 μg/kg body weight per day.

Thus problems of interpretations arise in comparing the levels of the various Aroclor® which have produced effects in animals, ranging from 125 μg in monkeys to the milligram levels in rodents, with the exposure estimates in humans from fish consumption. For example, estimates of the lifetime human risk of cancer and reproduction problems (Table 2) for exposure in the 90th percentile of fish eaters, i.e., 0.29, 0.21, and 0.13 μg PCB/kg body weight per day, indicate risk from exposure well below the average Michigan exposure and certainly well below the levels of exposure in the Yusho incident.

In light of the uncertainties upon which these risk estimates have been made, perhaps an equally compelling argument could be made for the establishment of either a 2-ppm or a 1-ppm tolerance. As suggested previously, the differences in risk between the two levels decreases only slightly even in the species of interest. From one point of view, this conclusion supports a rationale for proposing the 2-ppm tolerance based on the original calculation of an allowable daily intake resulting from the traditional use of the 1:10 safety factor which has been described in the previous section. In this case some similarities exist even in the face of uncertainty. To some, the rationale and logic for establishing 2 ppm can be justified from either the risk approach or the so-called safety approach. To others, nothing short of zero tolerance has any rationale or logic. For these, there may be no answer.

REFERENCES

1. Final Report, DHEW Subcommittee on the Health Effects of Polychlorinated Biphenyls and Polybrominated Biphenyls, U.S. Department of Health, Education, and Welfare, Washington, D.C., 1976.
2. **Veith, G. D.,** Baseline concentrations of polychlorinated biphenyls and DDT in Lake Michigan fish, 1971, *Pestic. Monit. J.,* 9, 921, 1975.
3. **Zitko, V., Hutzinger, O., and Chor, P. M. K.,** Contamination of the Bay of Fundy & Gulf of Marine Area with PCBs, PCTS, and chlorinated DBF and DBD, *Environ. Health Persp.,* 1, 47, 1972.
4. Seafood consumption study 1973—1974, National Purchase Diary, National Marine Fisheries Service — NOAA, U.S. Department of Commerce, Washington, D.C., 1976.
5. **Kuratsune, M., Masuda, Y., and Nagayama, J.,** Some of the recent findings concerning Yusho, National Conference on Polychlorinated Biphenyls, Chicago, Illinois, Publ. No. 560/6-75-004, Environmental Protection Agency, Washington, D.C., 1976
6. **Nagayama, J., Masuda, Y., and Kuratsune, M.,** Dibenzofurans in Kanechlors®, *Jpn. J. Hyg.,* 30, 126, 1975.
7. **Schwartz, L.,** Dermatitis from synthetic resins and waxes, *Int. J. Public Health,* 26, 586, 1936.
8. **Kimbrough, R. D.,** Toxicity of polychlorinated polycyclic compounds and related chemicals, *Crit. Rev. Toxicol.,* 2(4), 445, 1974.
9. **Kutz, F. W. and Yang, H. S.,** A note on polychlorinated biphenyls in air, National Conference on Polychlorinated Biphenyls, Chicago, Publ. No. 560/6-75-004, Environmental Protection Agency, Washington, D.C., 1976.
10. **Dennis, D. S.,** Polychlorinated biphenyls in the surface waters and bottom sediments of the major drainage basins of the United States, National Conference on Polychlorinated Biphenyls, Chicago, Publ. No. 560/6-75-004 Environmental Protection Agency, Washington, D.C., 1976.
11. **Humphrey, H. E. B., Price, H. A., and Budd, M. L.,** Evaluation of changes of the level of polychlorinated biphenyls (PCBs) in human tissue, Final Report of FDA Contract No. 223-73-2209, 1976.
12. **Wickizer, T. W., Brilliant, L. B., Copeland, R., and Tilden, R.,** *Am. J. Public Health,* 71(2), 132, 1981.
13. **Brown, J. M.,** Linearity vs. non-linearity of dose response for radiation carcinogenesis, *Health Phys.,* 31, 231, 1976.
14. **Crump, K. S., Hoel, D. G., Langley, C. H., and Peto, R.,** Fundamental carcinogenic processes and their implications for low dose risk assessment, *Cancer Res.,* 36, 2973, 1976.
15. **Guess, H., Crump, K., and Peto, R.,** Uncertainty estimates for low dose rate extrapolation of animal carcinogenicity data, *Cancer Res.,* 37, 3475, 1977.
16. **Crump, K. S., Guess, H. A., and Deal, K. L.,** Confidence intervals and test of hypotheses concerning dose response relations inferred from animal carcinogenicity data, *Biometrics,* 33, 437, 1977.
17. Chloroform environmental assessment, Environmental Studies Board, Commission on Natural Resources, National Research Council, National Academy of Sciences, Washington, D.C., 1978.
18. **Kimbrough, R. D., Squire, R. A., Linder, R. E., Strandbert, J. D., Mondali, R. J., and Bubse, V. W.,** Induction of liver tumors in Sherman strain female rats by polychlorinated biphenyl Aroclor® 1260, *J. Natl. Cancer Inst.,* 55, 1453, 1975.
19. **Kimbrough, R. D. and Linder, R. E.,** Induction of adenofibrosis and hepatomas of the liver in BALB c/J mice by chlorinated biphenyls (Aroclor® 1254), *J. Natl. Cancer Inst.,* 53, 547, 1974.
20. **Altman, N. H., New, A. E., McConnell, E. E., and Ferrell, T. L.,** A spontaneous outbreak of poly-chlorinated biphenyl (PCB) toxicity in rhesus monkeys (*Macaca mulatta*): clinical observations, *Lab. Anim. Sci.,* 29, 661, 1979.
21. **McConnell, E. E., Hass, J. R., Altman, N., and Moore, J. A.,** A spontaneous outbreak of polychlorinated biphenyl (PCB) toxicity in rhesus monkeys (*Macaca mulatta*): toxicopathology, *Lab. Anim. Sci.,* 29, 666, 1979.
22. Monograph on the carcinogenic risk of chemicals to humans, Vol. 18, International Agency on Research in Cancer, Polychlorinated biphenyls and polybrominated biphenyls, Lyon, France, 1978.
23. **Allen, J. R. and Norback, D. H.,** Pathobiological responses of primates to polychlorinated biphenyl exposure, Chicago National Conference on Polychlorinated Biphenyls, Publ. No. 560/6-75-004, Environmental Protection Agency, Washington, D.C., 1976.
24. National Cancer Institute, Bioassay of Aroclor® 1254 for possible carcinogenicity, DHEW Publ. No. NIH 78-838, Washington, D.C., 1978.
25. **Ito, N., Nagasaka, H., Ari, M., Makuira, S., Sugihara, S., and Hiro, K.,** Histopathologic studies on liver tumorigenesis induced in mice by technical polychlorinated biphenyl and its promoting effect on liver tumors induced by benzene hexachloride, *J. Natl. Cancer Inst.,* 51, 1637, 1973.
26. **Makuira, S., Aoe, H., Sugihara, S., Hirao, K., Ari, M., and Ito, N.,** Inhibiting effects of polychlorinated biphenyls on liver tumorgenesis in rats treated with 3'-methyl-4-dimethylaminoazo-benzene, *n*-2-fluoren-ylactamide, and diethylnitrosamine, *J. Natl. Cancer Inst.,* 53, 1255, 1974.

27. **Hendricks, J. D., Putnam, T. P., Bills, D. D., and Sinnhuber, R. O.,** Inhibiting effect of a polychlorinated biphenyl (Aroclor® 1254) on aflatoxin B_1 carcinogenesis in rainbow trout (*Salmo gairdneri*), *J. Natl. Cancer Inst.,* 59, 1545, 1977.
28. **Preston, B. D., Van Miller, J. P., Moore, R. W., and Allen, J. R.,** Promoting effects of polychlorinated biphenyls (Aroclor® 1254) and polychlorinated dibenzofurans-free Aroclor® 1254 on diethylnitrosamine-induced tumorigenesis in the rat, *J. Natl. Cancer Inst.,* 66(3), 509, 1981.
29. **Jelinek, C. F. and Corneliussen, P. E.,** Levels of PCBs in the U.S. food supply, National Conference on Polychlorinated Biphenyls, Chicago, Publ. No. 560/6-75-004, Environmental Protection Agency, Washington, D.C., 1976.
30. **McNulty, W. P.,** Primate study, National Conference on Polychlorinated Biphenyls, Chicago, EPA Publ. No. 560/6-75-004, Washington, D.C., 1976.
31. **Allen, J. B., Carstens, L. A., and Barsotti, D. A.,** Residual effects of short-term, low-level exposure of non-human primates to polychlorinated biphenyls, *Arch. Environ. Contam. Toxicol.,* 2, 86, 1974.
32. **Keplinger, M. L., Francher, O. E., and Calandra, J. C.,** Toxicologic studies with polychlorinated biphenyls, *Toxicol. Appl. Pharmacol.,* 19, 204, 1971.
33. **Linder, R. E., Gaines, T. B., and Kimbrough, R. D.,** The effects of PCB on rat reproduction, *Food Cosmet. Toxicol.,* 12, 63, 1974.

Chapter 11

SACCHARIN—A BITTER-SWEET CASE

D. L. Arnold and D. B. Clayson

TABLE OF CONTENTS

I. INTRODUCTION

Saccharin was discovered in 1879 by Constatin Fahlberg who was working in Ira Remsen's laboratory at Johns Hopkins University.[1] This discovery, similar to those for most other artificial sweeteners, was attributable more to serendipity than design.[2]

The controversy regarding the use of saccharin either as a food additive or table top sweetener was well underway by 1890[3] and is still continuing. This saga has involved members of the scientific community, legislative and regulatory authorities, the press, the public, trade associations, and even a president of the United States.[1] While much of saccharin's notoriety is attributable to its use as an artificial, noncaloric sweetener, it has also been used: (1) as an antiseptic and preservative to retard fermentation in food;[4] (2) for estimating the circulation time between the antecubital veins and capillaries of the tongue;[5] (3) as a brightener in nickel-plated automobile bumpers;[6] and (4) as a chemical for research interests completely unrelated to toxicology.[7]

From a toxicological point of view, saccharin is probably one of the most thoroughly investigated food additives ever consumed by man, but these findings have often been as contradictory as the discussions regarding its safety for consumption.[7]

Some of the first toxicological tests wherein saccharin was ingested by man reportedly occurred in France where diabetics received 5 g of saccharin per day for 5 months, with no harmful effects being observed.[8,9] However, less than 5 years after Fahlberg was awarded a patent for the "manufacture of saccharin compounds,"[4] France forbade its manufacture and importation.[10] During the next 8 years, Germany, Spain, Portugal, and Hungary limited the use of saccharin and expressly banned its use in food and beverages.[10]

In 1902 Dr. Harvey W. Wiley of the U.S. Department of Agriculture established his famous "poison squad," which consisted of young volunteers ingesting various food additives including saccharin. As a consequence Dr. Wiley attempted, without success, to persuade President Theodore Roosevelt to ban saccharin.[11] As a consequence of this discussion, Herter and Folin of the U.S. Department of Agriculture undertook a large number of studies to examine the effects of saccharin ingestion upon man.[13] While these studies did not implicate saccharin as a serious health problem for man, the use of saccharin in soft drinks and foods was eventually banned in the United States in 1912 during the presidency of W. H. Taft.[1]

During World War I, sugar supplies became so limited in Europe and North America that the ban on saccharin was temporarily lifted. The use of saccharin continued after the war and resulted in the U.S. Government attempting legal sanctions against the Monsanto Company in 1919 for violation of the Pure Food and Drugs Act. The case was dismissed in 1925 and the use of saccharin in various foods products continued to increase.[1]

II. ONE-GENERATION FEEDING STUDIES

The use of the chronic feeding study in the regulatory/evaluation process is a relatively recent development. Discussions regarding what constitutes a properly conducted chronic feeding study have been numerous.[13] and considerations pertaining to their more contentious aspects such as dose selection,[14-16] duration,[13] and the problems of geriatric pathology[17] often generate equal portions of heat and light. However, the chronic feeding study now constitutes a significant portion of the toxicological package submitted to regulatory agencies in support of the safety considerations put forth by the sponsor of any toxicological testing program.

The first chronic, one-generation feeding study to assess the toxicological effects of saccharin was reported in 1951 by Fitzhugh et al.[18] The authors concluded that saccharin was without effect at levels of 1.0% or less in the diet, and caused only slight toxic effects at 5.0%. However, true to the saccharin's history, the evaluation of the pathological changes observed in this study is still a matter for debate.[9,19]

Table 1
USE OF ARTIFICIAL SWEETENERS

1. Beverages	Soft drinks, powdered juices and drinks, wine
2. Sweets/desserts	Ice cream, popsicles, canned fruit, dessert toppings, cookies, candies, pudding
3. Condiments	Sauces and dressings
4. Miscellaneous	Chewing gums, pickled vegetables, jams, jellies, flavor enchancer
5. Table-top sweeteners	Tablets and solutions
6. Foods	Canned vegetables, ketchup, mint sauce, sweet ham and bacon
7. Pharmaceutical products	Pediatric liquid oral preparations, cough syrups, chewable tablets
8. Cosmetics	Toothpaste, dentrifice powders, lipstick, mouth wash
9. Tobacco products	Chewing tobacco, snuff, wrapper paper and mouth pieces

Adapted from References 1, 9, and 90.

The second chronic study employing saccharin was conducted by Lessel.[20] In the top dose group, whose diet contained 5% sodium saccharin, higher mortality and a slower growth rate were observed for both sexes; the latter in spite of a greater feed consumption than controls. It was also observed that one female and four male rats in the 5% group had urinary bladder calculi and another male had kidney calculi. The female rat with calculi had an extensive transitorial cell papilloma while another 5% female without calculi had hyperplasia and papillomata. As no evidence of nematodes was found, some have suggested that this study demonstrated that the feeding of diets containing levels of 5% saccharin produces bladder stones, which could eventually give rise to bladder tumors in a manner similar to other chemicals such as diethyleneglycol.[21]

Both of these studies presented problems to those attempting to analyze and interpret them,[16] but neither study resulted in any action by North America's regulatory agencies. During the next few years, the consumption of saccharin increased.[1] The ingestion of artificial sweeteners in North America escalated considerably after the U.S. Food and Drug Administration (FDA) in 1959, using new standards of identity allowed artificial sweeteners to be used in a variety of prepared foods which should be used by those who must restrict their intake of sugar. Such foods were initially marketed for those who either had to restrict their sugar consumption (diabetics) or should avoid excessive caloric intake (obesity), but subsequently they were promoted as "nonnutritive" or "noncaloric" sweetened foods for use by the general public.[1,10] This standard signaled the start of the diet soda "revolution" in North America as the consequence of a large marketing program which purportedly reached the 40 million U.S. dollar mark in 1978.[22]

In 1944, cyclamate was discovered[23] and was subsequently introduced into the market place as a second artificial sweetener. Eventually, cyclamate was added to saccharin to eliminate the unpleasant aftertaste of the latter. The commercial name of the cyclamate-saccharin (10:1) mixture is Sucaryl®[1] which was used in a variety of food products (Table 1) and as a table top sweetener. However, cyclamate was banned as a food additive in 1969[1] and this action resulted in an international effort to study further the toxicological effects of saccharin. Table II lists all of the one-generation chronic feeding studies, and their major findings which have been reported in the scientific literature. Many of these studies reported contradictory observations and several were either difficult to interpret, or led to markedly difficult evaluations due to early mortality, small numbers, or lack of a dose-response.[16,24,25] However, none of these studies caused any overt concern, resulting in regulatory action in North America.

Table 2
ONE-GENERATION SACCHARIN FEEDING STUDIES

(Ref.)	Species and strain	Dosages/dietary levels of saccharin (no. of males (M) and females (F) per group)		Study duration	Significant findings
18	Rat, Osborne-Mendel	Control 1.0% 5.0%	(7M;9F) (10M;10F) (9M;9F)	Until natural death or 2 years	Slight growth suppression in 5% group. Seven animals (sex unspecified) in 5% group had abdominal lymphosarcomas and 4 of these 7 has concurrent thoracic lymphosarcomas.
20	Rat, Boots-Wistar	Control 0.005% 0.05% 0.5% 5.0%	(20M;20F) (20M;20F) (20M;20F) (20M;20F) (20M;20F)	2 years	Mortality was higher in the 5% group, while growth was retarded despite greater feed consumption. One female and 4 males in the 5% group had bladder calculi and another male had kidney calculi. The female also had an extensive transitional bladder cell papilloma; and another female had bladder hyperplasia and a papillomata.
116	Rat, Sprague-Dawley	Control 0.2% 0.5%	(52M;52F) (52M;52F) (52M;52F)	Until natural death	The incidence of *Trichosomoides crassicauda* was 16% and was associated with a mild cystitis of urinary bladder.
117	Rat, Sprague-Dawley	Control 2.5 g/kg	(54M) (54M)	28 months	No significant findings reported.
16	Rat, Charles River (CD)	Control[a] 1% 5%	(20M;20F) (26M;26F) (26M;26F)	24 months	Only one bladder tumor was observed in this study; a female in the 5% study II group had a urinary bladder papilloma (OTA). While different sources had conflicting tumor incidence data (OTA), this study suggests that the female rats may have been more sensitive to the effects of saccharin than the males.
118	Rat, Wistar	Control[b] 1% 5%	(20M;20F) (26M;26F) (26M;26F)	28 months	No bladder tumors were observed, but a treatment related growth suppression was.
119	Rat, Sprague-Dawley	Control 10,000 ppm 50,000 ppm	(25M;1 group) (25M;2 groups) (25M;2 groups)	24 months	High dose group in one replicate had a low-grade, noninvasive transitional cell carcinoma of the bladder, and a bladder papilloma at the low dose. Only a transitional cell carcinoma of bladder in the low dose of the second replicate was found. A 30% incidence of *Trichosomoides crassicauda* was reported.

Table 2

ONE-GENERATION SACCHARIN FEEDING STUDIES

(Ref.)	Species and strain	Dosages/dietary levels of saccharin (no. of males (M) and females (F) per group)		Study duration	Significant findings
120	Rat, Sprague-Dawley	Control	(60M;60F)	26 months	A total of 5 bladder tumors were observed: 1-male control; 1M and 1F from 90 mg/kg group; 2M from the 810 mg/kg group. Three animals had grossly visible bladder calculi and 67 had calculi small enough to pass through the urethra. There was a body weight reduction in the 2430 mg/kg males and females, and a dose related decrease in the survivability of males.
		90 mg/kg	(60M;60F)		
		270 mg/kg	(60M;60F)		
		810 mg/kg	(60M;60F)		
		2430 mg/kg[b]	(60M;60F)		
117	Mice, dde	Control	(50M;50F)	21 months	The incidence of uterine cancers suggest an effect attributed to saccharin, but the incidence of all tumors was higher in the control group than the two lower dose groups.
		0.2%	(50M;50F)		
		1.0%	(50M;50F)		
		5.0%	(50M;50F)		
119	Mice, Charles River (CD)	Control	(25M;25F)	2 years	One transitional bladder cell carcinoma was observed in a control male and in one of the 50,000 ppm groups, two cases of papilloma hyperplasias of the bladder and two small papillomas were observed. However, the incidence of total tumors and vascular tumors suggest a relationship to saccharin treatment (OTA).
		10,000 ppm	(2 × 25M;25F)		
		50,000 ppm[c]	(2 × 25M;25F)		
121	Hamster Syrian Golden	Control[d]	(30M;30F)	80 weeks	The tumor types were similar in all groups and were within the range found for historical incidence of spontaneous tumors.
		0.156%	(30M;30F)		
		0.312%	(30M;30F)		
		0.625%	(30M;30F)		
		1.25%	(30M;30F)		

[a] Duplicate experiment, Study I.
[b] Study II; (54 to 56 males 2.5 g/kg per group).
[c] Approximately equivalent to a diet containing 5% saccharin.
[d] Drinking water study.

24
25
26
27

III. TWO-GENERATION FEEDING STUDIES

At the Eighth Annual Meeting of the U.S. Society of Toxicology in 1970, Dr. Leo Friedman suggested that "no-effect dose levels be specifically defined to include that period of an animal's lifetime from conception to sexual maturity."[26] This suggestion gave rise to the two-generation chronic/carcinogenic bioassays as currently conducted. From a pragmatic point of view, this suggestion should have provided the toxicologists with a very practical tool for ascertaining whether a substance may or may not be harmful; particularly when the substance being tested is a food additive to which one may be exposed from conception until death. However, due to the uncertainties in the extrapolation of the data from such studies to man, in part attributable to the dissimilarities between man and rodents and the lack of fetal exposure data, the usefulness of this bioassay is controversial.[16,27-29]

To date, a total of three studies, wherein saccharin has been incorporated in the feed and the two-generation bioassay protocol was employed, have been completed and reported in the literature. A fourth multidose study should be published in the near future.[30]

The first study was sponsored by industry and was conducted at the Wisconsin Alumni Research Foundation (WARF). In this study,[31] groups of 20 male and 20 female Sprague-Dawley rats were fed diets containing 0.0, 0.05, 0.5, or 5.0% sodium saccharin produced by the Remsen-Fahlberg procedure for 14 weeks prior to mating, and during mating, gestation, and lactation. Following weaning, the F_0 generation was killed and the F_1 generation was weaned onto their parent's diet, which they consumed until the study was terminated 100 weeks later. While saccharin had no long-term effects upon growth, feed consumption or hematological findings, the authors suggested that the incidence of tumors was apparently increased by the consumption of the 5.0% diet. Tumors such as squamous cell carcinomas of the uterus and transitional cell carcinomas of the urinary bladder were found only in the treated groups, with a higher incidence in the 5.0% groups.

Following a review of these results, the U.S. FDA removed saccharin from the "Generally regarded as safe" (GRAS) food additive list. This action would allow the FDA to ban saccharin if any further data raised concern about the safety of saccharin as a food additive.[32]

The second study was conducted by the FDA and it was a combination three-generation reproduction and two-generation chronic feeding study.[33,34] The F_0 generation consisted of groups of 10 male and 20 female Sprague-Dawley rats being fed diets containing 0.0, 0.01, 0.1, 1.0, 5.0, or 7.5% sodium saccharin. Another group was fed a diet containing 1.51% sodium carbonate, which resulted in a sodium ion concentration equivalent to the sodium in the 5% sodium saccharin group. The F_{1A} groups consisted of 48 males and 48 females, with interim kills of 4 per sex per group and 7 per sex per group at 14 and 18 months, respectively. The major findings included a depressed rate of body weight gain in the 5.0 and 7.5% saccharin groups with nine urinary bladder neoplasms in the 7.5% group (7 males; 2 females), one urinary bladder neoplasm in a male of the 5% group and one male in the control group. Since no histological evidence of the bladder parasite *Trichosomoides crassicauda* was found, the authors concluded that the parasites did not contribute to the neoplastic lesions and neither did the gross calculi they observed (the incidence of urinary bladder calculi in males was: control - 0/41; 0.01% saccharin - 0/43; 0.1% saccharin - 3/43; 1.0% saccharin - 2/43; 5.0% saccharin - 2/38; 7.5% saccharin -0/38).[34]

The lack of tumorigenic effects resulting from the feeding of saccharin during the one-generation studies was consistent with the findings of several epidemiological studies reported during the early seventies.[35-37] However, the tumorigenic effects in the two-generation studies led scientists in the Health Protection Branch to suggest that an impurity in sodium saccharin might be responsible for the apparently conflicting findings. Subsequent analytical studies found that the major impurity in all of the batches of sodium saccharin used in the various feeding studies was *ortho*-toluenesulfonamide (o-TS),[38-40] where levels of contamination

Table 3
INCIDENCE OF BLADDER TUMORS FOR RATS FED
DIETS CONTAINING o-TS OR SODIUM SACCHARIN

Treatment Groups	F_o generation Benign	Malignant	F_1 generation Benign	Malignant
Males				
Control	1	0	0	0
2.5 mg o-TS/kg/day	1	0	0	0
25 mg o-TS/kg/day	0	0	0	0
250 mg o-TS/kg/day	1	0	0	0
250 mg o-TS/kg/day + 1% NC₄Cl in drinking water	0	0	0	0
5% Sodium saccharin	4[a,b]	3[a]	4[c]	8[c,d]
Females				
Control	0	0	0	0
2.5 mg o-TS/kg/day	1	0	2	0
25 mg o-TS/kg/day	0	0	0	0
250 mg o-TS/kg/day	0	0	0	0
250 mg o-TS/kg/day + 1% NC₄Cl in drinking water	0	0	0	0
5% Sodium saccharin	0	0[e]	0	2

[a] Number of animals with tumors (i.e., benign plus malignant) was significantly higher ($p < 0.03$) than control. Statistics were based on numbers of animals surviving until first tumor was observed after 87 weeks on test (No. Control 38F, 36M; No. Saccharin 40F, 38M).

[b] Additional urinary tract tumors consisted of one (1) animal with a urethral tumor and one (1) animal with a malignant lesion of the kidney pelvis.

[c] Number of animals with malignant tumors ($p < 0.002$) and total tumors (benign plus malignant; $p < 0.01$) was significantly higher than control. Statistics were based on number of animals surviving until first tumor was observed 67 weeks on test (No. Control 45F, 42M; No Saccharin 49F, 45M).

[d] Additional urinary tract tumors consisted of one (1) animal with a urethral tumor.

[e] Additional urinary tract tumors consisted of two (2) animals with malignant lesions of the kidney pelvis.

From Arnold, D. L., Moodie, C. A., Grice, H. C., Charbonneau, S. M., Stavric, B., Collins, B. T., McGuire, P. F., Zawidzka, Z. Z., and Munro, I. C., Toxicol. Appl. Pharmacol., 5, 113, 1980. With permission.

ranged from 118 to 6100 ppm.[41] All of these samples were synthesized by the Remsen-Fahlberg method. Since o-TS is a known, though weak, inhibitor of carbonic anhydrase,[42,43] it was theorized that o-TS might be responsible for the tumors observed in the two-generation studies.[41,44] This hypothesis proposed an increase in urine alkalinity as a result of carbonic anhydrase inhibition,[45] a condition conducive to the production of bladder stones. A general correlation between bladder stones and bladder tumors has been suggested for rats and mice.[21]

To test this hypothesis, a third two-generation study was proposed wherein graded dosages of o-TS were added to the diet. In addition, ammonium chloride was to be added to the drinking water of one of the highest o-TS dosed groups to prevent the formation of an alkaline urine.[21,46] Prior to initiating this study, a preliminary gavage study was undertaken with o-TS which appeared to support the hypothesis to be tested in the chronic study.[47] Just before the start of the chronic study, some saccharin produced by the Maumee procedure became available. Since this method of saccharin synthesis does not result in detectable levels of o-TS[48] one group of saccharin fed rats was included in the experimental design as a control group.

This two-generation study[49] consisted of groups of 50 male and 50 female Sprague-Dawley rats (except where noted) receiving the following dietary treatments: control; 2.5 mg o-TS/kg/day; 25 mg o-TS/kg/day; 250 mg o-TS/kg/day; 250 mg o-TS/kg/day with 1% NH_4Cl in the drinking water (F_o = 40 males, 38 females; F_1 = 49 males, 50 females); or 5% sodium saccharin.

After 90 days on test, the animals were mated on a one-to-one basis within treatment groups. Following parturition 50 male and 50 female animals per group were randomly selected from the available animals to constitute the F_1 generation. Unlike the other two-generation studies, both the F_o and F_1 generation animals were kept on test for 142 weeks (3% of the initial animals still alive) and 127 weeks (20% of the initial animals still alive) post weaning, respectively. The incidence of tumors is indicated in Table III. Other significant findings included an increased urinary output in the saccharin groups and the urine was hypotonic; saccharin significantly increased the amount of urinary sodium and phosphorus excreted by male and female rats; no adults or ova of the parasite *Trichosomoides crassicauda* were found; growth rates were depressed in the 250 mg o-TS/kg/day, 250 mg o-TS/kg/day with 1% NH_4Cl in the drinking water and saccharin group; however, feed consumption was only decreased in the 250 mg o-TS/kg/day with 1% NH_4Cl in the drinking water groups; a total of ten animals had bladder or kidney stones, but no relationship between the stones or dietary treatment or sex or site of stone formation were apparent.

The fourth study previously mentioned is sponsored by the Calorie Control Council.[30] The major objective of this study is to ascertain the mechanism by which saccharin elicits bladder tumors in the second generation of Sprague-Dawley rats. Consequently, the F_o generation was terminated shortly after weaning. The study design is as follows:

Treatment groups	F_o Generation (No. animals)		F_1 Generation (No. animals)
	M	F	M
Control	125	250	350
1.00% Sodium saccharin	250	500	700
3.00% Sodium saccharin	209	418	500
4.00% Sodium saccharin	84	168	200
5.00% Sodium saccharin	52	104	125
6.25% Sodium saccharin	52	104	125
7.50% Sodium saccharin	52	104	125
3.50% Sodium hippurate			125
5.00% Sodium saccharin (only during mating and gestation)	52	104	125
5.00% Sodium saccharin (only after gestation)	52	104	125

A sodium hippurate group was included, for like saccharin, it is an organic acid; has a similar molecular weight, physical, chemical, and pharmacokinetic properties; and both are secreted and filtered into the urine; therefore the specificity of sodium saccharin to cause bladder tumors and the effects of sodium per se could be compared. It should also be pointed out that this is the first study in which diets containing concentrations of saccharin between

1 and 5% have been fed to rats. Additionally, the inverted group size design should greatly assist in any dose extrapolation exercises.

IV. DECISION TO BAN

When the data from all three of the two-generation assays were evaluated, there was little doubt that saccharin per se, and not an impurity, was the agent responsible for the bladder tumors observed in these studies. Consequently, based upon the Food and Drug Laws that exist in North America, some form of regulatory action was needed. It was also evident that additional research was needed; particularly regarding the mechanisms by which saccharin elicited bladder tumors in rats.

The best available data concerning the consumption of saccharin prior to March 9, 1977, when Canada announced the banning of saccharin use as a food additive[50] and the U.S. Food and Drug Administration announced their intent to do likewise,[51] showed that 70% of saccharin used as a "food" additive was consumed as diet soft drinks.[1,27] The average per capita consumption of diet soft drinks was less than one 8 oz bottle per day.[27] However, average consumption figures are somewhat misleading in this situation, since only 6% of the population represents the major consumers of diet sodas.[27]

The purported benefits of saccharin were its usefulness in: (a) diabetes; (b) weight control; (c) carbohydrate-restricted diets (e.g., hypertriglycerides and reactive hypoglycemia); (d) dental caries and periodontal disease; and (e) drug formulations. However, the literature available prior to March 9, 1977 and subsequent reviews of this data by others[16,52,53] has not provided any convincing documentation for such benefits.

The only epidemiological evidence that saccharin was a potential bladder carcinogen for humans in March of 1977 was a relatively small study conducted by the Canadian National Cancer Institute. This study, which was subsequently published, found that the risk of males exposed to saccharin vs. those who were unexposed was about 1:6.[54] Subsequent studies, some with significantly larger samples sizes, have not found such an increased risk attributable to saccharin consumption.[55-57] The design of an epidemiological study for food additives such as saccharin are fraught with many difficulties.[7] For example:

1. One cannot distinguish between the use of saccharin per se vs. the use of other artificial sweeteners.
2. Finding an unexposed population is practically impossible.
3. Accounting for known confounding factors such as cigarette smoking and occupational exposure can be accommodated, but all confounding factors may not be known for a food additive
4. Results obtained with diabetic populations may not be applicable to the general population even though diabetics are a subpopulation that would tend to consume greater amounts of artificial sweeteners.
5. The questioning procedures used to ascertain artificial sweetener consumption may be biased.
6. The use of death certificates, diagnostic procedures, and hospital populations other than those with bladder cancer may suffer from a lack of uniform criteria or unknown bias.
7. The size of the study populations used in epidemiology studies conducted to date may not be sufficiently large to ascertain an effect of saccharin which is a very weak animal carcinogen.
8. The urinary bladder may be an inappropriate organ for evaluating the toxicological effects of saccharin in man.

In addition to the toxicological and epidemiological data, other factors were given due consideration in the banning of saccharin, and included: (1) the health status of individuals consuming saccharin-sweetened products; (2) the potential health effects derived from consuming products containing saccharin; (3) the tendency of consumers to continue the use of products (i.e., tobacco) despite potential health hazards; (4) the economic impact; and (5) the use of warning labels to provide informed choice by consumers.[58]

V. MECHANISM STUDIES

A. Metabolism and Distribution

Saccharin does not appear to be metabolized by man or animals. Byard and colleagues[59] examined the excreta of each of four male volunteers who consumed 500 mg (C^{14})-saccharin. In the first 72 hr after dosing, 92.3% of the radioactivity was recovered in the urine and 5.8% in the feces for a total recovery of 98.1%. The radioactive material was shown by gas liquid chromatography to consist of unchanged saccharin. No further excretion occurred between 72 and 96 hr. In a study with six volunteers, they found that the overall recovery of unlabeled saccharin was about 94%. Again there was no evidence of metabolism.

Saccharin is not metabolized by rats according to Sweatman and Renwick[60] who showed that saccharin was not converted in vivo to its partial hydrolysis products: 2-sulfamoylbenzoic acid, 2-sulfobenzoic acid, and carbonate. The latter observations were an artifact induced by excessive pH changes during the work-up procedure; confirming previous observations of McChesney and Golberg[61] and Byard et al.[59]

Sweatman, Renwick, and Burgers studied the distribution and excretion of saccharin in man and in Sprague-Dawley rats.[62,63] For man, the intravenous injection of saccharin was followed by a rapid biphasic clearance with a terminal half-life of about 70 min. Following oral administration the clearance pattern was more complex with considerable subject to subject variation. Rats permitted a much more detailed investigation at the high levels (5%) that the sweetener was fed in the diet during various carcinogenic bioassays. A diurnal variation in excretion was noted with maximum excretion occurring at about six a.m. and the minimum at about six p.m. Urinary concentrations of saccharin were variable but on average about 20 times higher than those found in plasma or individual tissues. The urinary bladder and the kidney were the most highly exposed tissues but it is not clear whether these tissues were completely free of urinary contamination. The concentration of saccharin in plasma and urinary bladder wall fell at similar rates when saccharin feeding was discontinued. The fact that the urinary concentration of saccharin is so much higher than that in other tissues may help to account for the chronic toxicity of saccharin to the bladder epithelium in contrast to its relative lack of effect in other tissues.

B. Impurities

Saccharin contains several impurities whose concentrations depend largely on whether the sweetener was manufactured by the Remsen-Fahlberg or the Maumee process.[64] The most prevalent impurity in Remsen-Fahlberg manufactured saccharin is o-TS.[41] At one time its concentration in saccharin ranged up to 6,000 ppm (0.6%), but changes in manufacturing procedures have reduced its concentration to 25 mg/kg or less.[64]

The potential toxicity of o-TS was examined in detail because it is the most prevalent impurity in saccharin and because of its chemical-structural similarity to 4-ethylsulfonyl-naphthalene-1-sulfonamide.[65] *Ortho*-toluenesulfonamide, like saccharin, is of low toxicity (> 2g/kg)[66] and is neither teratogenic[20,67,68] nor carcinogenic.[49,66] The probable reason that o-TS is biologically inactive whereas 4-ethylsulfonyl-naphthalane-1-sulfonamide is a bladder carcinogen is that the latter chemical is a potent inhibitor of the enzyme carbonic anhydrase, which results in alkaline urine (pH > 8.5), bladder epithelial injury, and stone formation,[65,69] whereas o-TS lacks these strong effects. Clearly o-TS is unlikely to contribute to the carcinogenicity of saccharin.

No other saccharin impurities have been investigated for carcinogenicity or for other relevant forms of toxicity. This is reasonable since most of these impurities are at concentrations less than 10 mg/kg (0.01%) in the sweetener for the most part. The metabolic fate of 3-aminobenz(d)isothiazole-1, 1-dioxide[70] and benz (d) isothiasoline-1, 1-dioxide[71] have been studied in Sprague-Dawley rats.

C. Genotoxic Effects

In general, available tests for genotoxicity indicate that saccharin does not demonstrate genotoxic properties or clastogenicity.[64] Nevertheless, isolated experiments suggest saccharin may have some genotoxicity. For example, although tests using the Ames Salmonella-microsome system are negative[72,73] a system using a modified plate procedure and reisolated Salmonella tester strains appeared to give positive results.[74] Similarly, some workers have reported that saccharin may induce chromosome aberrations in mammalian cells.[75,76] Such effects were seen in intact mice fed 4 g/kg bw saccharin[77] but not in hamsters[78] fed 1.5 g/kg. Perhaps the most intensively investigated test has been Sister Chromatid Exchange in which many conflicting results have been obtained.[79-83]

There are two possible explanations for these conflicting series of results. Saccharin may possess a very low level of direct action on DNA that is expressed in these tests, or saccharin may act at relatively high concentrations in an indirect manner. With present information the second of these possibilities seems more likely. Saccharin does not form DNA adducts with rat liver or rat urinary bladder DNA[84] but it is an enzyme inhibitor that at the high levels permitted by its low level of toxicity,[85] may be highly effective. As discussed in the following section there is a need to determine whether saccharin at high concentrations is capable of inhibiting enzymes responsible for the different aspects of DNA replication and repair, and thus whether such inhibitions could account for the observed results of the various genotoxicity tests.

Trosko et al.[86] reported that saccharin inhibited cell to cell cooperation in 6-thioguanine-resistant and -sensitive V-79 cells. This ability to inhibit cooperation is a relatively novel technique that appears able to detect tumor-promoting agents such as phorbol esters.[87] Trosko's observations support studies conducted in vivo suggesting that saccharin promotes chemically induced bladder cancer.[88-93]

D. Mechanisms of Action on Urinary Bladder

The mechanism by which saccharin induces urinary bladder tumors is not known. The facts that saccharin: (1) does not undergo metabolism in mammalian systems; (2) possesses, at the most, marginal genotoxic activity; (3) does not contain detectable carcinogenic or genotoxic impurities; but (4) does induce urinary bladder tumors in rats in two-generation studies,[31,34,49] although it had negligible carcinogenicity in conventional one-generation bioassays,[16,24,25] strongly suggest saccharin-induced tumors arise by a nongenotoxic mechanism.[94]

It is therefore necessary to consider the biological properties of saccharin to delineate possible mechanisms for its tumorigenesis and to decide whether man at realistically determined levels of exposure is likely to develop cancer as a consequence of artificial sweetener ingestion.

Many nongenotoxic carcinogens induce cell proliferation in target tissues. Such proliferation may result from regeneration in response to cytotoxicity as with certain liver poisons[95] or to direct stimulation as with certain hormones.[96] Confusing results have so far been obtained with saccharin in various rat urinary bladder assays. Although Fukushima and Cohen et al.[97,98] described a dose-dependent increase in cell proliferation and a mild hyperplasia of the bladder epithelium of Fisher-344 rats fed 5% sodium saccharin, Lawson and Hertzog[99] failed to confirm this observation with Sprague-Dawley rats of the type used in the two-generation bioassays. Miyata et al.[100] failed to observe hyperplasia in Wistar rats fed 200 or 800 mg/kg saccharin. The proliferative response appears to be dependent on the type of rat used.

The luminal surface of the adult rat bladder epithelium is covered by a nearly unique asymmetric membrane that with the tight junctions between cells makes the epithelium relatively impermeable to ions, only slightly more permeable to water, but usually relatively permeable to lipophiles.[72,101] Lawson and Hertzog[122] showed that the urinary bladder epithelium of Sprague-Dawley rats fed 7.5% saccharin had the same degree of permeability to sodium ions as did untreated rats. That is, saccharin did not detectably affect the integrity of this barrier. Furthermore, Renwick and Sweatman,[102] also using Sprague-Dawley rats, demonstrated that contrary to earlier reports,[103] sodium saccharin did not appreciably pass across the urothelium despite the very high concentration gradient between the urine and the tissues. That is, in Sprague-Dawley rats, saccharin normally does not penetrate the urothelium. If saccharin is responsible for the observed bladder tumors in rats, it must either reach the tissue from the blood which implies a high level of "tissue susceptibility" to the agent, or it exerts its effect only in special circumstances.

There are two conditions in which the integrity of the luminal membrane is lost. The first occurs in rats at about the time of parturition when the two-cell-layer thick fetal bladder epithelium loses its superficial layer and during the next 20 days regenerates to produce the adult three-cell-layer thick urothelium.[104] Lawson and Hertzog[105] demonstrated that contrary to their negative results in the adult epithelium, 1% to 7.5% saccharin in the diet appeared to inhibit the level of bladder cell proliferation in the neonatal rat. The second example of loss of integrity of the epithelium is concerned with the systemic or local application of bladder toxic agents. Such agents, that "damage" urothelial cells, lead after a lag phase to regenerative proliferation and restoration of the integrity of the epithelium.[69] Bladder carcinogens generally have this effect.[106,107] It is possible therefore that the successful use of 5% sodium saccharin as a tumor-promoter[88,89] depends on the damage to the integrity of the epithelium caused by the carcinogen, a speculation supported by the observation that if the bladder is allowed time to recover from the initiating carcinogen, the promoting effect of saccharin is apparently reduced.[88]

The question therefore arises as to the effect of saccharin when it penetrates the urothelium. The exact mechanism is still speculative. The gross alteration of the ionic balance of the tissue provides one avenue of approach.[108] However, the high concentrations of saccharin present in rat urine after feeding saccharin at the 5% or 7.5% level in the diet are potent inhibitors of enzymes such as urease[109] and proteases.[85] The inhibition of proteases could be a factor in saccharin-mediated promotion of bladder carcinogenesis[88,89] since Kakizoe et al.[110] demonstrated that a specific protease inhibitor, leupeptin, enhanced the incidence of chemically induced bladder tumors. However, more work is required since these high concentrations of saccharin may inhibit enzymes other than those already investigated. Special attention should be paid to enzymes concerned in DNA repair or replication that might more directly explain the tumorigenic effects of saccharin.

VI. REGULATORY DILEMMA

The process of regulating food additives represents the general public's protection from the deliberate or inappropriate addition of toxic or otherwise harmful substances to food. Regulation is dependent on a wide variety of considerations other than the results of toxicological testing including, above all, public support of regulatory actions. Economic factors and the public perception of need for the additive, for example, the sensory properties of particular foodstuffs containing the chemical, the shelf-life of the foodstuff under normal wholesale and retail storage, and anticipated consumption patterns are among the considerations that the regulator must consider. Failure to take account of the toxicological evidence or any other of these considerations is likely to lead to less than acceptable regulation.

The demonstration by Oser et al.[111] that the ingestion by rats of a commercial mixture of saccharin and cyclamate resulted in cancer of the urinary bladder triggered regulatory action involving saccharin in North America. Despite the fact that it was not possible to decide whether saccharin, cyclamate, or the mixture of both substances was in fact responsible for tumorigenes in this experiment, the U.S. using the Delaney Clause, decided to prohibit the use of cyclamate. Several other countries also banned cyclamates. While this decision was more political than scientific in nature, there was a growing public disquiet about the ability of regulatory agencies to provide a food supply completely devoid of any toxicological risk. Following the banning of cyclamates, the U.S. F.D.A as well as other federal regulatory agencies called for further toxicological investigation into the safety of saccharin.

In subsequent carcinogenicity bioassays, cyclamate has proved to be devoid of carcinogenic activity with the apparent exception of a study of Friedman et al.[112] In this study, the few bladder tumors obtained were, in most cases, accompanied by bladder stones. It is highly probable that the stone rather than the chemical induced the tumors[21] and it should be noted that bladder stones do not appear to be associated with cancer in humans except in well defined, rare circumstances.[113]

Saccharin failed to give biologically and statistically significant incidences of bladder or other tumors on feeding to mice or rats from weaning to death.[16,24,25] However, three studies in which saccharin was fed to the mothers before and during pregnancy and was given to the offspring from parturition until death/termination of the study resulted in bladder cancer.[31,34,49] The discovery and confirmation that in such two-generation studies saccharin was carcinogenic led the Canadian Health Protection Branch to ban saccharin as a food additive[50] while the U.S. attempted to ban use of saccharin.[51,114] However, the public outcry against the removal of the only remaining permitted artificial sweetener from the marketplace resulted in over 100,000 items of mail protesting the action being delivered to the U.S. Congress, and to a moratorium on the regulation of saccharin that has been extended to 1983.[116] Industry, through the trade association of diet soft drink manufacturers (i.e., the Calorie Control Council), is using this respite to determine from a two-generation dose-response chronic/carcinogenic bioassay with rats[28] whether saccharin's tumorigenicity is confined to the dietary exposures of 5 or 7.5% that have previously been used, or whether a threshold might exist so that an appropriate safety factor might enable regulators to set a permissible level for the continued use of this sweetener.

ACKNOWLEDGMENTS

The authors wish to thank Mr. A. Peterkin for his assistance in preparing this manuscript, and Mrs. C. L. Clark for her typing assistance.

REFERENCES

1. **Munro, I. C., Stavric, B., and Lacombe, R.,** The current status of saccharin, in *Toxicology Annual 1974,* Winek, C. L., Ed., Marcel Dekker, Inc., New York, 1974, 71.
2. **Arnold, D. L. and Munro, I. C.,** Artificial sweeteners: their toxicological etiology is an interesting mix, presented at *Proc. Int. Conf. Environ. Aspects Cancer: The Role of Macro and Micro Components of Foods,* New York, 1982.
3. **deNito, G.,** Sull'ajeone tossica della saccarina. Richerche isto-patologiche. *Bull. Soc. Ital. Biol. Sper.,* 11, 934, 1936.
4. American Council on Science and Health, *Saccharin,* New York, 1979.
5. **Fishberg, A. M., Hitzig, W. M., and King, F. H.,** Measurement of the circulation time with saccharin, *Proc. Soc. Exp. Biol. Med.,* 30, 651, 1933.

6. The Sherwin-Williams Company, Products Directory Technical Bulletin 741, Chemicals Division, Cleveland, 1977.
7. **Arnold, D. L., Krewski, D. R., and Munro, I. C.,** Saccharin: a sweetener of historical and toxicological interest, *Toxicology,* 27, 179, 1983.
8. Saccharin ban goes beyond issue of cancer, *Chemical and Engineering News,* 57:17, 1977.
9. Saccharin and its salts. Proposed rule making, *Fed. Regist.,* 42, 19996, 1977.
10. **Hunter, B. T.,** *Consumer Beware! Your Food and What's Been Done to It,* Bantam Press, New York, 1977.
11. **Marine, G. and Van Allen, J.,** *Food Pollution: The Violation of Our Inner Ecology.* Holt, Rinehart and Winston, New York, 1972, 88.
12. **Herter, C. A. and Folin, O.,** *Influence of Saccharin on the Nutrition and Health of Man.,* Rep. No. 94, U.S. Department of Agriculture, Government Printing Office, Washington, D.C., 1911.
13. *Long-Term and Short-Term Screening Assays for Carcinogens: a Critical Appraisal,* Supplement 2, IARC Publication No. , International Agency for Research on Cancer, Lyon, 1980, 21.
14. **Munro, I. C.,** Considerations in chronic toxicity testing: the chemical, the dose, the design., *J. Environ. Pathol. Toxicol.,* 1, 183, 1977.
15. Proposed system for food safety assessment, Chronic toxicity testing, Food Safety Council, *Food Cosmet. Toxicol.,* 16 (Suppl. 2), 97, 1978.
16. *Cancer Testing Technology and Saccharin,* Stock No. 052-003-00471-2 Office of Technology Assessment, U.S. Congress, Washington, D.C., 1977.
17. **Burek, J. D.,** *Pathology of Aging Rats,* CRC Press, Boca Raton, Fla., 1980,
18. **Fitzhugh, O. G., Nelson, A. A., and Frawley, J. P.,** A comparison of the chronic toxicities of synthetic sweetening agents, *J. Amer. Pharm. Assoc.,* 40, 583, 1951.
19. *Proc. Toxicology Forum on Saccharin,* Toxicology Forum, Washington, D.C., 1977.
20. **Lessel, B.,** Carcinogenic and teratogenic aspects of saccharin, *Proc. 3rd Int. Congr., Food Science and Technology,* 505/70, Washington, D.C., 1970, 764.
21. **Clayson, D. B.,** Bladder carcinogenesis in rats and mice. Possibility of artifacts, *J. Natl. Cancer Inst.,* 52, 1685, 1974.
22. Saccharin Moratorium., Hearing Before the Subcommittee on Health and Environment of the CIFC, Committee on Interstate and Foreign Commerce, House of Representatives, 96th Congress, First Session on H.R. 4194, H.R. 1819, H.B. 4160, and H.R. 4172, Bills to amend the authority of the Food and Drug Administration respecting the availability of saccharin, Serial No. 96-15, May 23, 1979, U.S. Government Printing Office, Washington, D.C., 1979, 57.
23. **Lepkowski, W. C.,** Saccharin ban goes beyond issue of cancer, *Chem. Eng. News,* 55, 17, 1977.
24. **Crammer, M. F.,** *Saccharin: A report by Dr. Morris F. Crammer,* Scheer, G. H., Ed., American Drug Research Institute Inc., Tippecanoe, Indiana, 1980.
25. **Reuber, M. D.,** Carcinogenicity of saccharin, *Environ. Health Perspect.,* 25, 173, 1978.
26. **Friedman, L.,** Symposium on the evaluation of the safety of food additives and chemical residues. II. The role of the laboratory animal study of intermediate duration for evaluation of safety, *Toxicol. Appl. Pharmacol.,* 16, 498, 1970.
27. *Safety of Saccharin and Sodium Saccharin in the Human Diet,* Publication No. 238-137, Rep. to FDA National Research Council, Washington, D.C., 1974.
28. *Proc. Toxicology Forum on Saccharin,* Toxicology Forum, ibid, Washington, D.C., 1977.
29. **Grice, H. C., Munro, I. C., Krewski, D. R., and Blummenthal, H.,** *In utero* exposure in chronic toxicity/carcinogenicity studies, *Food Cosmet. Toxicol.,* 19, 373, 1981.
30. *Saccharin Working Group of the Toxicology Forum,* Toxicology Forum, Washington, D.C., 1980.
31. **Tisdel, M. O., Nees, P. O., Harris, D. L., and Derse, P. H.,** Long-term feeding of saccharin in rats, in *Symposium: Sweeteners,* Inglett, G. E., Ed., Avi Publishing Company, Inc., Westport, Conn., 1974, 145.
32. **Epstein, S. S.,** *The Politics of Cancer,* Sierra Club Books, San Francisco, 1978, 199.
33. **Taylor, J. M. and Friedman, L.,** Combined chronic feeding and three-generation reproduction study of sodium saccharin in the rat, *Toxicol. Appl. Pharmacol.,* 29 (Abstr. 200), 154, 1974.
34. **Taylor, J. M., Weinberger, M. A., and Friedman, L.,** Chronic toxicity and carcinogenicity to the urinary bladder of sodium saccharin in the *in utero*-exposed rat, *Toxicol. Appl. Pharmacol.,* 54, 57, 1980.
35. **Armstrong, B. and Doll, R.,** Bladder cancer mortality in diabetics in relation to saccharin consumption and smoking habits, *Br. J. Prev. Soc. Med.,* 29, 73, 1975.
36. **Kessler, I. I.,** Cancer mortality among diabetics, *J. Natl. Cancer Inst.,* 44, 673, 1970.
37. **Morgan, R. W. and Jain, M. G.,** Bladder cancer: Smoking, beverages and artificial sweeteners, *Can. Med. Assoc. J.,* 111, 1067, 1974.
38. **Stavric, B., Klassen, R., and By, A. W.,** Impurities in commercial saccharin. I. Impurities soluble in inorganic solvents., *J. Assoc. Off. Anal. Chem.,* 59, 1051, 1976.

39. **Stavric, B., Lacombe, R., Munro, I. C., By, A. W., Klassen, R., and Ethier, J.,** Studies on water soluble impurities in commercial saccharin, 167th Am. Chem. Soc. National Meeting, Los Angeles, March 31 — April 5, 1974, ANAL. 171, 1974.

40. **Stavric, B., Lacombe, R., Watson, J. R., and Munro, I. C.,** Isolation, identification and quantitation of O-toluenesulfonamide, a major impurity in commercial saccharins, *J. Assoc. Off. Anal. Chem.,* 57, 678, 1974.

41. **Grice, H. C.,** Commentary and discussion sections on toxicology: the report of the National Academy of Science, in Sweeteners: Issues and Uncertainties, Academy Forum, National Academy of Sciences, Washington, D.C., 1975, 143.

42. **Krebs, H. A.,** Inhibition of carbonic anhydrase by sulphonamides, *Biochem. J.,* 43, 525, 1948.

43. **Miller, W. H., Dessert, A. M., and Roblin, R. O., Jr.,** Heterocyclic sulfonamides as carbonic anhydrase inhibitors, *J. Am. Chem. Soc.,* 72, 4893, 1950.

44. **Lacombe, R., Moodie, C. A., Stavric, B., and Munro, I. C.,** Impurities in saccharin: orthotoluenesulfonamide toxicity study, presented at the Ann. Meet. Calorie Control Council, San Francisco, Calif., December 4—6, 1974.

45. **White, A., Handler, P., Smith, E. L., Hill, R. L., and Lehman, I. R.,** *Principles of Biochemistry,* 6th Ed., McGraw-Hill, New York, 1978, 1068.

46. **Flaks, A. and Clayson, D. B.,** The influence of ammonium chloride on the induction of bladder tumors by 4-ethylsulphonyl-naphthalene-1-sulphonamide, *Br. J. Cancer,* 31, 585, 1975.

47. **Arnold, D. L., Moodie, C. A., McGuire, P. F., Collins, B. T., Charbonneau, S. M., and Munro, I. C.,** The effect of *ortho*-toluenesulfonamide and sodium saccharin on the urinary tract of neonatal rats, *Toxicol. Appl. Pharmacol.,* 51, 455, 1979.

48. **Stavric, B.,** The Canadian saccharin study, in *Toxicology Forum on Saccharin,* Toxicology Forum, Washington, D.C., 1977, 73.

49. **Arnold, D. L., Moodie, C. A., Grice, H. C., Charbonneau, S. M., Stavric, B., Collins, B. T., McGuire, P. F., Zawidzka, Z. Z., and Munro, I. C.,** Long-term toxidity of *ortho*-toluenesulfonamide and sodium saccharin in the rat, *Toxicol. Appl. Pharmacol.,* 52, 113, 1980.

50. Canadian position on saccharin, News Release Health and Welfare Canada, 40, March 9, 1977.

51. The great saccharin snafu, Consumers Report, 42, 410, 1977.

52. **Rosenman, K.,** Benefits of saccharin: a review., *Environ. Res.,* 15, 70, 1978.

53. Saccharin: technical assessment of risks and benefits, National Academy of Sciences, Washington, D.C., 1978.

54. **Howe, G. R., Burch, J. D., Miller, A. B., Morrison, B., Gordon, R., Gordon, L., Weldon, L., Chambers, L. W., Fodor, G., and Winsor, G. M.,** Artificial sweeteners and human bladder cancer, *Lancet,* 2, 578, 1977.

55. **Hoover, R. N. and Strasser, P. H.,** Artificial sweeteners and human bladder cancer. Preliminary results, *Lancet,* 1, 837, 1980.

56. **Wynder, E. L. and Goldsmith, R.,** The epidemiology of bladder cancer, a second look, *Cancer,* 40, 1246, 1977.

57. **Wynder, E. L. and Stellman, S. D.,** Artificial sweeteners use and bladder cancer. A case-control study, *Science,* 207, 1214, 1980.

58. **Bradshaw, L., Arnold, D. L., and Krewski, D.,** A century of saccharin, in *Living with Risk: Environmental Risk Management in Canada,* Burton, I., Fowle, C. D., and McCullough, R. S., Eds., Institute for Environmental Studies, Toronto, 1982, 117.

59. **Byard, J. L., McChesney, E. W., Golberg, L., and Coulston, F.,** Excretion and metabolism of saccharin in man. II. Studies with ¹⁴C-labeled and unlabeled saccharin, *Food Cosmet. Toxicol.,* 12, 175, 1974.

60. **Sweatman, T. W. and Renwick, A. G.,** Saccharin metabolism and tumorigenicity, *Science,* 205, 1019, 1979.

61. **McChesney, E. W. and Golberg, L.,** The excretion and metabolism of saccharin in man. I. Methods of investigation and preliminary results, *Food Cosmet. Toxicol.,* 11, 403, 1973.

62. **Sweatman, T. W., Renwick, A. G., and Burgers, C. D.,** The pharmacokinetics of saccharin in man, *Xenobiotica,* 11, 531, 1981.

63. **Sweatman, T. W. and Renwick, A. G.,** The tissue distribution and pharmacokinetics of saccharin in the rat, *Toxicol. Appl. Pharmacol.,* 55, 18, 1980.

64. IARC Monographs on Evaluation of the Carcinogenic Risk of Chemicals to Humans. Some Non-Nutritive Sweetening Agents *IARC Sci. Publ.,* 22, 111, 1980.

65. **Clayson, D. B., Pringle, J. A. S., and Bouser, G. M.,** 4-Ethylsulphorylenaphthalene-1-sulphonamide: a new chemical for the study of bladder cancer in the mouse, *Biochem. Pharm.,* 16, 619, 1967.

66. **Schmähl, D.,** Experiments on the carcinogenic effect of *ortho*-toluolsulfonamid (OTS), *Z. Krebsforsch.,* 91, 19, 1978.

67. **Lorke, D.,** Untersuchungen von cyclamat und saccharin auf embryotoxische und teratogene wirkung an der maus, *Arzen.,* 19, 920, 1969.

68. **Fritz, H. and Hess, R.,** Prenatal development in the rat following administration of cyclamate, saccharin and sucrose, *Experientia,* 24, 1140, 1968.

69. **Clayson, D. B. and Cooper, E. H.,** Cancer of the urinary tract, *Adv. Cancer Res.,* 13, 271, 1970.

70. **Renwick, A. G.,** The fate of saccharin impurities: the metabolism and excretion of 3-amino[3-^{14}C] benz [d]-isothiazole-1,1-dioxide and 5-chlorosaccharin in the rat, *Xenobiotica,* 8, 487, 1978.

71. **Renwick, A. G. and Williams, R. T.,** The fate of saccharin impurities: the excretion and metabolism of [3-^{14}C] benz [d]-isothiazoline-1,1-dioxide (BIT) in man and rat, *Xenobiotica,* 8, 475, 1978.

72. **Ashby, J., Styles, J. A., Anderson, D., and Paton, D.,** Saccharin: an epigenetic carcinogen/mutagen? *Food Cosmet. Toxicol.,* 16, 95, 1978.

73. **Stoltz, D. R.,** *Proc. Toxicology Forum on Saccharin,* Toxicology Forum, Washington, D.C., 28, 101, 1977.

74. **Batzinger, R. P., Ou, S. Y. L., and Bueding, E.,** Saccharin and other sweeteners: mutagenic properties, *Science,* 198, 944, 1977.

75. **McCann, J.,** Cancer Testing Technology and Saccharin, Stock No. 052-003-00471-2, Office of Technology Assessment, U.S. Congress, Washington, D.C., 1977, 91.

76. **Yoshida, S. M., Masubueki, M., and Hiraga, K.,** Induced chromosome aberrations by artificial sweeteners in CHO K1 cells, *Mutat. Res.,* 54(Abstr. 45), 262, 1978.

77. **Masubachi, M., Yoshida, S., Hiraga, K., and Nawai, S.,** The mutagenicity of sodium saccharin (S-Na). II. Cytogenetic Studies, *Mutat. Res.,* 54, 219, 1978.

78. **van Went-De Vries, G. F. and Kragten, M. C. T.,** Saccharin: tack of chromosome-damaging activity in Chinese hamsters *in vivo, Food Cosmet. Toxicol.,* 13, 177, 1976.

79. **Fischman, H. K. and Rainer, J.,** Evaluation of sister chromatid exchanges in human leucocytes by exposure to saccharin, *Environ, Mutagens,* 1, 177, 1979.

80. **Abe, S. and Sasaki, M.,** Chromosome aberrations and sister chromatid exchanges in Chinese hamster cells exposed to various chemicals, *J. Natl. Cancer Inst.,* 58, 1635, 1977.

81. **Saxholm, H. J. K., Iverson, O. H., Reith, A., and Brogger, A.,** Carcinogenesis testing of saccharin. No transformation or increased sister chromatid exchange observed in two mammalian cell systems, *Eur. J. Cancer.,* 15, 509, 1979.

82. **Broegger, A., Ardito, S., and Waksock, V.,** Attempt to increase the sensitivity of the SCE method by means of caffeine, negative results with saccharin, *Mutat. Res.,* 74, 218, 1980.

83. **Wolff, S. and Rodinz, B.,** Saccharin — induced sister chromatid exchanges in Chinese hamster and human cells, *Science,* 200, 543, 1978.

84. **Lutz, W. K. and Schlatter, C. H.,** Saccharin does not bind to DNA of liver or bladder in the rat, *Chem. Biol. Interact.,* 19, 253, 1977.

85. **Lok, E., Iverson, F., and Clayson, D. B.,** The inhibition of urease and proteases by sodium saccharin, *Cancer Lett.,* 16, 163, 1982.

86. **Trosko, J. E., Dawson, B., Yotti, L. P., and Chang, C. C.,** Saccharin may act as a tumor promoter by inhibiting metabolic cooperation between cells, *Nature (London),* 285, 109, 1980.

87. **Yotti, L. P., Chang, C. C., and Trosko, J. E.,** Eliminations of metabolic cooperation in Chinese hamster cells by a tumor promoter, *Science,* 206, 1087, 1979.

88. **Cohen, S. M., Arai, M., Jacobs, J. B., and Friedell, G. H.,** Promoting effect of saccharin and DL-Tryptophan in urinary bladder carcinogenesis, *Cancer Res.,* 39, 1207, 1979.

89. **Fukushima, S., Friedell, G. H., Jacobs, J. B., and Cohen, S. M.,** Effect of L-tryptophan and sodium saccharin on urinary tract carcinogenesis initiated by N-[4-(5-nitro-2-furyl)-2-thiazolyl] formamide, *Cancer Res.,* 41, 3100, 1981.

90. **Hicks, R. M. and Chowaniec, J.,** The importance of synergy between weak carcinogens in the induction of bladder cancer in experimental animals and in humans, *Cancer Res.,* 37, 2943, 1977.

91. **Chowaniec, J. and Hicks, R. M.,** Response of the rat to saccharin with particular reference to the urinary bladder, *Br. J. Cancer,* 39, 355, 1979.

92. **Nakanishi, K., Hagiwara, A., Shibata, M., Imalda, K., Tatematsu, M., and Ito, N.,** Dose response of saccharin in induction of urinary bladder hyperplasia in Fischer 344 rats pretreated with N-butyl-N-(4-hydroxybutyl) nitrosamine, *J. Natl. Cancer Inst.,* 65, 1005, 1080.

93. **Nakanishi, K., Hirose, M., Ogiso, T., Hasegawa, R., Arai, M., and Ito, N.,** Effect of sodium saccharin and caffeine on the urinary bladder of rats treated with N-butyl-N-(4-hydroxybutyl)nitrosamines, *GANN.,* 71, 490, 1980.

94. International Commission for Protection against Environmental Mutagens and Carcinogens, Committee 2 Final Report: mutagen testing as an approach to carcinogeneus, *Mutat. Res.,* 99, 73, 1982.

95. **Peraino, C. R., Fry, J. M., Staffeldt, E., and Kisieleski, W. E.,** Effects of varying the exposure to phenobarbitol on its enhancement of 2-acetylaminofluorene-induced hepatic tumorigeneses in the rat, *Cancer Res.,* 33, 270, 1973.

96. **Nande, S.,** Hormonal carcinogenesis: a novel hypothesis for the role of hormones, *J. Environ, Pathol. Toxicol.,* 213, 1978.
97. **Fukushima, S. and Cohen, S. M.,** Saccharin-induced hyperplasia of the rat urinary bladder, *Cancer Res.,* 46, 734, 1980.
98. **Murasaki, G. and Cohen, S. M.,** Effect of dose of sodium saccharin on the induction of rat urinary bladder proliferation, *Cancer Res.,* 41, 942, 1981.
99. **Lawson, T. A. and Hertzog, P. J.,** The failure of chronically administered saccharin to stimulate bladder epithelial DNA synthesis in F_o rats, *Cancer Lett.,* 11, 221, 1981.
100. **Miyata, Y., Hagiwara, A., Nakatsuka, T., Murasaki, G., Arai, M., and Ito, N.,** Effects of caffeine and saccharin on DNA in the bladder epithelium of rats treated with N-butyl-N-(3-carboxypropyl)-nitrosamine, *Chem. Biol. Interact.,* 29, 291, 1980.
101. **Hicks, R. M.,** The mammalian bladder: an accommodating organ., *Biol. Rev.,* 50, 215, 1975.
102. **Renwick, A. G. and Sweatman, T. W.,** The absorption of saccharin from the urinary bladder, *J. Pharm. Pharmacol.,* 31, 650, 1979.
103. **Colburn, W. A.,** Absorption of saccharin from rat urinary bladder, *J. Pharm. Sci.,* 67, 1493, 1978.
104. **Firth, J. A. and Hicks, R. M.,** Membrane specialization and synchronized cell death in developing rat transitorial epithelium, *J. Anat.,* 113, 95, 1972.
105. **Lawson, T. and Hertzog, P.,** The apparent inhibition of urothelial DNA synthesis in neonatal rats by dietary saccharin, *Cancer Lett.,* 13, 159, 1981.
106. **Clayson, D. B., Lawson, T. A., Santana, S., and Bonses, G. M.,** Chemical induction of hyperplasia and of malignancy in the mouse bladder epithelium. *Brit. J. Cancer,* 19, 297, 1975.
107. **Lawson, T. A., Dawson, K. M., and Clayson, D. B.,** Acute changes in nucleic acid and protein synthesis in the mouse bladder induced by three bladder carcinogens, *Cancer Res.,* 30, 1586, 1970.
108. **Anderson, R. L.,** Response of male rats to sodium saccharin ingestion: Urine composition and mineral balance, *Food Cosmet. Toxicol.,* 17, 195, 1979.
109. **Anderson, R. L. and Kirkland, J. J.,** The effect of sodium saccharin in the diet of caecal micorflora, *Food Cosmet. Toxicol.,* 18, 353, 1980.
110. **Kakizoe, T., Esumi, H., Kawachi, T., Sugimura, T., Takeuchi, T., and Umezawa, H.,** Further studies on the effect of leupeptin, a protease inhibitor, on induction of bladder tumors in rats by N-butyl-N-(4-hydroxybutyl) nitrosamine, *J. Natl. Cancer Inst.,* 59, 1503, 1977.
111. **Oser, B. L., Carson, S., Cox, G. E., Vogin, E. E., and Sternberg, S. S.,** Chronic toxicity study of cyclamate: Saccharin (10:1) in rats, *Toxicology,* 4, 315, 1975.
112. **Friedman, L., Richardson, H. L., Richardson, M. E., Lethco, E. G., Wallace, W. C., and Sauro, F. M.,** Toxic response of rats to cyclamate in chow and semisynthetic diets, *J. Natl. Cancer Inst.,* 49, 751, 1972.
113. **Dodson, A. I.,** Urologeal Surgery, 4th Ed., C. V. Mosby, St. Louis, 1970.
114. Annual Report, U.S. Food and Drug Administration, 1977, 40.
115. Senate passes 2-year extension of saccharin ban moratorium, *Food Chem. News,* June 29, 1981, 70.
116. **Schmähl, D.,** Fehlen einer kanzerogenen. Wirkung von Cyclamat, Cytclohexylamin und Saccharin bei Ratten., *Arzneim. Forsch,* 23, 1973, 1466.
117. **Miyaji, T.,** cited in Publ. No. 238-137, Rep. FDA, National Research Council, Washington, D.C., 1974, 46.
118. **Furuya, T., Kawamata, K., Kaneko, T., Uchida, O., Horiuchi, S., and Ikeda, Y.,** Long-term toxicity study of sodium cyclamate and saccharin sodium in rats, *J. Pharmacol.,* 25(Suppl.) 55, 1975.
119. **Homburger, F.,** Negative lifetime carcinogen studies in rats and mice fed 50,000 ppm saccharin, in *Chemical Toxicology of Food,* Galli, C. L., Paoletti, R., and Vettorazzi, G., Eds., Elsevier/North-Holland Biomedical Press, Amsterdam, 1978, 359.
120. **Munro, I. C., and Moodie, C. A., Krewski, D., and Grice, H. C.,** A carcinogenicity study of commercial saccharin in the rat, *Toxicol. Appl. Pharmacol.,* 32, 513, 1975.
121. **Althoff, J., Cardesa, A., Pour, P., and Shubik, P.,** A chronic study of artificial sweeteners in Syrian golden hamsters, *Cancer Lett.,* 1, 21, 1975.
122. **Lawson, T. A., and Hertzog, P. J.,** unpublished observations.

Index

INDEX

I

N

O